BIOENERGY FOR SUSTAINABLE DEVELOPMENT AND INTERNATIONAL COMPETITIVENESS

The role of sugar cane in Africa

*Edited by Francis X. Johnson and
Vikram Seebaluck*

Routledge
Taylor & Francis Group

LONDON AND NEW YORK

First published 2012
by Routledge
2 Park Square, Milton Park, Abingdon, Oxon OX14 4RN

Simultaneously published in the USA and Canada
by Routledge
711 Third Avenue, New York, NY 10017

First issued in paperback 2018

Routledge is an imprint of the Taylor & Francis Group, an informa business

British Library Cataloguing in Publication Data
A catalogue record for this book is available from the British Library

Library of Congress Cataloging-in-Publication Data
Johnson, Francis X.
    Bioenergy for sustainable development and international competitiveness:
    the role of sugar cane in Africa/Francis X. Johnson and Vikram Seebaluck.
    p. cm.
    Includes bibliographical references and index.
    1. Sugar cane industry–Africa. 2. Biomass energy–Africa. 3. Renewable
    energy sources–Africa. 4. Sustainable development–Africa. I. Seebaluck,
    Vikram. II. Title.
    HD9117.A2J64 2012
    333.95′39–dc23
                                                          2011037177

ISBN 13: 978-1-138-54229-7 (pbk)
ISBN 13: 978-1-84971-103-6 (hbk)

Typeset in Bembo
by Wearset Ltd, Boldon, Tyne and Wear

CONTENTS

LIST OF FIGURES, TABLES AND PLATES

Figures

Tables

Plates

The plate section can be found between pages 260 and 261 of the text.

LIST OF CONTRIBUTORS

Goolam Houssen Badaloo is a Plant Breeder at the Mauritius Sugar Industry Research Institute, Mauritius and part-time Senior Lecturer in Crop Improvement/Genetics at the University of Mauritius and the Regional Training Centre. He has a Bachelor's degree in Agriculture, a Master's in Applied Genetics and a PhD in Quantitative Genetics. He has wide experience in plant breeding and genetics.

Bothwell Batidzirai is an International Bioenergy Consultant at the Fuels and Energy Department of the Chinhoyi University of Technology, Zimbabwe. He has 15 years of research experience in energy and development in Africa and has worked for various institutions and research groups, including SADC energy ministries and electricity utilities, the Southern Centre for Energy and Environment, the African Energy Policy Research Network, Forum for Energy Ministers of Africa, IEA Task 40 "Sustainable International Bioenergy Trade – Securing Supply and Demand" and Cane Resources Network for Southern Africa.

David Bauner has a Master's degree in Mechanical Engineering and a PhD in Industrial Economics. He has worked in the area of renewable fuels/energy and environmentally adapted vehicles in the corporate as well as governmental sector in Sweden. He has also been involved in projects in the EU, Latin America, Japan and Eastern Africa.

Gareth Brown is a Research Associate in the Centre for Energy Policy and Technology and the Centre for Environmental Policy at Imperial College London. His principal research interests are in modelling of bioenergy systems and assessments of their life-cycle environmental impacts.

Pontus Cerin is an Associate Professor at Umeå School of Business and a Guest Researcher at the Royal Institute of Technology, Sweden. His research includes eco-efficiency and innovation in environmentally and socially responsible investments and environmental policies in the auto industry, as well as the economics of green buildings. He is also Research Leader at Sustainable Investment Research Platform.

Maurizio Cocchi is a Manager at ETA Florence. He has conducted feasibility studies on biofuel and bioenergy production in Europe and developing countries and has managed European projects under the Sixth and Seventh Framework Programmes/Intelligent Energy for Europe Programme. He is also the Italian Representative for the IEA Bioenergy Task 40 "Sustainable Bioenergy Trade". He has Bachelor's and Master's degrees.

Kassiap Deepchand is a Consultant in the cane sugar industry. He has a Diploma in Agriculture and Sugar Technology, a Degree in Sugar Technology and a PhD that focused on systems study for the utilization of cane tops and leaves. He has worked as Head of the Sugar Technology Department of the Mauritius Sugar Industry Research Institute, and Technical Manager and Deputy Executive Director of the Mauritius Sugar Authority. He has also served as Coordinator of the World Bank/GEF Mauritian Sugar Energy projects and Principal Researcher in the African Energy Policy Research Network.

Rocio A. Diaz-Chavez, PhD, is a Research Fellow at the Centre for Environmental Policy of Imperial College London and Tutor at the Centre for Development, Environment and Policy, SOAS. She has over 15 years' experience in sustainability assessment and environmental management tools and methodologies in Latin America, Africa and Europe. She has a Bachelor's degree in Biology-Ecology.

Asha Dookun-Saumtally is a Principal Research Manager at the Mauritius Sugar Industry Research Institute, Mauritius. She has a Bachelor's degree in Microbiology and Biochemistry, a Master's degree in Microbiology and a Doctorate. She has wide experience in the fields of plant biotechnology, pathology and biosafety, and is currently the Chairperson of the Molecular Biology section of the International Society of Sugar Cane Technologists.

Tarryn Eustice is a Soil Scientist at the South African Sugarcane Research Institute. She has Bachelor's and Master's degrees in Agriculture and her research is focused on soil health and quality, with an emphasis on carbon relationships and their contribution to sustainable sugar cane production and greenhouse gas emissions.

Francesca Farioli is the Coordinator of the research unit "Energy, Environment and Development" at the Interuniversity Research Centre for Sustainable Development of the University of Rome, Italy. She has a PhD degree in the area of energy and environmental technologies for development and her research focuses on sustainability science. She is involved in the Global Bioenergy Partnership and Food Security Criteria and Indicators FAO project, and also serves as member of the Steering Committee of the International Conference of Sustainability Science.

Denilson Ferreira is the General Manager of Agroenergy at the Ministry of Agriculture, Livestock and Food Supply in Brazil, and Professor at São Paulo State Technological College. He has a Bachelor's degree in Business Management and a Master's degree in Energy.

Melinda Fones-Sundell is a Senior Research Fellow at the Stockholm Environment Institute. She is an Agricultural Economist with over 30 years' experience in agricultural development programmes in Africa and Latin America. In addition to university research and teaching, she worked for 18 years in the private sector, dealing primarily with environmental issues related to agriculture and with the environmental impact of biofuels in Southern and Eastern Africa.

Giuliano Grassi is Director of EUBIA, the European Biomass Association. He has worked in biofuel refineries and with hybrid town-vehicles. He has previously worked for the European Commission programme on research and development of bioenergy.

Antonio J. Gutiérrez-Trashorras is a Professor in the Department of Energy at the University of Oviedo, Spain. His PhD research focused on the energy consumption of petroleum products. His research interests include various aspects of renewable energy.

Luís Carlos Mavignier de Araújo Job is a specialist in public policies and governmental management and is currently on study leave for a Master's degree in Development Administration and Planning. He has worked at the Secretariat of Planning and Strategic Investment at the Ministry of Planning, Budget and Management, and also at the Secretariat of Production and Agroenergy at the Ministry of Agriculture, Livestock and Food Supply of Brazil.

Francis X. Johnson is a Senior Research Fellow at the Stockholm Environment Institute and visiting researcher in the Energy and Climate Studies Division at KTH Royal Institute of Technology in Stockholm, Sweden. He was previously a Senior Research Associate in Energy Analysis at the Lawrence Berkeley National Laboratory, USA. He was Scientific Coordinator of an EC-funded thematic research network on sugar cane in Southern Africa and has served as Evaluator, Advisor or Policy Analyst for FAO, UNIDO, EUROSTAT, the European Commission and European Parliament Environment Committee. He conducts research and analysis at the intersection of bioenergy, climate and development policies at local, national and global levels, focusing on both traditional and modern biomass use.

Lindsay Jolly is a Senior Economist at the International Sugar Organization, London, where he is responsible for the economic and policy analysis of world fuel ethanol, sweeteners and molasses markets. He studied Agricultural Economics and has worked for the Australian Bureau of Agricultural and Resource Economics in Canberra, and the Food and Agriculture Organization of the United Nations in Rome.

Dilip Khatiwada is a PhD Scholar in the division of Energy and Climate Studies of the Royal Institute of Technology, Sweden. He has a Bachelor's degree in Energy and Climate Studies. He has 12 years of professional experience and has served as Assistant Spokesperson at the Ministry of Environment of Nepal, Component Manager at the Danish International Development Agency, National Focal Person at the Regional Air Quality Project, Electro-Mechanical Specialist at the Asian Development Bank, and Under-Secretary at the Ministry of Industry in Nepal. His current research is focused on sustainability assessment of bioenergy systems.

Manoel Regis Lima Verde Leal is the Director of Sustainability of the Brazilian Bioethanol Science and Technology Laboratory. He has a Bachelor's degree in Aeronautical Engineering and a Master's and PhD in Mechanical Engineering. He has 25 years of experience of processing cane into sugar, ethanol and electricity at the Copersucar Technology Center/Sugarcane Technology Center, Center for Energy Planning of the University of Campinas and Brazilian Bioethanol Science and Technology Laboratory. He has also lectured at the Federal University of Ceará and Instituto Tecnologico de Aeronautica in Brazil.

Anna Lerner is part of the Climate Change Mitigation Team at the Global Environment Facility of the World Bank in Washington, DC, focusing on renewable energy and technology transfer. She has worked for the Deutsche Gesellschaft für Technische Zusammenarbeit (GTZ) as a Regional Advisor to the Southern African Development Community Secretariat on sustainability for biofuels and bioenergy.

Isaias Macedo is a Researcher for the Interdisciplinary Center of Energy Planning of the State University of Campinas in Brazil and Consultant for international and domestic public and private institutions. He has Bachelor's, Master's and PhD degrees in Mechanical Engineering as well as a postdoctorate degree. He has served as Head of the Sugarcane Technology Center and has held faculty positions at the Aeronautic Technological Institute and University of Campinas.

Maxwell Mapako is a Senior Researcher at CSIR in Pretoria, South Africa, and a PhD scholar at the Energy Research Centre, University of Cape Town. He has a Bachelor's degree in Biochemistry and Environmental Studies and a Master's degree in Energy Studies. His research interest is in sustainable energy options for poverty reduction.

Karoli Nicholas Njau is an Associate Professor at the Nelson Mandela African Institute of Science and Technology in Tanzania, where he studied Chemical Engineering. He has been an academic staff member at the University of Dar es Salaam for over 20 years and has led several multinational teams working on consultancy and research issues in environment and energy. His main areas of research are environment, bioenergy and food processing.

Alain Onibon is an Investment Officer at the FAO Subregional Office for West Africa of the UN Food and Agricultural Organization in Ghana. He has been an international Consultant on agricultural and rural development and has worked with the World Bank as an Agronomic Engineer. He was also the Director General of Land Use Management in Benin.

Henrique Pacini is a PhD scholar at the Royal Institute of Technology in Sweden and an Associate Economic Affairs Officer at the United Nations Conference on Trade and Development in Switzerland. He has a Bachelor's degree in Economics and a Master's degree in European Studies. His research focuses on energy systems, with an emphasis on the regulatory and market aspects of biofuels. His professional engagements have included involvement in regulatory affairs for the European Commission (DG ENER) and energy markets for the Brazilian government (SEAE/MF).

Brian S. Purchase is a Consultant and part-time Lecturer. He has a Bachelor's degree in Agriculture and his PhD focused on nitrification in tropical grasslands. He has worked as Lecturer at the University of Zimbabwe, Research Officer at the South African Sugar Milling Research Institute and Director at the Sugar Milling Research Institute.

Antonio Ramalho-Filho is an Agronomist, Soil Specialist and Consultant on land use planning, agro-ecological zoning and land evaluation. His PhD was about land use planning. He has been a Senior Researcher and Head of the Brazilian Agricultural Research Corporation. He has also served as President of the Brazilian Society of Soil Science.

Kishore Ramdoyal is a Principal Research Manager at the Mauritius Sugar Industry Research Institute, Mauritius. He has a diploma in Agriculture and Sugar Technology, a Bachelor's degree in Agriculture and a PhD in Agronomy/Crop Science. He has wide experience in the breeding and selection of sugar cane varieties and has headed the Sugar Cane Breeding Research and Development Programme of the Mauritius Sugar Industry Research Institute for the past ten years. He is a Member of the Germplasm and Breeding section of the International Society of Sugar Cane Technologists.

P.J. Manohar Rao is a Chemical Technologist with 52 years of experience in sugar mills, distilleries and co-product industries. He has worked for the Indian Ministry of Food and Agriculture for more than 30 years in various senior positions. He has also been the Managing Director/Advisor of the National Federation of Co-Operative Sugar Factories Ltd in New Delhi. He has published four books on sugar technology, co-products and energy conservation in sugar factories and distilleries.

Frank Rosillo-Calle is an Honorary Research Fellow in the Centre for Environmental Policy at Imperial College London. He has a PhD degree and over 30 years of research experience in bioenergy and related areas. He has published extensively on many aspects of biomass energy.

Joaquim Seabra is a Professor in the School of Mechanical Engineering at the University of Campinas, Brazil and joint appointment researcher at the Brazilian Bioethanol Science and Technology Laboratory. He has a Bachelor's degree in Food Engineering and a PhD in Energy Planning. His research experience and interests are in bioenergy and biorefineries, technology evaluation, life-cycle and sustainability assessment of energy systems, and GHG balances and accounting.

Vikram Seebaluck is a Senior Lecturer and Head of the Department of Chemical and Environmental Engineering at the University of Mauritius. He teaches and researches in the field of renewable energy technologies, with an emphasis on biomass energy. He has been a Doctoral Researcher at the Mauritius Sugar Industry Research Institute and Principal Researcher for an EC-funded thematic research network on sugar cane in Southern Africa. He has a Bachelor's degree in Chemical Engineering, with specialization in Sugar Engineering, and his PhD in Chemical Engineering focused on the study of cane bioenergy systems.

Semida Silveira is a Professor in Energy and Environmental Technology and Head of the division of Energy and Climate Studies at the Royal Institute of Technology, Sweden. She worked as Sustainability Expert at the Swedish Energy Agency, and as Energy and Climate Programme Manager at the Stockholm Environment Institute. Her research interests include bioenergy and climate change policy work, promotion of Swedish knowledge and technologies in development assistance and international business cooperation with corporate responsibility.

P.R.K. Sobhanbabu is President of Zenith Energy Services Private Limited, Hyderabad, India. He has worked as Programme Officer in the Energy and Environment division of Winrock International, New Delhi, and has over 11 years' experience in the fields of energy efficiency, renewable energy, and waste management. He has a Master's degree in Chemical Engineering.

Alexandre Betinardi Strapasson is a Doctoral Researcher at Imperial College London. He has a Bachelor's degree in Agronomy Engineering and a Master's degree in Energy. He worked at the Ministry of Agriculture, Livestock and Food

Supply of Brazil as Head of the Department of Sugarcane and Agroenergy and Chairman of the Chamber of Sugar and Ethanol. He also served as a UNDP Consultant at the Ministry of the Environment of Brazil.

Sergio C. Trindade is an international Consultant on sustainable energy, environment and technology businesses at SE2T International Ltd, and Director of Science and Technology at International Fuel Technology, Inc. He has served as the United Nations Assistant Secretary-General for Science and Technology and was Co-Laureate of the 2007 Nobel Peace Prize as a Member of the Intergovernmental Panel on Climate Change.

Rianto van Antwerpen, PhD, is a Senior Soil Scientist and Programme Manager at the South African Sugarcane Research Institute. He has worked for the Tobacco and Cotton Research Institute in Rustenburg. During his career he has worked on crops such as maize, peanuts, peas, wheat, cotton, tobacco, sugar cane and a range of green manure crops. He has a Master's degree in Agriculture.

José Nilton de Souza Vieira is the Deputy Director of the Department of Sugar Cane Agroenergy at the Ministry of Agriculture, Livestock and Supply in Brazil. He has a Bachelor's degree in Economics and a Master's degree in Industrial Economics and Technology. He has been a Specialist in public policy and government management at the Ministry of Planning, Budget and Management of Brazil. He has represented the Brazilian government for the Convention on Biological Diversity, Global Bioenergy Partnership, ISO and International Biofuels Forum.

Tom Walsh is the Chief Executive Officer of Renetech Bioresources Ltd. He has spent most of his international career in marketing, sales and management positions in the telecom, industrial materials and bioenergy sectors, before re-focusing to the area of renewable energy and co-founding Renetech in 2005. He is also currently the Secretary of the Irish Bioenergy Association. He has a Bachelor's degree in Commerce.

Arnaldo Walter is a Professor at the University of Campinas, Brazil and Researcher at the Brazilian Bioethanol Science and Technology Laboratory. He is an Executive Editor *of Energy* – the International Journal, is National Team Leader for the International Energy Agency task on Bioenergy Trade and is working with the ISO on a bioenergy sustainability standard.

Helen K. Watson, PhD, is a Senior Lecturer in the College of Agricultural, Earth and Environmental Sciences at the University of KwaZulu-Natal in Durban, South Africa. Her research and teaching focus on the use of remote sensing and geographic information systems to assess the influence of biophysiographic factors and the effects of land use practices and policies on land capability potential.

Jeremy Woods, PhD, is a Lecturer in Bioenergy at Imperial College London and Co-Director of the Porter Institute. His research focuses on development opportunities based on advanced bioenergy and bio-renewables. He has served as a Member of the Royal Society's working group on GHG emissions from agriculture and on biofuels, and on the advisory board of the UK Gallagher Review on the indirect land use change impacts of biofuels. He was involved in developing the international bioenergy platform and chairs the UK working group of the Scientific Committee on Problems of the Environment.

Francis D. Yamba is the Director of the Centre for Energy, Environment and Engineering, Zambia. He has a PhD degree and is a full Professor in Mechanical Engineering. He has wide experience in project management and research, having supervised and participated in over 40 projects in climate-related studies. He is among the IPCC scientists worldwide who have been awarded the 2007 Nobel Prize together with former US Vice President Al Gore.

ACKNOWLEDGEMENTS

This book resulted from more than ten years of collaboration across a network of researchers, policy analysts and consultants, as well as representatives from industry, national government agencies and international organizations. In the early years of that ten-year process, the significance of sugar cane as a renewable energy resource was appreciated by a relatively small audience outside of Brazil and a few other countries, but by the time this book was going to press, the situation is quite different and bioenergy from sugar cane is now recognized as a major success story, albeit one which still has quite some way to go to reach its potential in Africa.

The basic idea for this book arose specifically as a result of a policy research network that operated during 2002–2008: the Cane Resources Network for Southern Africa (CARENSA), which was supported as a Thematic Research Network (Contract ICA4-CT-2001-10103) with the European Commission. Additional collaborators joined the informal network that continued over the years and became co-authors. We wish to thank those organizations that hosted workshops of various kinds, including the UN Food and Agriculture Organisation (FAO), Imperial College London, the University of KwaZulu-Natal and the University of Mauritius, all of which helped to lay the groundwork for the eventual development of the book. Some partial funding for the conceptual design and editing of the book was also provided by the Swedish International Development Cooperation Agency (Sida) through the bioenergy component of the Climate for Development Programme of the Stockholm Environment Institute (SEI). Thanks are also extended to Richard Klein for his encouragement and support of this book project in his role as director of that programme during 2007–2010.

Many other persons and organizations gave their time above and beyond what was called for in order to advance the interdisciplinary analysis underlying this book and to build capacity around the role of sugar cane and other productive energy crops in contributing to sustainable development and global competitiveness in southern Africa and other regions where sugar cane is or can be grown. It is not possible to list all the persons and organizations that made some type of contribution and thus we apologise to those colleagues who could not be specifically mentioned.

Disclaimer

Neither the European Commission nor Sida nor the SEI had any role in the design or execution of the analysis and research contained in this book. The views expressed in this book are strictly those of the editors and authors and do not necessarily represent those of their respective organisations nor do they represent the views of the European Commission, the SEI or Sida.

PREFACE

Sugar cane has been used to make sugar (sucrose) in varying forms for thousands of years, but only in recent decades has its potential as a renewable energy source been exploited on any significant scale. As recently as a decade ago, in global terms it was still viewed mainly as a commodity crop for sugar, whereas today there is considerable interest in its use as a feedstock for many energy and non-energy products. Indeed, the long-term significance of biomass sources or feedstocks such as sugar cane is that they offer pathways for developing many bio-based products and materials as well as renewable energy in the form of ethanol, cogenerated electricity and biogas. Due to climatic factors, tropical biomass resources, such as sugar cane, are more productive than temperate biomass sources, offering many developing countries a comparative advantage over their northern counterparts in the emerging global market for bioenergy and bio-based materials. Improving that resource through careful breeding, agro-ecological zoning and agro-industrial infrastructure development creates the building blocks for low-carbon sustainable growth.

Long experience with sugar cane and good growing conditions in various areas dispersed throughout sub-Saharan Africa offer new opportunities to those countries and companies willing to make the significant long-term investment required. They can benefit directly from the experiences of the past few decades, including the pioneering effort to develop bioethanol from sugar cane in Brazil, the long experience in implementation of cogeneration systems in Mauritius and India and the experimentation with a wide range of bio-based products in many countries. The agricultural importance of sugar cane in Africa in combination with many African countries' vulnerability to external shocks highlights important linkages between agro-industrial development and the energy/climate forces that are driving new markets domestically and for export. Developing a highly productive and dispersed biomass resource such as sugar cane has strategic value throughout sub-Saharan Africa in addressing the triple challenge of energy insecurity, climate change and rural poverty.

This book aims to show that sugar cane is not only about agriculture, energy, climate and technology — important those these issues are — but also relates fundamentally to the choice of future pathways for economic growth and development in sub-Saharan Africa. The fossil fuel pathways followed by the developed countries are neither feasible nor desirable in the future: not only are these resources non-renewable but their socio-economic and environmental costs have begun to exceed their benefits. Highly productive and versatile biomass resources such as sugar cane can support a green economy that is better matched to local needs but also integrated into regional and global markets. In the fossil economy, poorer countries are vulnerable and powerless, whereas the bio-based economy of the future has the potential to link more closely to local resources and knowledge. Achieving this potential in Africa and elsewhere will be challenging: it will require a significant amount of well-targeted investment in infrastructure, technology deployment and human capacity, which in turn will require more effective policies and institutions at multiple scales. The physical building blocks are there, but the builders will need to be persistent and determined in order to succeed.

Francis X. Johnson and Vikram Seebaluck

ACRONYMS AND ABBREVIATIONS

ACP	African Caribbean Pacific
ACRU	Agricultural Catchments Research Unit
AEZ	agro-ecological zoning
BIG-CC	bagasse integrated gasification – combined cycle
BOD	biological oxygen demand
CDM	Clean Development Mechanism
CEST	condensing extraction steam turbine
CMC	carboxymethyl cellulose
CO_2eq	CO_2 equivalent
COB	chemical oxygen demand
CTCS	Centre Technique des Cultures Sucrières (Morocco)
CTL	cane tops and leaves
CV	coefficient of variation
DM	dry matter
DNA	deoxyribonucleic acid
DSCT	dehydrated sugar cane tops
EC	European Commission
EEZ	ecological-economic zoning
EtOH	ethanol
EU	European Union
FAO	Food and Agricultural Organization
FDA	Food Development Association
FFV	flexible-fuel vehicle
GCV	gross calorific value
GHG	greenhouse gas
GL	Gay-Lussac
GM	genetically modified

ha	hectare
HACCP	hazard analysis and critical control points
HTM	high-test molasses
ICUMSA	International Commission for Uniform Methods of Sugar Analysis
IGBP	International Geosphere-Biosphere Programme
IHTM	integral high-test molasses
IPP	independent power producer
ISO (1)	International Organization for Standardization
ISO (2)	International Sugar Organization
K_2O	Potassium (as fertilizer)
KESREF	Kenya Sugar Research Foundation
KSC	Kenana Sugar Company (Sudan)
kWh	kilowatt hour
L	litres
LCA	life-cycle assessment
LDCs	least developed countries
LHV	lower heating value
MAR	mean annual rainfall
Mbp	megabase pairs
MEG	monoethyleneglycol
$MJ\,kg^{-1}$	mega joules per kilogram
MJ	Mega Joule
MSIRI	Mauritius Sugar Industry Research Institute
N	nitrogen
P_2O_5	phosphorous (as fertilizer)
PCR	polymerase chain reaction
PGR	plant growth regulators
PHB	polyhydroxybutyrate
R$	Brazilian reals
SAR	sugar cane agricultural residues
SASRI	South African Sugar Research Institute
SCRI	Sugar Crops Research Institute, Giza (Egypt)
SMRI	Sugar Milling Research Institute
SUCEST	Sugar cane expressed sequence tags
t	tonne
tc	tonne of cane
US	United States
WISBEN	West Indies Sugar Cane Breeding and Evaluation Network
ZSAES	Zimbabwe Sugar Association Experiment Station

1

RENEWABLE RESOURCES FROM SUGAR CANE

The energy, environment and development context for Africa

Francis X. Johnson and Bothwell Batidzirai

Introduction

It is now widely accepted that the global consumption of non-renewable resources at the current rate is unsustainable and is likely to lead to serious social, economic and environmental problems. In this context, the desire to achieve continued human well being and meet growing demand for natural resources is an important challenge for the current generation. In particular, the twin challenges of energy insecurity and the threat of climate change need to be addressed through many approaches and in all world regions. For some countries and regions, bioenergy from sugar cane offers an option that is attractive in socio-economic and environmental terms as it draws on biophysical or climatic advantages and brings new economic opportunities to rural areas, while also facilitating production of the renewable resources that are needed to support a sustainable economy. Highly productive plants such as sugar cane are expected to play an important role in a knowledge-based bioeconomy of the future, being part of the portfolio of options to substitute non-renewable energy resources that are becoming costly and environmentally untenable.

In many respects, Africa represents a key testing ground for the future bioeconomy as it lies at the heart of the "biomass-poverty belt"; i.e. the tropical and subtropical regions of the world where extreme poverty coincides with great bioenergy potential. Africa is currently highly dependent on inefficient uses of biomass that deliver low-quality energy services (IEA, 2010), but at the same time has the greatest potential for bioenergy development of any world region (Smeets *et al.*, 2007). The exploitation of this potential can bring significant socio-economic benefits, especially to rural areas where jobs would be highly valued, as well as helping to stimulate advances in the agricultural and forestry sectors where better resource management and higher productivity are sorely needed (UNDESA, 2007). Among the various bioenergy options, the long experience and the considerable future potential for sugar cane in Africa makes it a valuable case study.

In this chapter, the overall background and context is provided for the role of sugar cane in Africa to support renewable energy expansion, economic growth and sustainable development. The key phases or sectoral dimensions are reviewed, including agriculture, industry, markets, impacts, climate, strategic issues and international comparisons. As sugar cane is a global industry, in some respects African regions are comparable to other world regions in terms of industrial structure and financing. In other respects, the context becomes highly location-specific, such as with agricultural conditions, or nationally specific, in terms of policies and institutions. A final section of this chapter summarizes the overall framework of analysis and provides a roadmap for the other chapters in the book.

Biomass conversion potential

The raw potential of biomass for energy depends primarily on photosynthetic efficiency, which varies considerably across plant species and varieties. There are two main photosynthetic pathways – C3 and C4, whose names derive from the fact that the first products of photosynthesis are the formation of 3-carbon and 4-carbon organic acids, respectively (El Bassam, 2010). The C4 pathway operates optimally at higher temperatures and is characterized by higher water use efficiency, greater CO_2 exchange, and greater biomass yields; sugar cane, sorghum and maize are the most relevant C4 crops. The C3 pathway plant species include many food crops such as potato, rye, wheat, barley, rice and cassava, but also many oil-bearing crops such as soya, sunflower and rape.

A summary of ecological requirements and yield-related information is provided in Table 1.1 for selected crops used for biofuel (ethanol) production, which are also used for production of food and fibre. The actual performance of the crops will vary considerably depending on factors such as the varieties chosen, location, climate and management practices. A full comparison of yields and other key parameters for energy crops is rather complex and location-specific and thus comparisons are best assessed for a specific region or agro-ecological zone (see Chapter 3 for a discussion of agro-ecological zones). As a C4 plant, sugar cane's geographical locus in tropical and subtropical climates endows it with one of the highest photosynthetic efficiencies among commercialized crops.

Although sugar cane is highly efficient in photosynthetic terms, it should also be noted that it is a rather demanding crop, being sensitive to frost and drought and having a high rainfall requirement; such characteristics are therefore among the key objectives of breeding programmes (Chapter 2). The high resource demands of sugar cane and the potential for water scarcity and other climate-related changes in some world regions has contributed to interest in alternative or complementary crops that can also provide food, fuel and fibre, such as sweet sorghum. The broader geographical suitability of sweet sorghum suggests that it may be preferred in the

Table 1.1 Characteristics and ecological requirements of selected feedstocks/crops

	Type	Photosynthetic pathway	Photosynthetic efficiency	Latitude range	Optimal temperature range[1]	Frost tolerance	Drought tolerance	Rainfall requirement[2]
Sugar crops								
Sugar cane	Perennial	C_4	Very high	37°N–31°S	24–26°C	Poor	Poor	High
Sweet sorghum	Annual	C_4	High	Adapted widely	27–30°C	Poor	Excellent	Low–moderate
Starch crops								
Cassava	Perennial	C_4	Moderate	30°N–30°S	25–29°C	Poor	Good	Moderate
Maize	Annual	C_4	Moderate–high	Adapted widely	18–33°C	Poor	Poor–moderate	Moderate
Wheat	Annual	C_3	Low–moderate	Adapted widely	15–23°C	Good	Moderate	Moderate
Lignocellulosic biomass								
Switchgrass	Perennial	C_4	Low–moderate	Adapted widely	20–32°C	Good	Excellent	Low–moderate
Miscanthus	Perennial	C_4	Moderate–high	Does well in temperate regions	Low temperatures	Excellent	Poor	Moderate

Source: adapted from El Bassam, 2010.

Notes

1 At most productive stage of growth.

2 Annual requirement, regardless of differences in growing periods and seasonal requirements.

future in locations where it is determined that the growing conditions for sugar cane are too close to borderline requirements (see discussion on sweet sorghum below).

Agricultural parameters

The sugar cane plant is thought to have originated in South and South East Asia more than 5,000 years ago, with its cultivation eventually spreading across all six continents. Sugar cane (*Saccharum officinarum*) is a genus of 6 to 37 species and a member of the grass or Poaceae family, which concentrates sucrose in its stalks. Its excellent capacity for energy binding with respect to surface area and time is a key factor in making it one of the most economically important crops in the world (El Bassam, 2010). It grows only in tropical and subtropical regions, with the result that over 90 per cent of sugar cane is grown in the developing world. Sugar cane re-emerges when cut, thus enabling several harvests to be obtained from the resulting "ratoon" crop. The number of required replantings increases with mechanization; since harvesting is still largely manual in much of Africa, the number of replantings is generally lower than elsewhere in the world, which saves considerable costs.

The potential to expand cane production in Africa depends on alternative land uses, vegetative cycles, soil properties and climate conditions, as well as the potential effect of such expansion on the availability of water and other key resource inputs. Under such considerations, the potential for expansion is quite limited in those countries which are currently the major producers – namely Mauritius, South Africa, Swaziland and Zimbabwe – that together currently account for about half of sub-Saharan African sugar cane production. Other countries such as Angola and the Democratic Republic of Congo that do have a considerable amount of suitable land and climate seem unlikely to expand significantly in the near term due to political instability.

The countries that appear to have considerable potential for expansion in Southern Africa include Malawi, Mozambique and Zambia (Watson, 2010). There is a fairly large region that is quite suitable in Central Africa but since much of this land is under forest cover, for reasons of biodiversity and carbon content it may be inappropriate for conversion. There are some suitable regions in West Africa and East Africa but water availability may limit the sustainability of expansion in some of those regions. It is also important to note that in addition to suitability, the *availability* of land must be assessed in relation to existing and potential uses, particularly in relation to food production. Application of detailed methodologies for evaluating bioenergy in relation to food security have shown that it is possible to match crops and management approaches to local needs, and thereby identify positive food–fuel synergies and avoid conflicts (FAO, 2010). A significant sugar cane project under way in Sierra Leone has demonstrated good practice methods to address both suitability and availability, thus illustrating development potential (see Chapter 17 for a brief discussion of this project).

In Africa, sugar cane is generally burnt before harvest to remove the extraneous residues, known as "cane trash". Harvesting and delivery systems are semi-mechanized in most African countries, through the use of loaders and related equipment (see Chapter 4). In other parts of the world, the practice of burning cane has been ended, generally in combination with a switch to mechanical harvesting, thereby creating the possibility to expand the cogeneration of heat and electricity by using cane trash. Mechanical harvesting of green cane results in lower sucrose recovery in the sugar factory, but on the other hand enhances energy generation in the boiler house by making additional fuel available (see Chapter 5). The growth of bioenergy markets and environmental concerns will eventually stimulate a similar transition in African regions.

Industrial configurations

Sugar cane processing infrastructure includes the sugar factory set-up, the configuration of the ethanol distillery (if any), the cogeneration plant (if any) and any other co-product facilities. Most sugar factories use a roller mill and associated equipment, which crush the cane and extract the juice, or alternatively they can use a diffuser where extraction is based more on "washing" the cane to release sucrose. The cogeneration plant uses bagasse from the sugar factory as its primary input fuel; bagasse is the fibrous residue left after juice extraction from cane. A cogeneration plant can be installed if the sugar factory processing capacity is adequate and efficient enough to generate sufficient surplus bagasse and if there are markets for surplus heat and electricity (Johnson *et al.*, 2007). As sugar cane becomes more valuable for its energy in comparison with sugar, there will be greater economic incentives to optimize sugar factories in energy terms, and this will enhance the opportunities for expanding renewable energy production from sugar cane (Chapter 5).

As the steam requirements are significant at sugar factories, improvements in steam economy at sugar factories can be obtained through higher calorific values for bagasse, energy conservation measures in the sugar mill department, boiling house and boiler house, optimization of boiler draft system, usage of instrumentation and control systems, and proper selection of boilers and auxiliaries. This enables surplus electricity to be generated. Simple energy conservation measures include replacement of turbo drives used for cane preparation and milling with hydraulic or electric drives; reduction of process steam consumption to 42–45 per cent by modifying the juice-heating and evaporation system; increasing the efficiency of boiler operations to a minimum of 70 per cent; introduction of more controls in power plant operations; and use of bagasse driers (Seebaluck *et al.*, 2007).

The current state-of-the-art system for bagasse cogeneration is the condensing extraction steam turbine (CEST), which can produce much more surplus electricity than the back pressure turbines commonly used at many existing factories in Africa. CEST systems are used extensively in India, Mauritius and Réunion and to some extent in Brazil, where they allow export of surplus electricity to the grid during the harvest season; a set of best-practice benchmarks is shown in Table 1.2. The surplus electricity obtainable is in some cases ten times what might be available from existing factory configurations that use back pressure turbines (Seebaluck *et al.*, 2008).

Table 1.2 Best-practice electricity production surplus using bagasse with CEST systems

Country	Pressure	Temperature	Surplus exportable electricity
Mauritius	82 bars	525 °C	130–140 kWh/tonne cane
India	67 bars	495 °C	90–120 kWh/tonne cane
	87 bars	515 °C	130–140 kWh/tonne cane
Brazil	67 bars	480 °C	40–60 kWh/tonne cane

Source: Seebaluck *et al.*, 2010.

Ethanol can be made directly from sugar cane juice or from molasses, a major by-product of sugar production. Fuel ethanol can be blended with gasoline at levels varying from 10 to 25 per cent with little or no effect on fuel economy and with relatively minor adjustments, if any (depending on the share) to the engine. Since typical ignition-spark engines are optimized for petrol, the substitution of ethanol results in lower output, generally between 65 and 80 per cent. The smaller scale of sugar production in African countries is among the reasons why it has been difficult to establish cost-competitive bioethanol production. The experience in Brazil suggests that economies of scale require a cane supply of between one million and two million tonnes per year; most African producers (with the exception of South Africa, Swaziland and Zimbabwe) operate near the lower end of this range (see also Chapters 5 and 6).

Although ethanol and cogenerated electricity are currently the most economically significant co-products from sugar cane, there are also a wide variety of other co-products whose commercial value and market share is likely to grow. Particle-board, corrugated boxes, and furfural are some potential fibre-based products. The sucrose stream of the cane also offers a wide range of co-product options, including various organic acids, monosodium glutamate, xanthan, dextran gums, yeasts and bioplastics (see Chapter 7). Through international investment and regional cooperation, the diversity and flexibility of sugar cane as a renewable resource can be employed as a powerful tool in support of sustainable industrial development, economic growth, and poverty reduction in sub-Saharan Africa.

Market scenarios and strategies

There are some key market issues that will impact the future direction of the sugar industry in Africa and its potential for becoming a major producer of renewable energy. The changes in the sugar and ethanol industries through reforms in the EU, USA and WTO will likely lead to some consolidation and geographical shifts; the lower sugar prices resulting from the removal of preferential markets will cause shifts to the regions with better conditions and higher agricultural productivity. At the same time, rising oil prices, energy security and climate change are providing incentives for expanded bioenergy production in the sugar sector. The emerging nexus between sugar policy reform and incentives for bioenergy production is important for sugar producers worldwide and not just in Africa (see Chapter 8).

Demand for electricity is expected to grow considerably in some areas in Africa, with growth rates of 3–5 per cent annually, and the need for new power capacity will become more acute by 2015. The sale of surplus electricity from sugar factories in the region would require installation of CEST systems as currently used in India, Mauritius and Brazil. Where economies of scale are reached and there are policies and regulations in place for independent power production, electricity cogenerated by CEST systems at sugar factories can be available at a reasonable price, generally in the range of 6–8 US cents per kWh (Johnson et al., 2007). Power from reliable and continuous sources would add diversity to the fuel mix for electric power in Southern Africa, which is currently overly dependent on hydro, or coal in the case of South Africa.

The potential for bioethanol production in those African countries with suitable land for sugar cane exceeds what would be required domestically for blending in all countries except South Africa, and thus indicates significant opportunities for export markets. The current global ethanol market is one in which Brazil exerts extreme market power, as the world's only major swing producer between sugar and ethanol. The expansion of the ethanol market could help to create many medium-size players and make the industry more competitive in the long term. The geographical diversification will contribute to a transition towards ethanol as a global bioenergy resource rather than a regional one (Johnson and Matsika, 2006).

The sugar industry can benefit from carbon finance under the CDM (Clean Development Mechanism) as well as other GHG emission markets or initiatives such as the Voluntary Carbon Standard. As of January 2010, there were 80 bagasse cogeneration projects registered under the CDM and another 72 at validation stage (UNEP, 2010). Ethanol, biogas and other projects associated with sugar cane can also benefit under the CDM, although additionality has to be proven, which is somewhat more difficult in the case of liquid and gaseous fuels due to low demand in many developing countries. To exploit carbon credit potential, the sugar industry must work within the regulatory environment for the carbon market and its associated institutions and procedures.

African countries can benefit from the experiences of Mauritius, India and Brazil with respect to electricity and ethanol from sugar cane through South-South technology transfer platforms (see Chapter 16). Significant investment will be needed to upgrade facilities and develop markets for ethanol, bagasse cogeneration and other sugar cane co-products. In addition to the high costs of the facilities, further costs for distribution, transport and storage would also be incurred due to the poor infrastructure in most regions of Africa. In spite of the difficult investment climate, there are now many different financial vehicles and ways of packaging new projects; consequently, renewable energy from a highly efficient crop such as sugar cane will entail lower risk and should attract investment from a broader range of sources (see Chapter 17).

Export of bioethanol to the US or the EU faces some trade barriers in the form of preferential markets, agricultural tariffs, quality requirements and difficulties in marketing channels (see Chapter 8). Although these barriers are being reduced, it will take time to implement reforms fully, so that least developed countries (LDCs) in Africa that have significant capacity will need to follow a dual strategy: developing domestic and export markets side by side (see Chapter 10). Since bioethanol is a finished product that does not need further refining (except for dehydration in some cases), there are little or no production costs associated with switching between export and domestic markets. For other energy and non-energy products from sugar cane, there may also be future export opportunities as international markets develop for bio-based products such as green chemicals (see Chapter 7).

Socio-economic impacts

The potential socio-economic impacts of medium or large-scale sugar cane and bioenergy initiatives can be quite significant, especially in rural areas where unemployment is high and infrastructure is poor (see Chapter 11). The cyclical nature of large-scale bioenergy feedstock harvesting can in some cases lead to reliance on temporary and/or seasonal workers, which might create other socio-economic and environmental pressures. Socio-economic impacts related to sugar cane and bioenergy can be broadly characterized as including the following aspects:

- land tenure and displacement risk for local residents;
- food security, income generation and distribution;
- employment, labour conditions and business structures and skills;
- socio-cultural and gender issues, including traditional lifestyles and governance systems;
- increased energy availability and access; and
- economics and infrastructure, including cross-sector effects.

Most of these issues are related to the agricultural side of the industry and are thus similar to the impacts encountered in most medium or large-scale agricultural initiatives. A key difference relates to the case when there is significant foreign investment and/or when the bioenergy products are intended solely or primarily for export. In such cases, additional care must be taken to avoid social conflicts over land rights and the displacement of local populations without appropriate compensation. Equity and sustainability conditions can be placed on foreign investment schemes (see Chapters 12 and 17). Even where land rights are formally granted, there can be differences as to how the government interprets definitions such as "idle" land and "productive use" requirements (Cotula et al., 2008). Clearer definitions will often be required to avoid allocation of lands on which local user groups depend for livelihoods (Nhantumbo and Salomão, 2010).

A key concern regarding the rapid expansion of large-scale bioenergy production is its potential impact on food security. There are fears that agro-resources could be shifted from food production to fuel production. However, the potential impacts of large-scale bioenergy production on food security are more complex, and include potential positive and negative feedbacks, which need proper policy guidance in implementation of the projects. Food security is generally measured in terms of food availability, access, utilization and stability. Based on these four issues, the socio-economic impacts of bioenergy on food security manifests in several ways (positive or negative), as follows:

1. changes in income levels for farmers and other affected groups;
2. changes in prices of other food crops or goods that affect food prices;
3. competition for inputs such as including land, water and nutrients;
4. land use prices/pressures/changing value for land;
5. social/rural stability; and
6. physical impacts on natural resources at the project site.

Whether the impacts are positive or negative depends on the reference group. Higher prices for food and non-food crops as well as increased value of the land itself are beneficial for farmers, but create additional costs for urban dwellers purchasing food. Nor is competition among uses and resources necessarily problematic, since competition can improve efficiency, such as in the use of water, fertilizers and other inputs.

Initially, the number of jobs would increase with sugar cane and bioenergy expansion, but as operations became more mechanized, the number of jobs on the agricultural side of the sugar cane sector would fall. However, mechanization would improve the overall quality of jobs and reduce the number of seasonal workers in favour of a more stable workforce that can help to build communities.

There can thus be some tension between the large-scale, centralized approach favoured by governments and large business vs. the possible benefits that could accrue to the rural poor from promoting a smaller-scale, decentralized approach to bioenergy and agriculture. In the case of sugar cane, the small-scale options can be implemented in two main ways. One is through outgrowers or contract farmers, who supply cane to the factory and are rewarded based on some objective measures of quality along with quantity. The other is for bioenergy or biofuel production itself to be geared towards smaller-scale markets, such as the use of bioenergy in decentralized energy production for heat and power (Lloyd and Visagie, 2007), or the promotion of household cooking with bioethanol (Stokes and Ebbeson, 2005). Where sugar cane is to be grown commercially for export in the form of sugar, ethanol and other products, the first option is more relevant and will require good organizational capacity for small farmers' groups.

There are also gender impacts of promoting small-scale options. The ratio of women to men employed by − or running − small-scale agricultural production appears to be greater than in large-scale production. According to previous feasibility studies, outgrower schemes can offer new opportunities for women (Cornland et al., 2001). Similarly, in the case of small-scale production of potable ethanol in the peri-urban areas of Harare, there was pronounced participation of women (see Chapter 11). This greater role for women was also exhibited during field trials of smallholder sweet sorghum production in Zambia (Woods et al., 2008).

The sugar cane agro-industry is mature and has fairly significant economies of scale, which in turn imply a large-scale, high-capital cost infrastructure. Such capital-intensive infrastructure has been costly to upgrade, particularly during the falling world sugar prices in the 1980s and 1990s. The region's industry is therefore saddled with some outdated factories that need improved efficiency so that surplus energy can be produced and sold. The investments in the physical infrastructure would have to be matched by various investments in soft infrastructure (e.g. education, health, training) and thus socio-economic benefits would accrue according to how well these investments proceed (Johnson et al., 2007).

Environmental impact

As with any agro-industrial activity, sugar cane production can have significant impacts on air, water, soil, flora, fauna and human activity as well as on the global climate. These impacts arise throughout crop production and industrial processing. As with most environmental impact assessments, local conditions will be quite important, since ecosystem functions, biodiversity, and other localized impacts vary widely. With the maturity of the sugar and sugar cane sector, the industrial side has become quite efficient, and therefore the impacts – as well as the areas for future improvements – are concentrated heavily on the agricultural side.

In general, the environmental impacts of new production are much greater than the marginal impacts due to modifications in the product mix (e.g. expanded production of bioethanol) based on existing sugar cane production. Extensive expansion of sugar cane entails a significant amount of land preparation and infrastructure provision, which is required for a permanent crop such as sugar cane. Land impacts in some African regions, especially Southern Africa, would be mitigated considerably by the low population density and the significant amount of pastureland available for agricultural expansion.

There are two types of impacts on the industrial side that are related to energy production and can be significant. First, the large wastewater stream resulting from ethanol production, known as vinasse or stillage, can contaminate water supplies if not discarded properly, or if used in the further production of substances like biogas or biofertilizers that can find their way back into the soil. Disposal regulations now exist in most countries, but their enforcement level varies. Second, the overall efficiency of the factory in terms of energy, water and other inputs determines the overall direction of sustainability in resource requirements, which is almost always negatively correlated with the age of the factory. An older, inefficient factory will use up much more bagasse, water, etcetera, and can result in much higher economic and environmental costs (Johnson et al., 2007).

The energy balance and climate impacts associated with sugar cane and bioenergy must be assessed on a life-cycle basis, and in relation to the baseline consumption of energy and materials. The comparison in the case of a sugar cane complex would include the associated substitution of ethanol for gasoline, cogenerated electricity for grid supply and any other products (e.g. biogas, green chemicals) in relation to the equivalent baseline (typically non-renewable) consumption. Consequently, energy balances and GHG balances will be location and facility-specific.

Among biofuel feedstocks that have been assessed for international standards, ethanol from sugar cane consistently ranks as the best performing among those commercially available, with reductions in GHG emissions of 70 per cent or more (EC, 2009; de Vries et al., 2010). There can be important location-specific issues that must be addressed in the GHG balances; for example, in the case of South Africa, irrigated sugar cane has higher yields, but has a much worse GHG balance because the electricity used for irrigation comes from coal (see Chapter 13). Similarly, unburnt cane frees up significant biomass for cogeneration, and where this can be substituted for coal, as is the case in South Africa, the GHG savings are significant.

The sweet sorghum alternative

An alternative crop that may serve as a valuable complement to sugar cane in expansion strategies is sweet sorghum, which is a sorghum variety with a higher relative sugar content than grain sorghum, making it suitable for ethanol production as well as being a source of fibre. It has a much shorter growing period than sugar cane – three or four months – and has lower water requirements and better tolerance of waterlogged soils (Dajue and Griffee, 2005). Sweet sorghum is more resistant to drought than most other crops on account of its superior water absorption and water-loss regulation abilities (Dercas and Liakatas, 1999). The crop grows successfully on a wide range of soils including heavy clays, calcareous and organic soils. It tolerates a pH between 5.5 and 8.5, as well as some degree of salinity (Dalianis et al., 2004). It does not have good properties for crystalline sugar production and consequently is generally not used for raw sugar production.

Sweet sorghum thrives in wet-dry tropical, subtropical, semi-arid and warm temperate regions, over a wider range of habitats than sugar cane, on account of its greater tolerance to both higher and lower temperatures, greater resistance to drought, and less particular soil requirements. Sweet sorghum achieves optimal growth where mean annual temperatures range from 27°C to 35°C. Growth is arrested below 8°C and above 40°C. While the average rainfall in areas where the crop is grown ranges from 450mm to 1,000mm per year, it achieves optimal growth where the mean annual rainfall is 800mm and is normally irrigated where the mean annual rainfall is less than 600mm (Dajue and Griffee, 2005).

A sweet sorghum crop is established from seed. The seed can be machine sown. Between 6kg and 8kg of seed is needed to establish a hectare of sweet sorghum (Dajue and Griffee, 2005). Under optimal conditions the seeds germinate and develop a root-and-canopy system completely covering the soil within a month. The crop is not photoperiod sensitive and reaches maturity in three to five months. During growth the plants accumulate sugars primarily in their stems. Towards maturity, the relative level of sucrose compared to total sugars increases. At maturity sucrose constitutes more than 70 per cent of the sugars (Woods, 2001). Since it is grown from seeds rather than plantings, it is a more flexible alternative than sugar cane and thus more suited to small-scale farming. There is little experience with sweet sorghum in Africa, but in Asia there has been considerable effort at commercializing the crop for joint production of fuel, food and fibre.

Sweet sorghum is more drought resistant, has lower rainfall requirements, and operates under a broader temperature range. Given the risks associated with climate change, there will be some clear advantages for sweet sorghum in water-scarce regions where expansion of sugar cane might threaten water needs for other uses and sectors. At the same time, given the long experience and economies of scale that have been developed for sugar cane, it seems likely that sweet sorghum would complement rather than replace sugar cane as an efficient source of fuel and fibre.

Synthesis and overview

Evaluating the suitability of land for sugar cane requires detailed geographical analysis, and the wide availability of specialized software and assessment approaches for soils and climates has facilitated better matching of crops to specific locations (Chapter 4). Agro-ecological zoning for sugar cane has been used extensively in Brazil and the same techniques could be applied in those African regions where data is available (Chapter 3). The desired plant characteristics will need to be adapted in the future to meet changing economic conditions as well as climatic changes that will affect temperatures, rainfall, pest incidence and periods of drought. Breeding of new varieties can be aimed at addressing such demands, although it must be recognized as a long process requiring many years of research and testing (Chapter 2). The harvesting and delivery of sugar cane to the factory is undergoing many changes as cane burning is phased out, mechanical harvesting is increasing and new transportation methods are devised (Chapter 4).

Sugar cane is arguably the most commercially important energy crop in the world, on top of its importance as the main source of crystalline sugar. Global sugar markets are undergoing considerable change as support schemes and preferential markets are gradually phased out (Chapter 8). In addition to ethanol and sugar, the existing or potential co-products of sugar cane are many and varied, ranging from cogenerated heat and electricity to bio-chemicals and agro-industrial feedstocks (Chapter 7). Indeed, a key attraction of sugar cane is precisely that it offers a rich array of energy and non-energy products and services. Cogeneration of heat and electricity from the fibrous residue of the sugar cane plant (bagasse) and field residues (cane trash) has already become commercially important in countries such as Brazil, India and Mauritius and there is scope for replication in most sugar cane-producing countries (Chapter 5). Supporting policies and the facilitation of international trade will open up new markets nationally and internationally (Chapters 9 and 10).

Ethanol from sugar cane has an excellent energy and GHG balance, unlike the use of corn or wheat as feedstock in the USA and Europe, respectively. The high efficiency results in lower costs and lower environmental impacts, especially when all the co-products are considered together (Chapter 6). In Brazil, ethanol is no longer directly subsidized and can compete economically with oil. However, even with the expanding production in Brazil, the supply of ethanol is insufficient to keep up with world demand. The expansion of ethanol production from sugar cane in other parts of the world therefore offers a variety of economic and environmental benefits. For oil-importing countries in Africa, the advantages are even more obvious, as an expansion in the production and use of ethanol can save foreign exchange, improve energy security, support global efforts on climate change and contribute to rural and agricultural development (Chapters 13 and 14).

There are naturally impacts and costs that must be weighed against the aforementioned benefits of expanding sugar cane as a renewable resource. As with any agro-industrial activity, a variety of environmental impacts arise, such as the potential for water pollution, the effects of burning green sugar cane before harvesting and the ecological drawbacks associated with a monoculture crop (Chapter 11). Socio-economic conflicts can also arise due to the capital-intensive nature of sugar cane, which tends to concentrate production and wealth without appropriate institutions and technical support. The use of manual harvesting entails risks for workers due to the arduous tasks involved in agricultural labour in warm climates. Recent sustainability schemes therefore may include social as well as environmental criteria (Chapter 12).

In order for sugar cane to become viable as both an energy and non-energy crop on a large scale in Africa, both financing and technological expertise must be more widely available geographically and across the entire life-cycle of production. Least developed countries (LDCs) generally face special challenges and opportunities in developing a sophisticated agro-industry like sugar cane. The experiences of LDCs in Asia and Latin America carry lessons for Africa, and furthermore they are all part of the new global development in biofuels that recognizes the higher value added in economic and environmental terms from substituting domestic for imported energy (Chapter 15). Financing is often particularly challenging for agricultural investments in LDCs and the financial packages will need to be tailored to the higher-risk scenarios and the common infrastructure gaps (Chapter 17). Similarly, the higher-risk environment for the technology transfer process and the management of technology requires careful integration of expertise with technical inputs and market channels, regardless of whether the markets are regional, national or international (Chapter 16).

References

Cornland, D.W., Johnson, F.X., Yamba, F., Chidumayo, E.N., Morales, M.M., Kalumiana, O. and Mtonga-Chidumayo, S.B. (2001) *Sugar Cane Resources for Sustainable Development: A Case Study in Luena, Zambia*, Stockholm Environment Institute, Stockholm.

Cotula, L., Dyer, N. and Vermeulen, S. (2008) "Fuelling Exclusion? The Biofuels Boom and Poor People's Access to Land", IIED, London. http://pubs.iied.org/12551IIED.html?k=fuelling exclusion.

Dajue, L. and Griffee, P. (2005) "Sweet Sorghum – the Old and the New – with Special Reference to China", unpublished report prepared for CARENSA.

Dalianis, C., Konstantinos, S. and Dercas, N. (2004) "An Assessment of the Suitability of Southern Africa for Sweet Sorghum Production", unpublished report prepared for CARENSA.

Dercas, N. and Liakatas, A. (1999) "Sorghum water loss in relation to water treatment", *Water Resources Management*, vol. 13, no. 1 (February), pp. 39–57.

de Vries, S.C., van de Ven, G.W.J., van Ittersum, M.K. and Giller, K.E. (2010) "Resource use efficiency and environmental performance of nine major biofuel crops, processed by first-generation conversion techniques", *Biomass and Bioenergy*, vol. 34, no. 5, pp. 588–601.

EC (2009) Directive 2009/28/EC of the European Parliament and of the Council of 23 April 2009 on the Promotion of the Use of Energy from Renewable Sources and Amending and Subsequently Repealing Directives 2001/77/EC and 2003/30/EC.

El Bassam, N. (2010) *Handbook of Bioenergy Crops: A Complete Reference to Species, Development and Applications*, Earthscan, London.

FAO (2010) "Bioenergy and Food Security – The BEFS Analytical Framework", *Environment and Natural Resources Series*, no 16, Rome. www.fao.org/docrep/013/i1968e/i1968e00.htm.

Hoogwijk, M.M. (2004) "On the Global and Regional Potential of Renewable Energy Sources", PhD thesis (NWS-E-2004-2), University of Utrecht, The Netherlands.

IEA (2010) *World Energy Outlook 2010*, International Energy Agency (IEA), Paris.

Johnson, F.X. and Matsika, E. (2006) "Bio-energy trade and regional development: The case of bio-ethanol in Southern Africa", *Energy for Sustainable Development*, 10, pp. 42–54.

Johnson, F.X., Seebaluck, V., Watson, H. and Woods, J. (2007) "Renewable resources for industrial development and export diversification: The case of bioenergy from sugar cane in southern Africa", *African Development Perspectives Yearbook*, University of Bremen Institute for World Economics and International Management, Lit-Verlag, Münster.

Lloyd, P.J. and Visagie, V.M. (2007) "The testing of gel stoves and their comparison to alternative cooking fuels", *Proceedings of the International Conference on the Domestic Use of Energy*, 11–12 (April), pp. 59–64, Cape Town.

Nhantumbo, I. and Salomão, A. (2010) "Biofuels, Land Access and Rural Livelihoods in Mozambique", IIED, London. http://pubs.iied.org/12563IIED.html?k=biofuels land access.

Seebaluck, V., Leal, M.R.L.V., Rosillo-Calle, F., Sobhanbabu, P.R.K. and Johnson, F.X. (2007) "Sugar Cane Bagasse Cogeneration as a Renewable Energy Resource for Southern Africa", *Proceedings of the 3rd International Green Energy Conference*, pp. 658–670.

Seebaluck, V., Mohee, R., Sobhanbabu, P.R.K., Rosillo-Calle, F., Leal, M.R.L.V. and Johnson, F.X. (2008) "Bioenergy for Sustainable Development and Global Competitiveness: The Case of Sugar Cane in Southern Africa, Thematic Report 2: Industry", Stockholm Environment Institute, Stockholm.

Smeets, E.M.W., Faaij, A.P.C., Lewandowski, I.M. and Turkenburg, W.C. (2007) "A bottom-up assessment and review of global bio-energy potentials to 2050", *Progress in Energy and Combustion Science*, 33, pp. 56–106.

Stokes, H. and Ebbeson, B. (2005) "Project Gaia: Commercializing a new stove and new fuel in Africa", *Boiling Point*, 50.

UNDESA (2007) "Small-Scale Production and Use of Liquid Biofuels in Sub-Saharan Africa: Perspectives for Sustainable Development", United Nations Department of Economic and Social Affairs (UNDESA), New York.

UNEP (2010) UNEP Risoe CDM/JI Pipeline Analysis and Database. http://cdmpipeline.org/.

Watson, H.K. (2010) "Potential to expand sustainable bioenergy from sugarcane in southern Africa", *Energy Policy*, doi:10.1016/j.enpol.2010.07.035.

Woods, J. (2001) "The potential for energy production using sweet sorghum in southern Africa", *Energy for Sustainable Development*, 10, pp. 31–38.

Woods, J., Mapako, M., Farioli, F., Bocci, E., Zuccari, F., Diaz-Chavez, R. and Johnson, F.X. (2008) "The Impacts of Exploiting the Sugar Industry Bioenergy Potential in Southern Africa – Options for Sustainable Development, Thematic Report 4 on Social and Environmental Impacts", Cane Resources Network for Southern Africa (CARENSA).

PART I

Agriculture

2

SUGAR CANE PHYSIOLOGY, BREEDING AND BIOTECHNOLOGY

Kishore Ramdoyal, Asha Dookun-Saumtally and Goolam Houssen Badaloo

Background

Sugar cane, an important industrial crop for the tropical and subtropical regions of the world, is known for its high photosynthetic efficiency. It accounts for about 80 per cent of the sugar produced in the world and about 8.5 million tonnes annually in Africa. Genetic improvement of sugar cane has led to an increase in both cane and sugar yields worldwide, and has enabled expansion of the crop into a larger range of environments. Much effort has been directed towards the development of new sugar cane varieties that meet the specific requirements of growers as well as addressing emerging uses of the crop. Future advances will require a synergy across several fields of research, including traditional breeding, genetics, physiology and biotechnology. The goal is to tap the enormous potential of the crop in both food and energy production, thereby transforming conventional sugar factories into sugar cane processing complexes for multiple products, especially electricity and ethanol, thus helping to sustain the long-term viability of the industry. The potential of using sugar cane as a biofactory for the accumulation of value-added products is yet to be fully exploited. A combination of conventional and molecular breeding approaches can enhance breeding efficiency for introgression breeding for biomass options. This chapter reviews the crop's physiology, breeding and selection principles for developing new and improved varieties for sugar and biomass and the application of biotechnology for enhancing breeding programmes.

Introduction

Sugar cane (*Saccharum* spp. Linnaeus, Poaceae) holds a prominent place among tropical plants and together with sugar beet (*Beta vulgaris* spp. Linnaeus, Amaranthaceae) they form the basis of the sugar industry worldwide. Sugar is produced in more than 100 countries, with global production of 174 million tonnes in 2008 (FAOSTAT, 2008); sugar cane accounts for about 80 per cent and sugar beet for about 20 per cent of total sugar produced (see Chapter 8). Africa accounted for about 8.54 million tonnes of sugar produced from about 1.48 million hectares harvested (Table 2.1). Cane yield varies widely depending on geographical location, husbandry practices, soil fertility, water availability and irrigation practices. Many African countries, particularly those with no breeding programme, cultivate old varieties; there is excellent potential for increasing productivity through improved varieties or genetic improvement. Sugar cane is remarkable for its production of biomass: historically the crop has been exploited for its capacity to store sugar while only more recently has its renewable energy potential become widely recognized.

Genetic improvement of sugar cane has enabled significant increases in cane and sugar yields worldwide and has facilitated expansion of the crop's growing area. In Queensland, Australia, the average cane yield increased from under 50 tonnes ha^{-1} to over 95 tonnes ha^{-1} between 1942 and 1998, a rate of increase of 0.75 tonnes ha^{-1} per year (Cox *et al.*, 2000). In Mauritius, new varieties helped to boost average sugar yield from 6.8 tonnes ha^{-1} to 9.8 tonnes ha^{-1} between 1948 and 1988. Increases

TABLE 2.1 Area harvested, total cane produced and average cane yield for selected African countries in 2008 (only those with more than 20,000 ha are listed)

African countries	Area harvested (ha)	Total cane produced (tonnes)	Average yield (tonnes/ha)
Cameroon	145,000	1,450,000	10
Côte d'Ivoire	25,875	1,630,000	63
Democratic Rep. of Congo	40,000	1,550,000	39
Egypt	135,962	16,469,947	121
Ethiopia	21,482	2,300,000	107
Kenya	54,465	5,112,000	94
Liberia	26,000	265,000	10
Madagascar	82,000	2,600,000	32
Malawi	23,000	2,500,000	109
Mauritius	62,011	4,533,000	73
Mozambique	180,000	2,451,170	14
Nigeria	62,000	1,500,000	24
Réunion	24,528	1,773,411	72
South Africa	314,000	20,500,000	65
Swaziland	52,233	5,000,000	96
Uganda	35,000	2,350,000	67
United Rep. of Tanzania	23,000	2,370,000	103
Zambia	24,000	2,500,000	104
Zimbabwe	39,000	3,100,000	79
Others	108,929	5,437,772	50
Africa – Total	**1,478,485**	**85,392,300**	**58**

Source: FAOSTAT, 2008.

in sucrose content have contributed significantly to the increased sugar yield. Between 1930 and 1990, the sucrose content of cane in Louisiana increased from 9.7 per cent to 13.6 per cent (Legendre, 1995). A yield plateau or decline associated with the monocropping system has been experienced in some regions over a specific period of time (Burnquist et al., 2010)[1]; nevertheless it is believed that genetic improvement in conjunction with sound management practices have mitigated yield declines (Garside et al., 1997; Ramdoyal et al., 2009).

The high genotype × environment (G×E) interaction and the differential response of genotypes with respect to environment that has been observed for sugar cane have compelled most of the sugar cane industries to set up their own breeding programmes to meet their specific industry requirements. Most of the programmes are financed by the planting community or the sugar industry through different modalities. Owing to the high costs involved in operating breeding programmes, regional breeding networks have been established, as for example the West Indies Sugar Cane Breeding and Evaluation Network (WISBEN), which services the Caribbean countries. In the African/Indian Ocean regions, cane research institutes in Mauritius (MSIRI) and South Africa (SASRI) and Réunion (eRcane) are active in providing varieties and fuzz (seeds) to the African countries or private companies under special contracts that entail the payment of royalties when varieties achieve commercial status. Breeding and selection programmes have also been established in Kenya (KESREF), Egypt (SCRI), Morocco (CTCS), Sudan (KSC), Zimbabwe (ZSAES), while Ethiopia is planning to expand its cane area and embark on breeding activities. Sugar cane germplasm is maintained in two world collections in Florida, USA, and Kannur, India (Berding and Roach, 1987).

The sugar cane plant

Sugar cane belongs to the Poaceae family of the tribe andropogoneae, abundant in tropical and subtropical regions and of which both maize (*Zea mays* Linnaeus) and sorghum (*Sorghum bicolor* Linnaeus) are members. The plant is a large perennial grass generally grown within 30° of the equator. Sugar cane is endowed with a C4 photosynthetic pathway and is among the most efficient converters of solar energy to chemical energy, with a conversion efficiency of 2.24–2.29 per cent compared to switch grass at 0.22–0.56 per cent (Klass, 2004); sugar cane carbon fixation rates are as high as $28\,mg\,CO_2\,m^{-1}\,s^{-1}$ (Irvine, 1975).

Sugar cane is propagated using cuttings, and the first crop, the "plant cane" crop, is generally harvested 8–24 months after planting. The portion of the stalk that is left underground gives rise to the succeeding crop known as the stubble or "ratoon" crop, usually harvested at 12-month intervals. A whole cycle may last between four and ten years (Fauconnier, 1993). Differential tillering ability and tiller survival are factors that explain variation in ratooning. The sugar cane crop also experiences ratoon decline, attributed to reduced growth, vigour and death of the stools.

Stalk and leaf

Sugar cane stalks are unbranched and differentiated into nodes and internodes with adventitious root primordia, a growth ring above each root band and an axillary bud. The lateral buds, inserted alternately along the stalk (Artschwager, 1940), are capable of sprouting into new plants. The shoot roots arise from underground nodes, and the axillary buds at these nodes give rise to tillers. The leaf consists of a sheath that overlaps itself, tightly enveloping the stem and a blade. Leaves are differentiated into long (1–2 m) blades and shorter (0.2–0.3 m), stalk-clasping sheaths (Artschwager, 1940). The blade is widest at its midpoint and tapers towards a narrowed base and pointed tip. A prominent midrib projects from the basal two-thirds of the lower surface of the blade. The green leaves are continuously renewed to maintain a green leaf area, in order to maximize the proportion of incident radiation intercepted for the production of dry matter (DM). The extent to which the canopy intercepts the available radiation depends on leaf area per unit soil surface area (LAI), leaf angle and canopy architecture. The establishment of a canopy depends on the rate at which leaves are produced and the longevity of the individual leaves (Van Dillewijn, 1952; Inman-Bamber, 1994). The leaf production rate is temperature sensitive. The time taken between the appearances of two successive fully expanded leaves (phyllochron) is cultivar-specific (Singels *et al.*, 2005) with a base temperature between 7.8 °C and 11 °C (Inman-Bamber, 1994).

Root

Sugar cane is traditionally propagated using vegetative cuttings, usually consisting of three nodes or in some areas whole stalks are planted. They are placed horizontally; rootlets arise from the root bands and subsequently an aerial shoot develops from the single bud. Normally the aerial shoot feeds upon the parent stalk cutting for a period of one to two months, and during this period the cutting develops roots so that the aerial shoot receives its mineral food through the cutting. After a month or two, depending on environmental conditions, the aerial shoot normally forms its own roots and the seed cutting gradually loses its function in supporting the aerial shoot.

Flower and flowering in sugar cane

Sugar cane usually flowers at the age of 10–12 months. The inflorescence (also known as the tassel or arrow) is an open panicle and is on average about 20–50 cm long. The flowers have both male and female organs, but not all produce fertile pollen.

Flowering in sugar cane is a complex physiological process consisting of multiple stages of development, each having specific environmental and physiological requirements (Julien, 1972), such as day length, temperature and moisture. The time and intensity of flowering are influenced by the environment, latitude, day length, altitude, temperature, moisture, stress and nutrition. Flowering is not desirable in commercial conditions as it signals the end of vegetative growth and it can be detrimental to both cane productivity and sucrose content when harvested late.

Growth and sugar storage

Growth and tillering

The sugar cane cycle comprises four distinct phenological phases, from planting or ratooning until harvest – germination, tillering, stalk elongation and ripening (Soopramanien, 1979). The sugar cane sett contains one or more buds, which, under favourable conditions, germinate and give rise to a primary stalk, under the control of enzymes and growth-regulating substances (Plate 7). Maximum germination and shoot vigour result in a temperature range of 27–33 °C (Keating *et al.*, 1999).

Growth is very sensitive to temperature; optimum growth is reached at temperatures ranging between 30 °C and 33 °C (Fauconnier, 1993); it is very slow below 15 °C and ceases when temperature exceeds 35 °C (Van Dillewijn, 1952). Tillering comprises tiller production, tiller death and stable population. The number of tillers increases exponentially until peak tiller-density stage, whilst the rate of canopy closure increases. Once the canopy closes in there is no further production of tillers. Light intensity and duration, temperature and moisture affect the production of tillers. During the tiller death phase, the tillers compete for nutrients, water and light, and the less vigorous or late-formed ones are eliminated leading to a stable phase (Van Dillewijn, 1952) that normally remains unchanged until harvest. Avoidance of water stress and provision of nitrogenous fertilizer increase tiller survival, but establishment of a high density of primary shoots remains the most effective method of maximizing crop yield (Bell and Garside, 2004). Stalk elongation, characterized by phytomer production and the rapid elongation of the internodes, is influenced by temperature, solar radiation, moisture supply and variety (Van Dillewijn, 1952).

Sugar accumulation, ripening and partitioning

The ripening and maturation phase in a 12-month crop lasts for about three months, starting from 270–360 days after planting; sugar synthesis and rapid accumulation take place during this phase. In general, maturation proceeds in two phases, characterized by increased sucrose accumulation in basal internodes; further increments of sucrose accumulation of whole stalks depend mostly on ripening of distal internodes (Inman-Bamber *et al.*, 2002). A major part of the biomass is deposited as sucrose in the internodes under specific conditions such as low temperature and low moisture, since there is no growth and the demand for dry matter is reduced (Richard, 2003; Nayamuth *et al.*, 2005). Consequently, sucrose content rises so that a "massive" accumulation of sucrose is induced, which is termed "ripening". The most recently expanded internodes near the top of the stem stop elongating and photosynthates are channelled into storage as sucrose (Inman-Bamber *et al.*, 2002).

Enzymatic control of sucrose accumulation

Extensive work carried out to determine the physiological processes of sucrose accumulation in sugar cane was reviewed by Moore (1995). Sucrose storage is the composite result of a cycle of sucrose synthesis and sucrose hydrolysis, and thus involves the coordinated activity of several enzymes (Whittaker and Botha, 1997). In hybrid cultivars that store high quantities of sucrose, the level and its rate of accumulation is a function of the environment and crop development throughout the growing season. The accumulation of sugars includes sucrose and simple sugars (fructose and glucose). Sucrose content is regulated by a cytoplasm localized acid invertase (AI) and a vacuole localized neutral invertase (NI). Re-mobilization of photosynthate occurs when sucrose is needed for energy in other parts of the plant. The AI converts sucrose to glucose and fructose (reducing sugars). When no more energy is required for growth the NI synthesizes the sucrose back to storage. Hatch and Glasziou (1963) demonstrated a linear regression between AI activity and the rate of elongation of immature internodes, as well as between NI activity and sucrose storage.

Partitioning within the stalk

Sugar production depends directly on partitioning of crop biomass to the stalk and then to sucrose stored in stalk parenchyma. The stalk may be considered as a two-sink system, namely sucrose and stalk structure (essentially fibre). Partitioning of dry matter (DM) to the two sinks is a dynamic process that varies with crop age, cultivar, water stress, temperature, time of harvest and variety (Singels *et al.*, 2005). Cooler temperatures favour partitioning away from the leaves towards the stalk such that sucrose content is higher in winter because of a reduction in the sink strength for structural growth. Singels *et al.* (2004), using the Canegro model, observed marked differences between cultivars in the partitioning of assimilate to stored sucrose.

Factors affecting sucrose accumulation

The factors affecting sucrose accumulation are still poorly understood and no consistent pattern has emerged which pinpoints certain enzyme activities as important controlling steps (Rohwer and Botha, 2001). Sucrose accumulation patterns differ between varieties; early-ripening varieties produce more sugar per tonne of cane at the start of the season than late varieties, but accumulate less sugar during the subsequent ripening phase (Herbert and Rice, 1972). Nayamuth and colleagues (2005) have followed sucrose accumulation at the exponential growth phase and after the onset of the ripening phase, and reported that parent varieties had a different sucrose accumulation pattern; namely, precocious, early, middle and late-maturing types. Reduced irrigation and/or the development of mild water stress, referred to as a drying-out period prior to harvest, usually enhances sucrose accumulation, while lack of nitrogen late in the season reduces growth and favours sucrose accumulation. Low temperatures during cool sunny winters slow down growth rates and carbon consumption, while photosynthesis may continue, thereby favouring sucrose accumulation. Immature internodes grown at 13–16 °C had higher sucrose content than those grown at 27–29 °C (Alexander and Samuels, 1968).

Artificial ripeners

Under commercial conditions, artificial ripeners are used under specific circumstances to enhance ripening (Donaldson and Van Staden, 1993). Most chemical ripeners are plant growth regulators (PGRs) or herbicides that are applied in sublethal doses to induce and/or increase ripening. The PGRs restrict growth and allow the plant to accumulate sucrose. However, the response of a particular crop to artificial ripeners depends on the nutritional status of the crop, the age, and overall health, climate, juice quality and variety. The successful development of a universally satisfactory chemical has not yet been achieved (O'Shea, 2000).

Origin and species

The term "*Saccharum* complex" denotes a closely interrelated group composed of the genera *Saccharum*, *Erianthus* (sect. *Ripidium*), *Sclerostachya*, *Narenga* and *Miscanthus* that are involved in the origin of sugar cane. The taxonomy, evolution, distribution and characteristics of the genera of the *Saccharum* complex and the species of the genus *Saccharum* have been reviewed by Daniels and Roach (1987). Three species were cultivated in the past, *Saccharum officinarum*, *S. barberi* (India) and *S. sinense* (China); the latter two species were subsequently confirmed to be hybrids (D'Hont *et al.*, 2002). New Guinea is the centre of diversity of *S. officinarum* L.; clones of this species were referred to as "noble", a term coined by Dutch scientists in Indonesia. They have thick stalks, a high sucrose content and high juice purity; they are low in fibre, have a low starch content, and are of low vigour, with low adaptability to environmental stresses and low resistance to diseases. Two wild species, *S. spontaneum* and *S. robustum*, have played a major role in the development of modern sugar cane hybrids. The more important of the two, *S. spontaneum* L., is distributed widely (Daniels and Roach, 1987; Sreenivasan *et al.*, 1987). Clones are highly polymorphic, low in sucrose and very fibrous. *S. robustum* Brandes and Jesweit ex Grassl is indigenous to New Guinea; stalks are generally thicker than those of *S. spontaneum*, being hard and woody, with little juice (Stevenson, 1965), and occur in two cytotypes. Among the associated genera, *Erianthus* michx Sect. *Ripidium* Henrard (Daniels and Roach, 1987) and *Miscanthus* Anderss are assuming increasing importance (Chen *et al.*, 1986) in broadening the genetic base and introducing useful traits into new varieties. Clones of *Erianthus* are tall with slender stalks of good diameter; they are highly vigorous, with good ratooning ability and tolerance to moisture stress. The genus *Miscanthus* occurs from sea level in Indonesia (Berding and Koike, 1980) to 3,300 m in Taiwan (Lo *et al.*, 1978); clones vary from small, wiry types to tall reeds.

Modern sugar cane varieties

Genetic improvement of sugar cane has been marked by two important events: the discovery in 1888 that sugar cane produced fertile seeds (Stevenson, 1965); and the onset of interspecific hybridization when the breeders at the Proefstation, Java made use of the wild *S. spontaneum*, which not only produced disease-resistant offspring but brought about an unexpected improvement in vigour, cane and sugar yields, ratooning ability, and adaptability to stress conditions. This spectacular work was realized with the use of a naturally occurring F1 hybrid, Kassoer, and one or two successive backcrossing of its progeny to a noble variety. Jeswiet coined the term "nobilization" for this process of incorporating a wild species genome into a noble cane genetic background (Stevenson, 1965). This work culminated, in 1921, with the production of the "wonder cane", POJ 2878, that within eight years came to occupy 90 per cent of the area under cane in Java. Introduced into other countries across the world, the variety has been a valuable parent in generating commercial varieties in many breeding programmes, including one in Mauritius, where one of its progeny, M 134/32, occupied 90 per cent of the cane area in 1952 and generated many valuable varieties in turn (Ramdoyal and Badaloo, 2002). Concurrently in India, in 1912, the nobilization of wild *Saccharum* species using a 64-chromosome clone led to another source of varieties, including Co 205, and intercrossing with other species produced tri-species hybrids that were useful in subtropical India and other parts of the world (Roach and Daniels, 1987).

Cytogenetics and genetics of sugar cane

Modern sugar cane varieties have a complex genomic structure involving more than one species (aneuploids) with chromosome numbers ranging between 100 and 130 (Sreenivasan *et al.*, 1987). First-generation crosses (F1) between *S. officinarum* and *S. spontaneum* and the first-generation backcrosses (BC1) between *S. officinarum* (recurrent female parent) and the F1 show a particular chromosome transmission of $(2n+n)$. *S. officinarum* transmits its whole chromosome complement $(2n)$ when crossed with *S. spontaneum*, which in turn transmits its haploid (n) chromosome number (Roach, 1968). In modern programmes commercial parents are often used as the recurrent female parent to attenuate the drop in vigour in BC1 and BC2 progenies (Ramdoyal and Badaloo, 2002); chromosome transmission involving commercial hybrids is strictly of the $n+n$ type (Burner and Legendre, 1993). The proportion of *S. spontaneum* chromosomes in commercial hybrids varies between 10 per cent and 20 per cent, while the proportion of recombinant chromosomes between the two parental species is relatively low (5–17 per cent) (Piperidis and D'Hont, 2001; Cuadrado *et al.*, 2004). The peculiar genetic structure and its meiotic irregularities render the genetic improvement of sugar cane laborious.

Breeders and quantitative geneticists are interested to study the transmission of traits of importance and the repeatability of traits to derive the best breeding approach and increase the efficiency of selection programmes. Phenotypic variation for a trait can be partitioned into genetic and non-genetic (environmental) components. Heritability of a trait is the proportion of the total variation displayed by a trait, which is due to gene effects and ranges from 0 (non-heritable) to 1 (highly heritable). Quantitative genetic experiments enable the estimates of the genetic component of variances useful in deriving the heritability of traits. Furthermore, genetic effects of traits are contributed by two components of the gene action. The additive portion is transmitted to progenies and is therefore important in hybridization programmes, while the non-additive portion is the interaction between the two parents in determining the expression of characters and is not predictable.

Generally, sugar cane geneticists agree that non-additive genetic variance is more important for the productivity parameters such as cane and sugar yields (Miller, 1977; Badaloo *et al.*, 1997). Breeding efforts to improve productivity parameters exploit specific combining ability (positive association of genes between two parents) from cross evaluation trials and family selection, which mitigate the environmental effects. Both additive and non-additive genetic variance are important for yield components such as stalk number, stalk height and cane diameter, whereas additive genetic variance and relatively higher heritability can be expected for the cane-quality components, Brix (total dissolved solids of juice), pol (measure of sucrose content), fibre content, dry matter percentage, juice purity (ratio of pol:Brix) (Hogarth *et al.*, 1981; Mamet *et al.*, 1996; Ramdoyal and Badaloo, 1998; Badaloo *et al.*, 2005). Moderate heritability is also reported for milling characters, starch content and reducing sugars per cent extract (Brown *et al.*, 1968), whereas Hogarth and Kingston (1984) reported a high component of genetic variance for ash per cent juice (mineral content of juice), indicating that parental choice would be effective in improvement programmes.

Knowledge of the mode of inheritance is also significant when breeding for disease resistance, and has been mentioned by various authors in relation to the common diseases: *smut* (Wu *et al.*, 1983); *rust* (Comstock *et al.*, 1992; Ramdoyal *et al.*, 1999b); *Fiji leaf gall* (Hogarth, 1977); and *yellow spot* (Ramdoyal *et al.*, 2001).

Genetic analyses also enable the estimate of genotypic and phenotypic correlations between the traits and indicate whether the selection for a trait will result in a positive, neutral or negative influence for another trait. In commercial populations, cane yield is more closely correlated with stalk number than with weight per stalk or stalk diameter (Mariotti, 1972). No negative correlation between cane yield and sucrose content on a fresh-weight basis is evident, such that selection for cane yield may not have negative consequences on sucrose content (Brown *et al.*, 1968; Mariotti, 1972). However, in commercial populations as well as in interspecific crosses (Jackson, 1994; Santchurn, 2010), genetic and phenotypic correlations between sucrose and fibre content of cane stalk are negative and point to the difficulty of increasing both characters simultaneously. In early generations of interspecific derived populations, high stalk number is associated with low stalk diameter, low sugar content and high fibre content. The relationships among attributes correspond to the association within the species genetic stock and point to the difficulty of breaking down linkage groups, particularly when homozygous pairing between chromosomes of the same species occurs and the frequency of intergenomic recombination between different species is low (D'Hont *et al.*, 1996).

Principles of sugar cane improvement

Sugar cane, an out-breeding plant, is endowed with both a sexual and an asexual mode of reproduction. Breeding principles comprise the creation of a new population of heterogeneous seedlings, through the hybridization of parent varieties each year; the evaluation, selection and differential propagation of selected genotypes through several clonal stages (Plate 1). The time taken from the seedling stage, the selection of clones, their multiplication and the eventual release of a variety may range from eight to 15 years (Mamet and Travailleur, 1998). The evaluation of varieties for disease resistance in disease trials and/or pest tolerance is mandatory to any breeding programme. Sugar cane differs from other crops in its long crop cycle and perennial nature, and the need for evaluation across a number of ratoon crops to ensure its economic exploitation. Additionally, the relatively large stature of the plant and its flowering stalk require impressive crossing facilities, important experimental areas, high labour requirements and costly mechanized harvesting machinery and weighing equipments in programmes where mechanized harvest is practised. With the improvement of analytical technology, the classical cane analysis for measuring Brix, pol, fibre and conductivity has evolved into one with automatic data acquisition from bar-coded labels and near infrared-spectroscopy technology using fibrated cane samples for the automatic measurement of cane quality (Cox *et al.*, 2000).

It is important that breeding objectives are set correctly as per the requirements of the industry. Traditional breeding programmes aim to increase cane and sugar yield per hectare, sucrose content, ratooning ability, disease resistance, pest tolerance, and suitability to harvest dates and mechanized harvest. Specific objectives include breeding for abiotic stresses such as tolerance to frost, drought, waterlogging or salinity. Successful breeding programmes should possess as wide a collection of parents, developed locally and imported from other breeding programmes, as can be managed in order to provide the breeder with a high level of genetic variability. Parent nursery collection should be localized in an environment conducive to flowering. However, breeding stations situated in subtropical and temperate zones – such as the South African Sugar Research Institute (SASRI), Bureau of Sugar Experiment Station, Australia (BSES), United States Department of Agriculture – Agricultural Research Service (USDA-ARS), Canal Point, Florida and Houma, Louisiana – provide photoperiodic facilities to induce flowering in shy and non-flowering varieties (Nuss and Berding, 1999). Under more tropical climates, as in Mauritius, night-break treatments are imposed to delay flowering in the early-flowering *S. spontaneum* clones and early-generation interspecific hybrids to enable flower synchrony with the later-flowering noble and commercial varieties (Julien, 1972).

Generally, flowering is restricted to a few weeks. The choice of parents is determined by the priorities of the programme, flower synchronization and sexing of parent varieties, the ability of parents and crosses to produce high selection rates or elite progenies (Heinz and Tew, 1987; Domaingue *et al.*, 1988; Badaloo *et al.*, 1999), breeding value of parents and the genetic value of crosses that include economic selection indices (Wei *et al.*, 2007). Computing systems are widely used to generate crossing lists and process breeding and selection data (Wu, 1987; Nuss *et al.*, 1989; Sun-Yuan Hsu *et al.*, 1991). The IT system for managing germplasm and its utilization in computer-aided crossing in Mauritius has been described by Ramdoyal and colleagues (1999a) and Mundil and colleagues (2005).

Sugar cane crossing techniques have been reviewed by Heinz and Tew (1987) and hybridization procedures practised in many breeding stations have been documented by Ramdoyal and colleagues (2003). Crossing is conducted in covered sheds or under controlled conditions in glasshouses. Whole flowering stalks maintained in a preservative and nutritive solution are isolated in special units or crossing lanterns for pollination, which may last between seven and 15 days (Ramdoyal *et al.*, 1995a). Two types of crosses are made: the biparental crosses (identity of both female and male parents are known); and the polycross system (female parents are crossed with different male parents) for generating more variability in the progenies. Seed-setting and maturation proceed for a further three to four weeks before seeds are harvested (Ramdoyal *et al.*, 1995b).

Selection methods

Depending on programmes, from 30,000 to over 500,000 seedlings are produced for selection (Skinner *et al.*, 1987; Mamet and Travailleur, 1998). At the seedling stage, breeders either use mass selection or select the best genotypes from all families; they may also use a combination of family selection and mass selection in evaluating the families as a whole for sugar yield in replicated trials and selecting the best genotypes within the best families (Cox *et al.*, 2000).

The number of clonal stages varies between three and five (Mamet and Travailleur, 1998). During the early stages of selection, a large population of seedlings are evaluated in small plot sizes, their numbers being reduced at each stage. The selected ones are evaluated more precisely on larger plots in replicated trials in which environmental variation can be controlled by more efficient experimental designs. Initially, on account of large environmental variation, selection is based on visual appreciation of the genotypes, or on yield-correlated characters (e.g. stalk number, stalk diameter); in subsequent clonal stages, cane weight is determined along with quality characters. The interaction of the genotype with the environment (G × E) interactions are complicating factors, and it is important to evaluate genotypes in a wide range of environments over several years to gauge both the differential response of the genotype in different years (G × Y) and performance across ratoon crops (Bissessur *et al.*, 2010). A simplified flowchart of the Mauritian sugar cane selection programme is given in Figure 2.1.

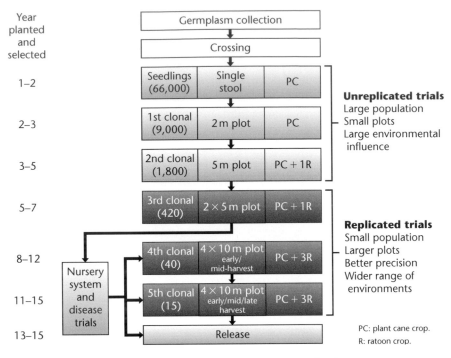

FIGURE 2.1 Simplified flowchart of the MSIRI sugar cane selection programme

Variety distribution and exchange

Promising varieties are multiplied in a stepwise manner so that enough planting material is available at the time of a variety's release. The modalities for releasing a variety involve extensive consultation between the variety improvement team, extension officers and stakeholders. In Mauritius, all data pertaining to a candidate variety are compiled and presented to a "Cane Release Committee" that includes different stakeholders. This committee makes recommendation for the release of a variety to be ratified by the presiding Ministry before the variety is registered in an official list for commercial exploitation. In some countries, such as Australia, Brazil, South Africa and the USA, varieties are protected under their respective plant breeders' right (PBR).

Breeding for biomass

Above-ground productivity

Sugar cane yields vary widely depending on the environmental conditions and the length of the crop cycle. Osgood (2003) reported cane yield in Hawaii for a 24-month crop to be as high as 380 tonnes ha^{-1} (190 tonnes ha^{-1} yr^{-1}) while recoverable sugar yield was 24< tonnes ha^{-1} yr^{-1}, and an equivalent amount of fibre was expected with a 50:50 partitioning of dry matter to sugar and fibre. A harvest of 100 tonnes ha^{-1} yr^{-1} representing about 30 tonnes ha^{-1} yr^{-1} of total solids (soluble solids and fibre) is considered normal under Hawaiian conditions (Tew and Cobill, 2008). The world average fresh-weight yield is about 65 tonnes>ha^{-1}, representing around 17 tonnes ha^{-1} of total solids. Breeding of sugar cane for biomass requires an appreciation of the contribution of the different plant parts to the production of above-ground biomass (AGB). Globally, it is estimated that sugar cane produces roughly 80 per cent of total AGB and 20 per cent below ground. In Mauritius, of the above-ground biomass, about 69 per cent is constituted of the mature cane stalks destined for milling, 21 per cent of the green cane tops and leaves (CTL) left in the field after harvest, and 10 per cent of dry leaves or sugar cane trash, amounting to a ratio of stalk to non-stalk of 70:30 (Beeharry, 1996). This is close to what has been reported in Puerto Rico – 69 per cent cane stalk, 17 per cent CTL, 14 per cent trash (Alexander, 1985). Mature cane stalks consist of approximately 25 per cent total solids partitioned between soluble solids, 13 per cent (>90 per cent sucrose) and fibre, 12 per cent, and 75 per cent water. CTL contains about 20–25 per cent fibre and 75–80 per cent water. Dry leaves or trash contain 80 per cent fibre and 20 per cent water.

Cane as an energy feedstock

Depending on the variety of cane cultivated, one tonne of above-ground biomass yields about 69 kg (6.9 per cent) of sugar, 207 kg (20.7 per cent) of bagasse at 50 per cent moisture, 21 kg (2.1 per cent) of molasses after sugar exhaustion, 35 kg (3.5 per cent) of scum and 357 kg (35.8 per cent) of water and impurities. In addition, CTL and trash amount to 124 kg and 155 kg of bagasse (at 50 per cent moisture), respectively, thus yielding total fibrous biomass of 486 kg for every tonne of above-ground cane biomass (Beeharry, 1996). Bagasse is used in boilers for generating the internal energy (steam and electricity) requirements of the sugar cane processing plant and any surplus electricity can be exported to the grid. CTL and trash constitute a potential feedstock that can be baled and transported to the mill for energy production independent of the cane harvest (see Chapter 5). As a biomass crop, sugar cane fares well with respect to energy balance, with an output/input (O/I) ratio varying between 1.8 and 3.3 for sugar cane cultivated for sugar. An O/I ratio in excess of 8:1 has been reported in Brazil when sugar cane is processed for ethanol production (Tew and Cobill, 2008).

Breeding strategies for biomass options

Sugar cane breeding and selection has traditionally been geared towards maximizing sugar production per unit area and has resulted in generating varieties with reduced fibre content. High fibre content has been considered an undesirable characteristic, as it reduces the mill throughput and increases sugar losses in the bagasse. However, a minimum quantity of bagasse is necessary for mills to process canes efficiently, rendering the system energy self-sufficient.

Large potential gains in both sugar and fibre production can be achieved from sugar cane if breeding programmes are concentrated on total biomass production and not simply increasing the sugar concentration in the stem. Tew and Cobill (2008) describe three different types of canes as regards fibre : sugar ratio and water content:

a. Traditional sugar cane – 12 per cent fibre : 13 per cent sugar, 75 per cent water, grown primarily for sugar.
b. Type 1 canes – 17 per cent fibre : 13 per cent sugar, 70 per cent water, where the vegetative biomass is an important product.
c. Type 2 energy canes – 30 per cent fibre : 5 per cent sugar, 65 per cent water, where the vegetative biomass is the main product.

In Mauritius, multivariate data analyses of sucrose content, fibre content and biomass based on genotypes belonging to different nobilized generations defined four groups of canes (Santchurn, 2010), as summarized in Table 2.2. The thresholds for pol and fibre are relative to each other and are based on the values obtained from current analyses of cane samples under local conditions with the current analytical methodology adopted. Thresholds can vary depending on the geographical conditions, growth stage and crop age.

Type 1 canes are the conventional varieties, with more or less equal partitioning of dry matter between pol and fibre – 13 per cent : 12 per cent. Type 2 canes – pol 13 per cent : fibre 14–16 per cent – are the enhanced-fibre commercial types, where sucrose content is not affected by an increase in fibre. Type 3 canes – pol > 8–13 per cent : fibre 16–20 per cent – are multipurpose varieties that compromise on sucrose concentration to the advantage of fibre and can be used for electricity/ethanol/sugar in a socio-economic environment where bio-energy from sugar cane would be more profitable than or equally profitable to sugar. Type 4 canes – pol <8 per cent : fibre >20 per cent – are energy canes, which are used for electricity generation. A selection index has been developed accordingly for selecting different types of canes. A breeding programme should cater for the

TABLE 2.2 Classification of varieties based on multivariate analyses for biomass traits of a population of interspecific derived clones of different nobilized generations

Types of canes	Pol % cane	Fibre % cane	Remarks
Type 1	13	12	Conventional commercial varieties
Type 2	13	14–16	Commercial varieties with enhanced fibre content
Type 3	>8–13	16–20	Multipurpose canes, compromise on pol % to the advantage of fibre for multipurpose use – sugar/ethanol/electricity
Type 4	<8	>20	Energy canes for cogeneration with little juice for fermentation

different options, with emphasis on a main core programme. The benefit that can be derived from higher-fibre canes also depends on the operating conditions (Lau et al., 2003).

Conventional breeding programmes can routinely generate Type 2 canes (Tew and Cobill's commercial types) with selection pressure operating jointly for pol and fibre (Domaingue, 1993). In Mauritius, variety M 1672/90 (Type 2), released in 2007, produced cane and sugar yields equal or superior to the conventional varieties, with a fresh-weight fibre content superior to the standards by 15 per cent to 25 per cent (Ramdoyal et al., 2007). To generate Type 3 and Type 4 canes, introgression breeding using S. spontaneum or Erianthus and Miscanthus should be resorted to. The classical interspecific programme described under nobilization breeding would generate the different types of varieties when the appropriate selection pressure for the main biomass traits is applied within the respective nobilized generations (Ramdoyal and Badaloo, 2007). Generally, one would expect a higher proportion of Type 4 fibrous canes from F1 populations and more Type 2 and Type 3 canes from the BC1 and, to a lesser extent, from the BC2 generations, the latter producing more of the Type 1 canes. Morphologically, Types 1 and 2 are similar, whereas Types 3 and 4 are characterized by abundant, taller stalks of smaller diameter. The MSIRI breeding programme devotes about 10 per cent of its annual total seedling population to biomass breeding involving the wild species. The interspecific derived seedlings (F1, BC1, BC2) are selected visually at the "seedling" stage and planted directly into 5 m plots at the first clonal stage (thus skipping the intermediate 2 m plot stage), sampled for fibre and pol, and assessed for biomass yield for evaluation at the first replicated stage and final stages (Figure 2.1).

Three "high-fibre" varieties, L 79–1002, HoCP 91–552 and Ho 00–961, were released for the US biofuels industry (Cobill, 2007). In Australia, special "Hi-Energy Canes" are being bred from the wild relatives, as well as high-value, smut-resistant varieties for production of both sugar and energy. Similar ventures in China are aimed at utilizing S. spontaneum for biomass composition and yield components using DNA markers-assisted methodology (Wang et al., 2008) while in Japan, Terajima and colleagues (2005, 2007) have reported developing high-biomass sugar cane for multipurpose use from S. spontaneum, Sorghum bicolor, Miscanthus spp. and Erianthus spp. SASRI have experimented with test plots for high-fibre cane varieties (Plate 2).

The Barbados Agricultural Management Company (BAMC) proposed the conversion of the Barbados sugar industry into a cane industry, where electrical energy would be produced from very high-fibre canes, the "Fuel Cane Project" (Albert-Thenet et al., 2004). These varieties contain 18–27 per cent fibre and 8–10 per cent pol, compared to 15 per cent fibre and 17 per cent pol for the standard variety. They produce a 26–62 per cent higher cane yield, 24–74 per cent more dry matter (Brix + fibre), and 24–61 per cent more biomass (cane + tops) than the control (Rao et al., 2007). The commercial-level experience is reported by Harm de Boer (2009).

Sugar cane biotechnology

In view of its complex genetic makeup and the long time interval required to develop new varieties by conventional breeding methods, the sugar cane crop is a good candidate for biotechnological applications that can speed up variety develop-ment. In practice, these new technologies have brought about notable progress during the last two decades and have contributed to the understanding of the complex sugar cane genome, as well as in the production of transgenic lines with improved traits. The potential of using sugar cane as a biofactory for the accumula-tion of value-added products has also been demonstrated, but is yet to be exploited.

Tissue culture

Tissue culture techniques are increasingly utilized in the propagation of sugar cane and are the basis for cryopreservation of germplasm, elimination of pathogens and gene transfer. Several sugar cane-growing countries have invested in tissue culture laboratories for *in vitro* mass propagation of new varieties. This technique has a number of advantages and includes a high rate of multiplication within a short period, clonal propagation from a small stock of plant material, and the ability to produce disease-free plantlets. However, the support of a disease-diagnostic labora-tory is necessary to ensure that the production is free from pathogens, which is also important for the safe movement of germplasm. Plantlets free of the diseases *sugar cane mosaic virus*, *Fiji leaf gall virus* and *sugar cane yellow leaf virus* have been produced (Odda *et al.*, 1992; Wagih *et al.*, 1995; Parmessur *et al.*, 2002). Apart from explants such as axillary bud, apical bud and meristem used to initiate *in vitro* plants, leaf roll or disc culture is taking more importance (Geijskes *et al.*, 2003). The development of an immersion system for mass propagation of sugar cane has also been investi-gated (Lorenzo *et al.*, 2001).

Progress in transgenic sugar cane

Although research in the genetic transformation of sugar cane has been ongoing for over 20 years, there is in 2011 no commercially grown transgenic sugar cane any-where in the world. Progress has been made in this field following the successful production of a transgenic sugar cane callus expressing the gene for resistance to an antibiotic via electroporation of protoplasts (Chen *et al.*, 1987). Thereafter, Bower and Birch (1992) produced the first transgenic sugar cane, again expressing an anti-biotic resistance, using microprojectile bombardment. A number of genes have subsequently been inserted after much effort devoted to overcome the difficulties in transforming the high polyploid genome. Successful transgenic sugar cane has been produced for herbicide (Gallo-Meagher and Irvine, 1996; Leibbrandt and Synman, 2003), viral (Gilbert *et al.*, 2005), bacteria (Zhang *et al.*, 1999) and insect tolerance (Smith *et al.*, 1999). Modification for sucrose metabolism (Rossouw *et al.*,

2007) has also been given much attention, although most of the transgenic lines have been produced through glasshouse trials, while work carried out by Groenewald and Botha (2008) to downregulate an endogenous gene by anti-sense expression were evaluated in the field in one ratoon crop.

The first field experiments with genetically modified (GM) sugar cane commenced in 1996 (Birch, 1997). In Colombia and the US *sugar cane yellow leaf virus*-resistant clones have been evaluated (Victoria *et al.*, 2005; Gilbert *et al.*, 2009), while Butterfield and colleagues (2002) have obtained progeny after crossing transgenic herbicide and viral-resistant sugar cane lines as male parents with non-transgenic clones. Stable inheritance of transgenes was demonstrated in this study while Leibbrandt and Snyman (2003) reported that transgenes for herbicide were stable in multiple cycles of vegetative propagation. Sugar cane lines transformed for resistance to *Sugar cane mosaic virus* strain E were evaluated in the field in Florida (Gilbert *et al.*, 2005).

The long-awaited commercial transgenic plants are likely to be released by 2015. Herbicide and insect-tolerant sugarcane (Round-up-ready and Bt varieties) from Monsanto are expected to be amongst the first. In view of commercialization of genetically modified canes, a number of countries are already preparing regulations to that end.

Sugar cane as a biofactory for high-value products

The particularity of sugar cane to produce a large biomass favours the crop as a biofactory to accumulate high-value molecules. In Hawaii, researchers have developed transgenic sugar cane lines that produce the human cytokine granulocyte macrophage colony-stimulating factor (GM-CSF) (Wang *et al.*, 2005). The GM-CFS protein was accumulated up to 0.02 per cent of total soluble protein, and field trials showed stable protein accumulation. Despite considerable attempts through conventional and molecular breeding, it has not been possible significantly to increase the sugar content of the crop. In Australia, Wu and Birch (2007) managed to express a bacterial sugar isomerase (SI) gene with the potential to accumulate the high-value sugar isomaltulose in sugar cane tissues, without any decrease in the storing capacity of sucrose concentration. Researchers have been successful in expressing the genes from the bacterium *Ralstonia eutropha* in sugar cane plants to produce bioplastics, polyhydroxybutyrate (PHB) (Brumley *et al.*, 2007). So far, PHB has been accumulated in the leaves to a maximum of 1.88 per cent of dry matter. Efforts are presently targeted to increase the accumulation of PHB to a higher percentage. Woodard and colleagues (2009) have reported the production of transgenic sugar cane expressing an antimicrobial bovine lysosyme in both the stalk and the leaves.

Prospects of cellulosic ethanol production from sugar cane bagasse are being investigated by the Syngenta Centre for Sugar Cane Biofuels Development in Australia (Sainz *et al.*, 2010). More energy and value could be derived from bagasse by converting it into biofuel and biogas. However, the cell wall in plants makes the production of cellulosic ethanol a challenge. Cellulose exists within a matrix of other polymers, primarily hemicelluloses and lignin. Pre-treatment of bagasse with heat, enzymes or acids can remove these polymers from the cellulose before hydrolysis, but this is an expensive process. The expression of enzymes in sugar cane represents a promising approach to reducing the costs of enzyme production for next-generation biofuel and preliminary results obtained by Syngenta show promising avenues for this new approach.

DNA marker systems in sugar cane and genetic mapping

Molecular marker technologies can be useful for marker-assisted selection, particularly for pests, diseases and other economic traits, and much work is currently being undertaken in this field. These techniques can provide information on fingerprints and genetic diversity/relatedness on varieties, predict location of genes, and also determine DNA markers linked to specific phenotypic traits, as well as allow genetic map construction by analysing populations derived from crosses of genetically diverse parents, and estimating the recombination frequency between the genetic loci.

The main constraint in mapping sugar cane resides in its highly complex genome, commercial sugar cane having a genome size of approximately 3,000 mega base pairs (Mbp) compared to rice, with a genome size of 430 Mbp (Tomkins *et al.*, 1999). The breakthrough in sugar cane genetic mapping became evident when it was realized that single-dose and double-dose markers, segregating in a $1:1$ and $3:1$ ratio respectively in an F1 progeny, could be mapped (Wu *et al.*, 1992). Following this a number of genetic maps have been constructed using different types of markers. The first sugar cane genetic maps were constructed from arbitrary primed PCR (Al-Janabi *et al.*, 1993) and restriction fragment length polymorphism – RFLP (Da Silva *et al.*, 1993, 1995), while the most recent maps have involved markers such as simple sequence repeats – SSRs, amplified fragment length polymorphisms – AFLPs, and expressed sequence tags – ESTs (Hoarau *et al.*, 2001; Aitken *et al.*, 2005; Oliveira *et al.*, 2007; Al-Janabi *et al.*, 2007; Parmessur *et al.*, 2010). Although these recent maps contain more than 1,000 markers, they are still incomplete and need to be further saturated. A number of these maps have been used together with phenotypic data to identify quantitative trait loci – QTLs (Sills *et al.*, 1995; Daugrois *et al.*, 1996; Guimarães *et al.*, 1997a; Ming *et al.*, 2001; Hoarau *et al.*, 2002; Al-Janabi *et al.*, 2007).

Recent advances in sugar cane genomics

The complete genome of sugar cane is not available, but significant progress is being made in this area. In 2003, DNA sequence information became available through the release of some 238,000 expressed sequence tags (ESTs) assembled through the work carried out under the sugar cane EST project (SUCEST) (after more than 260,000 cDNA clones were partially sequenced (Vettore *et al.*, 2003)).

These ESTs were assembled into 43,141 putative transcripts. But previously, researchers from South Africa and Australia had produced a small collection of ESTs (Carson *et al.*, 2002; Casu *et al.*, 2001). Through the transcriptome information genes involved in biotic and abiotic stress response, disease resistance and sucrose accumulation have been identified (Arruda and Silva, 2007; Nishiyama *et al.*, 2010). In addition, cDNA microarrays have been used to identify genes determining important traits (Casu *et al.*, 2003). The high colinearity that exists between the sorghum and sugar cane genomes allows information from one grass to be used by another (Guimarães *et al.*, 1997b). Comparative mapping with sorghum, the closest cultivated relative of sugar cane, has allowed the evaluation of corresponding QTLs affecting different traits such as flowering and plant height (Aitken *et al.*, 2009; Ming *et al.*, 2002). Sorghum ESTs are also being mapped on sugar cane (Parmessur *et al.*, 2010).

Biotechnology to detect sugar cane pests and diseases

Some 90 sugar cane diseases have been reported. Molecular diagnostics are increasingly becoming the preferred method for the diagnosis of diseases. They are essential in the production of clean seed, play a vital role in surveillance of pathogens, in germplasm exchange and in the selection of cultivars resistant to diseases. Molecular diagnostics are available for most sugar cane pathogens of bacterial, viral, phytoplasmal and fungal origin. Xie and colleagues (2009) reported the simultaneous detection of four viruses using one-step RT-PCR. Efficient PCR methods have been developed for the quantification and detection of major bacterial pathogens of sugar cane (Rott and Davis, 2004) while molecular tools have been instrumental in improving the detection of phytoplasmas (Dookun-Saumtally *et al.*, 2008). The genome of the two sugar cane bacteria, namely *Leifsonia xyli subsp xyli* and *Xanthomonas albilineans* have been sequenced, providing useful information for the identification of pathogenicity genes (Monteiro-Vitorello *et al.*, 2004; Pieretti *et al.*, 2009). Recently, researchers have also described the application of DNA barcodes to distinguish species of sugar cane insect pests such as armyworms (*Mythimna* spp.) (Joomun *et al.*, 2010).

Prospects for the future include microarray or DNA chips and processors that will be made available to characterize unknown DNA samples, enabling the identification of sugar cane pathogens.

Outlook

Recognizing the enormous biomass potential of sugar cane and the various renewable bioenergy options, the African region, with its tropical and subtropical climate, is well suited to expand its cane production with the transformation of its sugar industries into a sugar cane industry. Enormous potential exists for the development of sugar-producing units into a "sugar cane cluster" from which sugar, bioenergy mainly in the form of ethanol and electricity, and high-value products can be produced. Additionally, the high resilience of the crop to adverse climatic factors, its impact in mitigating environmental degradation and its socio-economic multifunctionality place this crop in a key position towards a sustainable agro-food security and bioenergy development strategy for the region. The cultivation of sugar cane for its main product, sugar, will continue as long as the world market sugar price is attractive enough, but a shift in its usefulness to varying proportions of sugar and energy is expected as the availability of fossil fuel dwindles and oil prices increase. Cultivation of the crop in marginal areas for its total biomass potential beyond its ecological suitability can further be envisaged. New molecular technologies can provide useful tools for genetic fingerprinting, genome mapping, marker-assisted selection and the safe movement of healthy material that can enhance breeding efficiency for sugar, energy and value-added products. A combination of conventional and molecular breeding approaches is expected to enhance breeding efficiency for introgression breeding for biomass options. The development of new cropping systems, viewed from a holistic angle, that include management practices, harvesting, transportation and processing, would pave the way to a new era in managing the new sugar cane crop for total biomass production.

Note

1 Monocropping system: The agricultural practice of growing the same crop year after year on the same land, without crop rotation using other crops.

References

Aitken, K.S., Jackson, P.A. and McIntyre, C.L. (2005) "A combination of AFLP and SSR markers provide extensive map coverage and identification of hom(oe)ologous linkage groups in sugarcane", *Theoretical and Applied Genetics*, 110, pp. 789–801.

Aitken, K.S., McNeil, M.D., Hermann, S., Bundock, P.C., Eliott, F.G., Heller-Uszynska, K., Kilian, A., Henry, R.J. and Jackson, P. (2009) "Comparative Mapping of a Sugarcane Genetic Map to the Sorghum Genome", paper presented to the 14th Australasian Plant Breeding Conference (APBC) held in conjunction with the 11th Society for the Advancement of Breeding Research in Asia and Oceania (SABRAO) Congress, Cairns, Queensland, 10–14 August.

Albert-Thenet, J.R., Simpson, C.O., Rao, P.S. and Gardner, M. (2004) "The BAMC Fuel Cane Project", Proceedings of the West Indies Society of Sugar Cane Technologists, Barbados.

Alexander, A.G. and Samuels, G. (1968) "Controlled-temperature studies of growth, enzymology, and sucrose production by two sugarcane varieties in Puerto Rico", *Journal of Agriculture of the University of Puerto Rico*, PR 52, pp. 204–217.

Alexander, A.G. (1985) *The Energy Cane Alternative*, Elsevier, Amsterdam.

Al-Janabi, S.M., Honeycutt, R.J., McClelland, M. and Sobral, B.W.S. (1993) "A genetic linkage map of *Saccharum spontaneum* L. '*SES 208*'", *Genetics*, 134, pp. 1249–1260.

Al-Janabi, S.M., Parmessur, Y., Kross, H., Dhayan, S., Saumtally, S., Ramdoyal, K., Autrey, L.J.C. and Dookun-Saumtally, A. (2007) "Identification of a major quantitative trait locus (QTL) for yellow spot (*Mycovellosiella koepkei*) disease resistance in sugarcane", *Molecular Breeding*, vol. 19, no.1, pp. 1–14.

Arruda, P. and Silva, T.R. (2007) "Transcriptome analysis of the sugarcane genome for crop improvement", in R.K. Varshney and R. Tuberosa (eds), *Genomics Applications in Crops*, vol. 2, Swedish, Springer, pp. 483–494.

Artschwager, E. (1940) "Morphology of the vegetative organs of sugar cane", *Journal of Agricultural Research*, 60.

Badaloo, M.G.H., Kearsey, M.J., Domaingue, R. and Julien, M.H.R. (1997) "Cross prediction in sugar cane, I Evaluation of biparental seedling populations in two contrasting environments", in "Quantitative Genetics of Sugar Cane: Cross Evaluation of Major Agronomic and Morphological Trait", PhD Thesis, University of Birmingham, UK.

Badaloo, G.H., Kearsey, M.J., Domaingue R. and Julien M.H.R. (1999) "The application of cross prediction in sugar cane breeding in Mauritius", *Proceedings of the International Society of Sugar Cane Technologists*, 3, pp. 496–503.

Badaloo, M.G.H., Ramdoyal, K. and Nayamuth, A.R.H. (2005) "Variation and inheritance of sucrose accumulation patterns and related agronomic traits in sugarcane families", *Proceedings of the International Society of Sugar Cane Technologists*, 25, pp. 430–441.

Beeharry, R.P. (1996) "Extended sugarcane biomass utilisation for exportable electricity production in Mauritius", *Biomass Bioenergy*, 11, p. 441–449.

Bell, M.J. and Garside, A.L. (2004) "Shoot and stalk dynamics, dry matter partitioning and the yield of sugarcane crops in tropical and subtropical Queensland, Australia", in "Review of Knowledge of Sugarcane Physiology and Climate–Crop–Soil Interactions", Sugar Research and Development Corporation Final Report No. CSE006, pp. 44–45.

Berding, N. and Koike, H. (1980) "Germplasm conservation of the Saccharum complex: A collection from the Indonesian Archipelago", *Hawaii Plant Records*, 59, pp. 87–176.

Berding, N. and Roach, B.T. (1987) "Germplasm collection, maintenance and use", in D.J. Heinz (ed.), *Sugar Cane Improvement through Breeding*, Elsevier, Amsterdam, pp. 143–210.

Birch, R.G. (1997) "Transgenic sugarcane: opportunities and limitations", in B.A. Keating and J.R. Wilson (eds), *Intensive Sugarcane Production: Meeting the Challenges Beyond 2000*, CAB International, Brisbane, Qld, Ch. 8, pp. 125–140.

Bissessur, D., Ramnawaz, C. and Ramdoyal, K. (2010) "Use of historical multi-location sugarcane variety trials data to identify relationships among environments in terms of genotype response", *Proceedings of the International Society of Sugar Cane Technologists*, 27, p. 9.

Bower, R. and Birch, R.G. (1992) "Transgenic sugarcane via microprojectile bombardment", *Plant Journal*, 2, pp. 409–416.

Brown, A.H.D., Daniels, J. and Latter, B.D.H. (1968) "Quantitative genetics of sugarcane I. Analysis of variation in a commercial hybrid commercial population", *Theoretical and Applied Genetics*, 38, pp. 361–369.

Brumley, S.M., Purnell, M.P., Petrasovits, L.A., Nielsen, L.K. and Twine, P.H. (2007) "Developing the sugarcane biofactory for high-value biomaterials", *International Sugar Journal*, 109, pp. 5–14.

Burner, D.M. and Legendre, B.L. (1993) "Chromosome transmission and meiotic stability of sugar cane (*Saccharum* spp.) hybrid derivation", *Crop Science*, 33, pp. 600–606.

Burnquist, W.L., Redshaw, K. and Gilmour, R.F. (2010) "Evaluating sugarcane R&D performance: evaluation of three breeding programs", *Proceedings of the International Society of Sugar Cane Technologists*, 27, pp. 1–14.

Butterfield, M.K., Irvine, J.E., Valdez Garza, M. and Mirkov, T.E. (2002) "Inheritance and segregation of virus and herbicide resistance transgenes in sugarcane", *Theoretical and Applied Genetics*, 104, pp. 797–803.

Carson, D.L., Huckett, B.I. and Botha, F.C. (2002) "Sugarcane ESTs differentially expressed in immature and maturing internodal tissues", *Plant Science*, 162, pp. 289–300.

Casu, R., Dimmock, C., Thomas, M., Bower, N., Knight, D., Grof, C., Mcintyre, L., Jackson, P.D., Whan, V., Drenth, J., Tao, Y. and Manners, J. (2001) "Genetic and expression profile in sugarcane", *Proceedings of the International Society of Sugar Cane Technologists*, 24, pp. 542–546.

Casu, R.E., Grof, P.L., Rae, A.L., McIntyre, L.C., Dimmock, C.M. and Manners, J.M. (2003) "Identification of a novel sugar transporter homologue strongly expressed in maturing stem vascular tissues of sugarcane by expressed sequence tag and microarray analysis", *Plant Molecular Biology*, 52, pp. 371–386.

Chen, W.H., Gartland, K.M.A., Davey, M.R., Sotak, R., Gartland, J.S., Mulligan, J., Power, J.B. and Cocking, E.C. (1987) "Transformation of sugarcane protoplasts by direct uptake of a selectable chimaeric gene", *Plant Cell Reporter*, 6, pp. 297–301.

Chen, W.H., Huan, Y.J., Shen, I.S. and Shih, S.C. (1986) "Utilisation of *Miscanthus* germplasm in sugar cane breeding in Taiwan", *Proceedings of the International Society of Sugar Cane Technologists*, 18, pp. 641–648.

Cobill, R.M. (2007) "Development of energy canes for an expanding biofuels industry", *Sugar Journal*, vol. 70, no. 5, pp. 6–7.

Comstock, J.C., Wu, K.K. and Schenell, R.J. (1992) "Heritability of resistance to sugarcane rust", *Sugar Cane*, 6, pp. 7–10.

Cox, M., Hogarth, D.M. and Smith, G. (2000) "Cane breeding and improvement", in P.G. Allsopp (ed.), *Manual of Cane Growing*, BSES, Indooroopilly.

Cuadrado, A., Acevedo, R., Moreno Dias de la Espina, S., Jouve, N. and de la Torre, C. (2004) "Genome remodelling in three modern *S. officinarum* × *S. spontaneum* sugarcane cultivars", *Journal of Experimental Botany*, 55, pp. 847–854.

Da Silva, J., Honeycutt, R.J., Burnquist, W., Al-Janabi, S.M., Sorrells, M.E., Tanskley, S.D. and Sobral, W.S. (1995) "*Saccharum spontaneum* L. '*SES 208*' genetic linkage map combining RFLP- and PCR-based markers", *Molecular Breeding*, 1, pp. 165–179.

Da Silva, J., Sorrells, M.E., Burnquist, W. and Tanskley, S.D. (1993) "RFLP linkage map and genome analysis of *Saccharum spontaneum*", *Genome*, 36, pp. 782–791.

Daniels, J., and Roach, B.T. (1987) "Taxonomy and evolution", in D.J. Heinz (ed.) *Sugarcane Improvement through Breeding*, Elsevier, Amsterdam, pp. 7–84.

Daugrois, J.H., Grivet, L., Roques, D., Hoarau, J.Y., Lombard, H., Glaszmann, J.C. and D'Hont, A. (1996) "A putative major gene for rust resistance linked with a RFLP marker in sugarcane cultivar 'R570' ", *Theoretical and Applied Genetics*, 92, pp. 1059–1064.

D'Hont, A., Grivet, L., Feldmann, P., Rao, P.S., Berding, N. and Glaszmann, J.C. (1996) "Characterization of double genome structure of modern sugarcane cultivars (*Saccharum* spp.) by molecular cytogenetics", *Molecular and General Genetics*, 250, pp. 405–413.

D'Hont, A., Lu, A.Y.H., Paulet, F. and Glaszmann, J.C. (2002) "Oligoclonal interspecific origin of 'North Indian' and 'Chinese' sugarcanes", *Chromosome Research*, 10, pp. 253–262.

Domaingue, R. (1993) "Selection of sugar cane varieties with higher fibre content", *Revue Agricole et Sucrière de l'Ile Maurice*, vol. 71, nos 2–3 (May–December), pp. 210–217.

Domaingue, R., Ramdoyal, K., Rivet, L. and Mamet, L.D. (1988) "Vers un choix judicieux de parents dans le programme d'amélioration de la canne à sucre à l'Ile Maurice", 3e Congrès international de l'ARTAS, 16–23 October, Réunion, pp. 326–331.

Donaldson, R.A. and Van Staden, J. (1993) "Effects of Fusilade Super on the growth and yields of the sugar cane variety NCo 376", *Proceedings of the South African Sugar Technology Association*, 67, pp. 44–48.

Dookun-Saumtally, A., Saumtally, S., Parmessur, Y., Joomun, N. and Al-Janabi, S. (2008) "Phytoplasmas infecting sugarcane", in N.A. Harrison, G.P. Roa and C. Marcone (eds), *Characterization, Diagnosis and Management of Phytoplasmas*, Studium Press LLC, Houston, TX, Ch. 12.

FAOSTAT (2008) www.fao.org/economic/ess/publications-studies/statistical-yearbook/ fao-statistical-yearbook-2007–20008/b-agricultural-production/en/.

Fauconnier, R. (1993) "Sugarcane", in *The Tropical Agriculturalist*, Macmillan, London.

Gallo-Meagher, M. and Irvine, J.E. (1996) "Herbicide resistant transgenic plants containing the bar gene", *Crop Science*, 36, pp. 1,367–1,374.

Garside, A.L., Smith, M.A., Chapman, L.S., Hurney, A.P. and Magarey, R.C. (1997) "The yield plateau in the Australian sugar industry: 1970–1990", in B.A. Keating and J.R. Wilson (eds), *Intensive Sugarcane Production, Meeting the Challenges Beyond 2000*, CAB International, Brisbane, Ch. 7.

Geijskes, R.J., Wang, L.F., Lakshmanan, P., McKeon, M.G., Berding, N., Swain, R.S., Elliott, A.R., Grof, C.P.L., Jackson, J. and Smith, G.R. (2003) "SmartsettTM seedlings: tissue culture seed plants for the Australian sugar industry", *Sugarcane International*, May/ June, pp. 13–17.

Gilbert, R.A., Gallo-Meagher, M., Comstock, J.C., Miller, J.D., Jain, M. and Abouzid, A. (2005) "Agronomic evaluation of sugarcane lines transformed to resistance to sugarcane mosaic strain E", *Crop Science*, 45, pp. 2,060–2,067.

Gilbert, R.A., Glynn, N.C., Comstock, J.C. and Davis, M.J. (2009) "Agronomic performance and genetic characterization of sugarcane transformed for resistance to sugarcane yellow leaf virus", *Field Crop Research*, 111, pp. 39–46.

Groenewald, J.-H. and Botha, F.C. (2008) "Down-regulation of pyrophosphate: fructose 6-phosphate 1-phosphotransferase (PFP) activity in sugarcane enhances sucrose accumulation in immature internodes", *Transgenic Research*, 17, pp. 85–92.

Guimarães, C.T., Honetcutt, R.J., Sills, G.R. and Sobral, B.W.S. (1997a) "Genetic maps of *Saccharum officinarum* L. and *Saccharum robustum* Brandes & Jesw Ex Grassl", *Genetic Molecular Biology*, 22, pp. 125–132

Guimarães, C.T., Sills, G.R. and Sobral, B.W.S. (1997b) "Comparative mapping of Andropogoneae: *Saccharum* L. (sugarcane) and its relation to sorghum and maize", *Proceedings of the National Academy of Science* (USA), 94, pp. 14261–14266.

Harm de Boer, G. (2009) "Experience with High Fibre Cane – A Review", National Agricultural Conference, Barbados, July, pp. 17–18.

Hatch, M.D. and Glasziou, K.T. (1963) "Sugar accumulation cycle in sugar cane. II. Relationship of invertase activity to sugar content and growth rate in storage tissue of plants grown in controlled environments", *Plant Physiology*, 38, pp. 344–348.

Hebert, L.P. and Rice, E.R. (1972) "Maturity studies of commercial sugarcane varieties in Florida", *Proceedings of the International Society of Sugar Cane Technologists*, 14, pp. 137–143.

Heinz, D.J. and Tew, T.L. (1987) "Hybridisation procedures", in D.J. Heinz (ed.), *Sugarcane Improvement Through Breeding*, Elsevier, Amsterdam.

Hoarau, J.Y., Grivet, L., Offmann, B., Robin, L.-M., Diorflar, J.-P., Payet, J., Hellmann, A., D'Hont, A. and Glaszmann, J.C. (2002) "Genetic dissection of a modern cultivar (*Saccharum* spp.). II. Detection of QTLs for yield components", *Theoretical and Applied Genetics*, 105, pp. 1027–1037.

Hoarau, J.Y., Offmann, B., D'Hont, A., Risterucci, A.M., Roques, D., Glaszmann, J.C. and Grivet, L. (2001) "Genetic dissection of a modern cultivar (*Saccharum* spp.). I. Genome mapping with RFLP", *Theoretical and Applied Genetics*, 103, pp. 84–97.

Hogarth, D.M. (1977) "Fiji inheritance studies", *Proceedings of the Queensland Society of Sugar Technologists*, 44, pp. 89–91.

Hogarth, D.M. and Kingston, G. (1984) "The inheritance of ash in juice from sugarcane", *Sugar Cane*, 1, pp. 5–9.

Hogarth, D.M., Wu, K.K. and Heinz, D.J. (1981) "Estimating genetic variance in sugar cane using a factorial cross design", *Crop Science*, vol. 21, no. 1, pp. 21–25.

Illovo (2010) www.illovo.co.za/Libraries/2010 Annual Report/Annual Report 2010 Part 6.sflb.ashx.

Inman-Bamber, N.G. (1994) "Temperature and seasonal effects on canopy development and light interception of sugarcane", *Field Crop Research*, 36, pp. 41–51.

Inman-Bamber, N.G., Muchow, R.C. and Robertson, M.J. (2002) "Dry Matter Partitioning of Sugarcane in Australia and South Africa", *Field Crop Research*, 76, pp. 71–84.

Irvine, J.E. (1975) "Relations of photosynthetic rates and leaf and canopy characters to sugarcane yield", *Crop Science*, 15, pp. 671–676.

Jackson, P. (1994) "Genetic relationships between attributes in sugarcane clones closely related to *Saccharum spontaneum*", *Euphytica*, vol. 79, nos 1–2, pp. 101–108.

Joomun, N., Ganeshan, S. and Dookun-Saumtally, A. (2010) "Identification of three armyworm species (Lepidoptera: Noctuidae) using DNA barcodes and restriction enzyme digestion", *Proceedings of the International Society of Sugar Cane Technologists*, 27, p. 11.

Julien, R. (1972) "The photoperiodic control of flowering in *Saccharum*", *Proceedings of the International Society of Sugar Cane Technologists*, 55, pp. 389–399.

Keating, B.A., Robertson, M.J., Muchow, R.C. and Huth, N.I. (1999) "Modelling sugarcane production system 1: development and performance of the sugarcane module", *Field Crops Research*, 61, pp. 253–271.

Klass, D.L. (2004) "Biomass for renewable energy and fuels", in *Encyclopedia of Energy*, Elsevier, USA.

Lau, A.W.A.F., Kong Win Chang, K.T.K.F. and Gunness, D. (2003) "Factors influencing surplus electricity production in co-generation plants", ISSCT Co-Products Workshop: Ethanol Production and Use, 14–18 July, Piracicaba, São Paulo (programme and abstracts: CD-ROM, Paper).

Legendre, B.L. (1995) "Potential for increasing sucrose content of sugar cane: an assessment of recurrent selection in Louisiana", *Sugar Cane*, 3, pp. 4–8.

Leibbrandt, N.B. and Synman, S.J. (2003) "Stability of gene expression and agronomic performance of a transgenic herbicide-resistant sugarcane line in South Africa", *Crop Science*, 43, pp. 671–677.

Lo, C.C., Chia, Y.H., Chen, W.H., Shang, K.C., Shen, I.S. and Shih, S.C. (1978) "Collecting *Miscanthus* in Taiwan", *Proceedings of the International Society of Sugar Cane Technologists*, 16, pp. 59–69.

Lorenzo, J.C., Ojeda, E., Espinosa, A. and Borroto, C. (2001). "Field performance of temporary immersion bioreactor-derived sugarcane plants", *In Vitro Cellular and Development Biology-Plant*, 37, pp. 803–806.

Mamet, L.D., Galwey, N.W. and Julien, M.H.R. (1996) "Earliness of ripening in sugar cane (*Saccharum* spp. L.) in Mauritius: variation and inheritance studies", *Sugar Cane*, 4, pp. 3–11.

Mamet, L.D. and Travailleur, A. (1998) "International survey of sugar cane selection schemes", Abstract, 4th ISSCT Breeding and Germplasm Workshop, 4–9 May, Mauritius.

Mariotti, J.A. (1972) "Associations among yield and quality components in sugarcane hybrid components", *Proceedings of the International Society of Sugar Cane Technologists*, 14, pp. 297–302.

Miller, J.D. (1977) "Combining ability and yield component analyses in a five-parent diallel cross in sugar cane", *Crop Science*, 17, pp. 545–547.

Ming, R., DelMonte, T., Moore, P.H., Irvine, J.E. and Paterson, A.H. (2002) "Comparative analysis of QTLs affecting plant height and flowering time among closely-related diploid and polyploidy genomes", *Genome*, 45, pp. 794–803.

Ming, R., Liu, S.C., Moore, P.H., Irvine, J.E. and Paterson, A.H. (2001) "QTL mapping in a complex autoploidy: genetic control of sugar content in sugarcane", *Genome Research*, 11, pp. 2075–2084.

Monteiro-Vitorello, C.B., Camargo, L.E., Van Sluys, M.A., Kitajima, J.P., Truffi, D., do Amaral, A.M., Harakava, R., de Oliveira, J.C., Wood, D., de Oliveira, M.C., Miyaki, C., Takita, M.A., da Silva, A.C., Furlan, L.R., Carraro, D.M., Camarotte, G., Almeida N.F Jr, Carrer, H., Coutinho, L.L., El-Dorry, H.A., Ferro, M.I., Gagliardi, P.R., Giglioti, E., Goldman, M.H., Goldman, G.H., Kimura, E.T., Ferro, E.S., Kuramae, E.E., Lemos, E.G., Lemos, M.V., Mauro, S.M., Machado, M.A., Marino, C.L., Menck, C.F., Nunes, L.R., Oliveira, R.C., Pereira, G.G., Siqueira, W., de Souza, A.A., Tsai, S.M., Zanca, A.S., Simpson, A.J., Brumbley, S.M. and Setúbal, J.C. (2004) "The genome sequence of the gram-positive sugarcane pathogen Leifsonia xyli subsp xyli", *Molecular Plant Microbe Interaction*, vol. 17, no. 8, pp. 827–883.

Moore, P.H. (1995) "Temporal and spatial regulation of sucrose accumulation in the sugarcane stem", *Australian Journal of Plant Physiology*, 22, pp. 661–679.

Mundil, D., Ramdoyal, K. and Rivet, L., See Cheong, F.M.N. and Chintaram, E. (2005) "Information technology as a tool to improve the utilisation and management of sugar cane germplasm at the MSIRI", *Proceedings of the 7th Meeting of Agricultural Scientists*, 4–6 May, MSIRI (Abstract only).

Nayamuth, A.R., Mangar, M., Ramdoyal, K. and Badaloo, M.G.H. (2005) "Early sucrose accumulation, a promising characteristic to use in sugarcane improvement programs", *Proceedings of the International Society of Sugar Cane Technologists*, 25, pp. 421–429.

Nishiyama Jr, M.Y., Vicente, F.F.R., Lembke, C.G., Sato, P.M., Dal-Bianco, M.L., Fandiño, R.A., Hotta, C.T. and Souza, G.M. (2010) "The SUCEST-FUN regulatory network database: Designing an energy grass", *Proceedings of the International Society of Sugar Cane Technologists*, 27.

Nuss, K.J. and Berding, N. (1999) "Planned recombination in sugar cane breeding: artificial initiation of flowering in sugar cane in sub-tropical and tropical conditions", *Proceedings of the International Society of Sugar Cane Technologists*, vol. 23, no. 2, pp. 504–508.

Nuss, K.J., Harding, L. and Blose, M.J. (1989) "The use of a computer in a sugarcane breeding programme", *Proceedings of the International Society of Sugar Cane Technologists*, 20, pp. 963–967.

Odda, M., Nadif, A., Madrane, A. and Hesse, F.W. (1992) "In Vitro Culture of the Apex for Cleaning Sugar Cane Seeds from Mosaic Virus", Société de Technologie Agricole de Sucrière de l'Ile Maurice. 9ème Congrès International. 12–16 October, MSIRI (Abstract only).

Oliveira, K.M., Pinto, L.R., Marconi, T.G., Margarido, G.R.A., Pastina, M.M., Teixeira, L.H.M., Figueira, A.V., Ulian, E.C., Garcia, A.A. and Souza, A.P. (2007) "Functional integrated genetic map based on EST-markers for a sugarcane (*Saccharum* spp.) commercial cross", *Molecular Breeding*, vol. 20, no. 3, pp. 194–220.

Osgood, R.V. (2003) "Cane planter, sugarcane yield and record yield", *Sugar Journal*, 66, p. 7.

O'Shea, M. (2000) "Sucrose accumulation", in D.M. Hogarth and P.G. Allsopp (eds), *Manual of Cane Growers*, BSES, Brisbane, pp. 82–88.

Parmessur, Y., Al-Janabi, S. and Dookun-Saumtally, A. (2010) "Mapping of sugarcane variety M 134/75 using EST-SSR markers", *Proceedings of the International Society of Sugar Cane Technologists*, 27, Molecular Biology Section.

Parmessur, Y., Al-Janabi, S., Saumtally, S. and Dookun-Saumtally, A. (2002) "Sugar cane yellow leaf virus and sugar cane yellow phytoplasma: detection and elimination by tissue culture", *Plant Pathology*, 51, pp. 561–566.

Pieretti, I., Royer, M., Barbe, V., Carrere, S., Koebnik, R., Cociancich, S., Couloux, A., Darrasse, A., Gouzy, J., Jacques, M.-A., Lauber, E., Manceau, C., Mangenot, S., Poussier, S., Segurens, B., Szurek, B., Verdier, V., Arlat, M. and Rott, P. (2009) "The complete genome of *Xanthomonas albilineans* provides new insights into the reductive genome evolution of the xylem-limited *Xanthomonadaceae*", *BMC Genomics*, 10, pp. 616.

Piperidis, G. and D'Hont, A. (2001) "Chromosome composition analysis of various *Saccharum* interspecific hybrids by genomic *in situ* hybridisation (GISH)", *Proceedings of the International Society of Sugar Cane Technologists*, 11, pp. 565–566.

Ramdoyal, K. and Badaloo, G.H. (1998) "Inheritance of agronomic traits in commercial hybrid sugar cane populations in contrasting environments and in different crop cycles", *Journal of Genetics and Breeding*, 52, pp. 361–368.

Ramdoyal, K. and Badaloo, G.H. (2002) "Prebreeding in sugar cane with an emphasis on the programme of the Mauritius Sugar Industry Research Institute", in J.M.M. Engels, V. Ramanatha Rao, A.H.D. Brown and M.T. Jackson (eds), *Managing Plant Genetic Diversity*, IPGRI, CABI Publishing, Wallingford, Ch. 29, pp. 307–321.

Ramdoyal, K. and Badaloo, M.G.H. (2007) "An evaluation of interspecific families of different nobilized groups in contrasting environments for breeding novel sugarcane clones for biomass", *Proceedings of the International Society of Sugar Cane Technologists*, 26, pp. 625–632.

Ramdoyal, K., Badaloo, G.H. and Goburdhun, P. (2003) "An international survey of sugar cane crossing programmes", *Sugar Cane International*, July–August, pp. 3–11.

Ramdoyal, K., Badaloo, G. and Mangar, M. (1995a) "Effect on seed setting of potassium metabisulphite in sugar cane crossing solution", *Sugar Cane*, 2 (March–April), pp. 3–7.

Ramdoyal, K., Badaloo, G. and Mangar, M. (1995b) "Improvement in maturation techniques in cane hybridization", *Sugar Cane*, 6, pp. 12–16.

Ramdoyal, K., Bissessur, D., Badaloo, G.H. and Mungur, H. (2007) "M 1672/90, a high yielding sugar cane variety with increased fibre content for cogeneration released in 2007 for the subhumid zone of Mauritius", *Revue Agricole et Sucrière de l'Ile Maurice*, vol. 86, no. 1 (January–April), pp. 27–37.

Ramdoyal, K., Mamet, L.D., Rivet, L., Badaloo, G. and Domaingue, R. (1999a) "Sugar cane hybridization procedures and computer-aided crossing at the Mauritius Sugar Industry Research Institute", *Sugar Cane*, September, pp. 5–10.

Ramdoyal, K., Santchurn, D., Badaloo, G. and Bissessur, D. (2009) "Comparative analysis of commercial productivity of new and older sugarcane varieties in Mauritius", 2009 ISSCT Breeding and Germplasm Workshop: Sugarcane Crop Improvement – Opportunities and Challenges, 17–21 August, Novotel Rockford Palm Cove, Cairns, Qld, pp. 43 (Abstract only).

Ramdoyal, K., Sullivan, S., Badaloo, G.H., Dhayan, S., Saumtally, S. and Domaingue, R. (2001) "Inheritance of resistance to yellow spot in segregating populations of sugarcane", *Proceedings of the International Society of Sugar Cane Technologists*, vol. 24, no. 2, pp. 422–429.

Ramdoyal, K., Sullivan, S., Lim Shin Chong, L.C.Y., Badaloo, G.H., Saumtally, S. and Domaingue, R. (1999b) "The genetics of rust resistance in sugar cane seedling populations", *Theoretical and Applied Genetics*, 100, pp. 557–563.

Rao, P.S., Davis, H. and Simpson, C. (2007) "New sugarcane varieties and year round sugar and ethanol production with bagasse-based cogeneration in Barbados and Guyana", *Proceedings of the International Society of Sugar Cane Technologists*, pp. 1169–1176.

Richard, C. (2003) "Understanding cane maturity", *Sugar Journal*, vol. 66, no. 6, p. 6.

Roach, B.T. (1968) "Quantitative effects of hybridisation in *Saccharum officinarum* × *Saccharum spontaneum* crosses", *Proceedings of the International Society of Sugar Cane Technologists*, 13, pp. 939–954.

Roach, B.T. and Daniels, J. (1987) "A review of the origin and improvement of sugar cane", Copersucar International Sugar Cane Breeding Workshop, Copersucar Technology Center, Piracicaba-SP, Brazil, pp. 1–31.

Rohwer, J.M. and Botha, F.C. (2001) "Analysis of sucrose accumulation in the sugar cane culm on the basis of *in vitro* kinetic data", *Biochem Journal*, 358, pp. 437–445.

Rossouw, D., Bosch, S., Kossmann, J.M., Botha, F.C. and Groenewald, J.H. (2007) "Down-regulation of neutral invertase activity in sugarcane cell suspension cultures leads to increased sucrose accumulation", *Functional Plant Biology*, 34, pp. 490–498.

Rott, P. and Davis, M.J. (2004) "Identification of sugarcane bacterial pathogens and assessing their genetic diversity using molecular techniques", in G.P. Rao, A.S. Saumtally and P. Rott (eds), *Sugarcane Pathology. Bacterial and Nematode Diseases*, Science Publishers, Enfield, pp. 175–183.

Sainz, M.B., Chaudhuri, S., Jepson, I. and Dale, J. (2010) "Prospects of cellulosic ethanol from sugarcane bagasse", *Proceedings of the International Society of Sugar Cane Technologists*, 27, Molecular Biology Section.

Santchurn, D. (2010) "Evaluation and selection of different types of sugarcane varieties for multi-purposes use from a population of interspecific derived clones in Mauritius", MSc Thesis, University of Free State, South Africa.

Sills, G., Bridges, W., Al-Janabi, S. and Sobral, B.W.S. (1995) "Genetic analysis of agronomic traits in a cross between sugarcane (*Saccharum officinarum* L.) and its presumed progenitor (*S. robustum* Brandes, Jesw, Ex.Grassl)", *Molecular Breeding*, 1, pp. 355–363.

Singels, A., Donaldson, R.A. and Smit, M.A. (2004) "Opportunities for improved biomass production and partitioning in sugarcane: theory and practice", Review of Knowledge of sugarcane physiology and climate–crop–soil interactions, Sugar Research and Development Corporation Final Report No. CSE006, pp. 61–62.

Singels, A., Donaldson, R.A. and Smit, M.A. (2005) "Sugarcane physiology: Integrating from cell to crop to advance sugarcane production", *Field Crops Research*, vol. 92, nos 2–3, pp. 291–303.

Singels, A. and Inman-Bamber, N.G. (2002) "The responses of sugarcane to water stress: preliminary results from a collaborative project", *Proceedings of the South African Sugar Technologists Association*, 76, pp. 240–244.

Skinner, J.C., Hogarth, D.M. and Wu, K.K. (1987) "Selection methods, criteria and indices", in D.J. Heinz (ed.), *Sugarcane Improvement through Breeding*, Elsevier, Amsterdam.

Smith, G.R., Joyce, P.A., Nutt, K.A., McQualter, R.B, Taylor, G.O. and Allsopp, P.G. (1999) "Evaluation of transgenic sugarcane engineered for resistance to mosaic and canegrubs", *Proceedings of the International Society of Sugar Cane Technologists*, 23, pp. 291–297.

Soopramanien, G.C. (1979) "The physiological basis of sucrose yield variation in four sugarcane varieties planted and harvested on four different dates under three contrasting environments", PhD Thesis, University of Reading.

Sreenivasan, T.V., Ahloowalia, B.S. and Heinz, D.J. (1987) "Cytogenetics", in D.J. Heinz (ed.), *Sugar Cane Improvement through Breeding*, Elsevier, Amsterdam, pp. 211–253.

Stevenson, G.C. (1965) *Genetics and Breeding of Sugar Cane*, Longman, London.

Sun-Yuan, H., Lin, C.J. and Lo, C.C. (1991) "Personal computer as an aid in sugarcane crossing program", *Report of the Taiwan Sugar Research Institute*,132, pp. 1–11.

Terajima, Y., Matsuoka, M., Irei S., Sakaigaichi, T., Fukuhara, S., Ujihara, K., Ohara, S., Tatsuhiro, H. and Sugimoto, A. (2007) "Breeding for high-biomass sugarcane and its utilisation in Japan", *Proceedings of the International Society of Sugar Cane Technologists*, 26, pp. 759–763.

Terajima, Y., Matsuoka, M., Ujihara, K., Irei, S., Fukuhara, S., Sakaigaichi, T., Ohara, S., Tatsuhiro, H. and Sugimoto, A. (2005) "The simultaneous production of sugar and biomass ethanol using high-biomass sugarcane derived from inter-specific and inter-generic cross in Japan", Proceedings of Biomass – Asia Workshop 2 (Abstract), Bangkok, p. 65.

Tew, T.L. and Cobill, R.M. (2008) "Genetic improvement of sugarcane (*Saccharum* spp.) as an energy crop", in W. Vermerris (ed.), *Genetic Improvement of Bioenergy Crops*, Springer Science and Business Media, New York, pp. 249–272.

Tomkins, J.P., Yu, Y., Miller, S.H., Frisch, D.A., Woo, S.S. and Wing, R.A. (1999) "A bacterial artificial chromosome library for sugarcane", *Theoretical and Applied Genetics*, 99, pp. 419–424.

Van Dillewijn, C. (1952) *Botany of Sugarcane*, The Chronica Botanica, Waltham, MA.

Vettore, A.L., da Silva, F.R., Kemper, E.L., Souza, G.M., da Silva, A.M., Ferro, M.I.T., Silva, F.H., Giglioti, E.A., Lemos, M.V.F., Coutinho, L.L., Norbrega, M.P., Carrer, H., França, S.C., Bacci, M. Jr, Goldman, M.H.S., Gomes, S.L., Nunes, L.R., Carnago, L.E.A., Siqueira, W.J., Van Sluys, M.-A., Thiemann, O.H., Kurarnae, E.E., Santelli, R.V., Marino, C.L., Targon, M.L.P.N., Ferro, J.A., Silveira, H.C.S., Marini, D.C., Lemos, E.G.M., Tambor, J.H.M., Carraro, D.M., Roberto, P.G., Martins, V.G., Goldman, G.H., de Oliveira, R.C., Truffi, D., Colombo, C.A., Rossi, M., de Araujo, P.G., Sculaccio, S.A., Angella, A., Lima, M.M.A., de Rosa, V.E. Jr, Siviero, F., Coscrato, V.E., Machado, M.A., Grivet, L., Di Mauro, S.M.Z., Nobrega, F.G., Menck, C.F.M., Braga, M.D.V., Telles, G.P., Cara, F.A.A., Pedrosa, G., Meidanis, J. and Arruda, P. (2003) "Analysis and functional annotation of an expressed sequence tag collection for tropical crop sugarcane", *Genome Research*, vol. 13, no. 12, pp. 2725–2735.

Victoria, J.I., Avellaneda, M.C., Angel, J.C. and Guzman, M.L. (2005) "Resistance to sugarcane yellow leaf virus in Colombia", *Proceedings of the International Society of Sugar Cane Technologists*, 25, pp. 664–669.

Wagih, M.E., Gordon, G.H., Ryan, C.C. and Adkins, S.W. (1995) "Development of an axillary bud culture technique for Fiji disease virus elimination in sugar cane", *Australia Journal of Botany*, 43, pp. 135–143.

Wang, L.P., Jackson, P.A., Lu, X., Fan, Y.-H., Foreman, J.W., Chen, X.-K., Deng, H.-H., Fu, C., Ma, L. and Aitken, K.S. (2008) "Evaluation of Sugarcane × *Saccharum spontaneum* progeny for biomass composition and yield components", *Crop Science*, 48, pp. 951–961.

Wang, M.L., Goldstein, C., Su, W., Moore, P.H. and Albert, H.H. (2005) "Production of biologically active GM-CSF in sugarcane: a secure biofactory", *Transgenic Research*, 14, pp. 167–178.

Wei, X., Stringer, J., Jackson, P. and Cox, M. (2007) "Maximising Whole Industry Benefits from the Australian Sugarcane Improvement Programme through an Optimal Genetic Evaluation System", Final Report – SRDC project BSS267.

Whittaker, A. and Botha, F.C. (1997) "Carbon partitioning during sucrose accumulation in sugarcane internodal tissue", *Plant Physiology*, 115, pp. 1651–1659.

Woodard, S.L., White, S.G., Damaj, M.B., Gonzalez, J., Mirkov, E.T. and Nikolov, Z.L. (2009) "Evaluation of processing options for transgenic sugarcane tissue expressing bovine lysozyme", American Society of Agricultural and Biological Engineers, Reno, NV, 21–24 June (Abstract only).

Wu, K.K. (1987) "Computer applications in sugarcane improvement", in D.J. Heinz (ed.), *Sugarcane Improvement through Breeding*, Elsevier, Amsterdam.

Wu, K.K., Burnquist, W., Sorrells, M.E., Tew, T.L., Moore, P.H. and Tanksley, S.D. (1992) "The detection and estimation of linkage in polyploids using single-dose restriction fragments", *Theoretical and Applied Genetics*, vol. 83, pp. 294–300

Wu, K.K., Heinz, D.J. and Meyer, H.K. (1983) "Heritability of sugarcane smut resistance and correlation between smut grade and yield components", *Crop Science*, 23, pp. 54–56.

Wu, L. and Birch, R.G. (2007) "Doubled sugar content in sugarcane plants modified to produce a sucrose isomer", *Plant Biotechnology Journal*, 5, pp. 109–117.

Xie, Y., Wang, M., Xu, D., Li, R. and Zhou, G. (2009) "Simultaneous detection and identification of four sugarcane viruses by one-step RT-PCR", *Journal of Virological Methods*, 162, pp. 64–68.

Zhang, L., Xu, J. and Birch, R.G. (1999) "Engineered detoxification confers resistance against a pathogenic bacterium", *Nature Biotechnology*, 17, pp. 1021–1024.

3

AGRO-ECOLOGICAL ZONING AND BIOFUELS

The Brazilian experience and the potential application in Africa

Alexandre Betinardi Strapasson, Antonio Ramalho-Filho, Denilson Ferreira, José Nilton de Souza Vieira and Luís Carlos Mavignier de Araújo Job

Introduction

This chapter presents the Brazilian experience of agro-ecological zoning for bio-fuels production and considers how the zoning approach could be adapted and applied in African countries in order to promote the sustainable production of food and fuels. Agro-ecological zoning is a platform for countries interested in starting or expanding biofuels production in harmony with food production and environment conservation. This zoning offers a key primary source of technical support for land use planning aimed at improving the sustainability of agriculture.

The Brazilian ethanol and biodiesel programmes represent the largest and argu-ably the most successful examples for incorporating biofuels into a national energy mix. Moreover, they have led to improved use of zoning as a guide to overall agri-cultural production strategies. The experience with ecological-economic and climate risk zonings paved the way to develop a pioneering model – agro-ecological zoning (AEZ) – which has been an important and innovative tool for the biofuels industry. "Agro" is used here in a broad sense to mean all aspects of plant and animal husbandry, including forestry, cattle, crops and environment sustainability. Sugar cane was the first biofuel crop to be evaluated and expanded using AEZ, fol-lowed by oil palm. Both zonings were coordinated by the Brazilian Ministry of Agriculture, Livestock and Food Supply (MAPA) and carried out by its research board, the Brazilian Agricultural Research Corporation (Embrapa), together with other research centres.

Brazil is a leading country for renewable energies, which represent 47 per cent of its energy mix and around 85 per cent of the electricity generation (EPE, 2010). It is the world's largest producer of sugar and ethanol from sugar cane, consuming already more ethanol than gasoline at a national level. Although it must be acknowledged that the Brazilian biofuels programmes cannot be simply copied or transposed onto other countries biofuels' strategies, the zoning experiences are more widely applicable and could strengthen the crucial role of agriculture in many African countries.

Zoning as a land use planning tool

Land use planning must be part of a strategic management policy to set guidelines for sustainable development. In the case of agriculture, this challenge is even greater, because of the need to increase feedstock production to meet the demand for biofuels. Given that land is a limited resource and land use for agriculture has significant environmental impacts, the identification of the productive potential of different areas is mandatory in order to limit these impacts to the lowest possible level.

A first step in developing any sort of tool for land use planning is to have a thorough database of technical information on different aspects, such as: soil, climate, topography, water resources and land use. Additionally, it is necessary to identify the areas not recommended for economic uses, including those areas prioritized for environmental preservation to protect sensitive ecosystems and maintain areas of high biodiversity. Areas recommended and not recommended for economic uses are identified through so-called ecological and economic zoning (EEZ). Once completed, the second step is to identify the activities most favourable for the areas considered suitable for economic use, and this is done through agro-ecological zonings, based on soil and climate surveys. The zoning concepts are discussed below.

Ecological-economic zoning

The goal of EEZ is to assist policy makers in developing programmes, focusing on the identification of priority areas for environmental preservation, as well as those with great potential for economic uses, especially for agriculture and mining. Thus, there is the need to involve different areas of government, especially those responsible for environmental policy and economic policy for agricultural production.

EEZ has tactical, social, legal and political features, and can be used by governments at different levels to define programmes and plans. EEZ is premised on the sustainable development of a given large area or region. Its main objective is to protect the environment. The approach defines zones based on a diagnosis of natural and socio-economic resources, creates scenarios and provides policy guidance. Hence, EEZ must be complemented by an agro-ecological zoning, which has a technical and operational characteristic, aimed at determining the best use of land on a local basis.

Zoning is an instrument for the development of spatial planning policies. Therefore, it is also important to conduct a dialogue with the local population, who will be affected by the implementation of the policy. This becomes evident when the zoning indicates the need to either restrict the land occupation or protect certain areas, for example by creating environmental reserves. The areas considered suitable for economic uses can be classified according to the type of investment appropriate to them, including constructing cities, agricultural use or exploitation of water resources and minerals. Therefore, the ecological and economic zoning is the first step towards promoting more efficient use of natural resources in harmony with environmental conservation goals.

Agricultural zoning

Agriculture has changed considerably in recent decades. The use of more sophisticated technology extends to different fields: genetics, machinery and equipment, fertilizers and agrochemicals. However, these new technologies only produce good results when they are used rationally. Hence it is essential to have information on the different conditions of soil and climate, in order to make appropriate choices about which crop to grow in which area.

This set of information can be obtained through agricultural zoning. The zoning has two main segments: pedological (identification and survey of different soil types, and interpretation in each region) and climatic (temperature range, rainfall and distribution of rainfall throughout the year). These two parts form what is called pedo-climatic zoning. Additional information, such as current land use, topography, logistics and land tenure should be considered. In the case of climate, it is important to have a long historical data series, especially because there can be wide variations from one year to another. Lack of such technical information can be an obstacle for the development of a mapping project. On the other hand, it is recommended to start up the study with available data and make future updates, rather than waiting for all required information.[1]

The gathering and organization of all these data will lead to the zoning for different crops. Basically it is a process where local pedological and climate information are compared with technical requirements for different crops and different varieties of a same crop. The results are shown in a set of maps, identifying the suitable areas to grow each one and the appropriate planting season (climatic risk zoning).

The agricultural zoning and EEZ form agro-ecological zoning (AEZ), which indicates the suitability or non-suitability of different areas. Information from this zoning is then analysed by an interdisciplinary group of policy makers responsible for developing public policies. The results of AEZ analyses should be available to farmers to help them in making decisions about what to grow on their land. AEZ also provides useful tools to help private investors in choosing or evaluating project locations.

The zoning experience in Brazil

The available zoning systems will be discussed below in reference to the Brazilian experience with the three types of zoning: ecological-economic zoning, agricultural zoning and, more recently, agro-ecological zoning.

Ecological and economic zoning

Brazil, as the biggest Latin American country, has many different biomes, soils and climatic conditions. It is rich in biodiversity and natural resources, but these abundant resources must be used efficiently and responsibly. The Brazilian government has therefore adopted various measures to promote socio-economic development with respect to the environment, publishing a vast legal framework on this issue: of special importance is the Brazilian Forest Code, established by Federal Law 4.771 in 1965, and the new Constitution of 1988, which provided for the establishment of ecological-economic zoning for all Brazilian states. According to the legislation, the Ministry of the Environment is responsible for coordinating the EEZ at the national level, but the zoning must be implemented autonomously by each Brazilian state. As it requires a long process that depends on many technical studies and public consultations, as well as the political environment for approval by state legislatures, only a few states have concluded their EEZs.[2]

Therefore, state governments have some autonomy to develop their policies on land use, but they must observe certain restrictions imposed by the Federal Constitution and national environmental policy. For example, in the case of the Brazilian Amazon, which constitutes 42 per cent of the country, there are important restrictions on land use. On the other hand, until the approval of the EEZ, farmers must preserve 80 per cent of the area of their land; i.e. they can use only 20 per cent of the land for agriculture or livestock. The EEZ can suggest areas with high potential for agriculture and low environmental impacts, where the preserved area can be reduced to 50 per cent, but the environmental protection of sensitive areas may exceed 80 per cent of the land. Additionally, in the Amazon region, there is macro-ecological-economic zoning, named the Macro-EEZ project, which is coordinated by the Brazilian Ministry of the Environment. The Macro-EEZ aims to contribute to the Sustainable Amazon Plan, establishing policy alignments towards a sustainable development model for the region.

Climate risk zoning

In the 1970s, the MAPA established a programme to support agricultural activities, by providing technical information to farmers, with guidelines for planting and crop management in order to reduce the risk of production failure. Following these guidelines, farmers could more easily access loans from banks and obtain agricultural insurance. Although the MAPA promoted important advances through this programme, major changes occurred only in the late 1990s, with the upgrade and integration of different databases, to allow the development of so-called agricultural zoning of climate risk, involving a variety of institutions and expertise.

Many sorts of information are used by this zoning system. There is a large database, which contains information on natural resources for all Brazilian municipalities owned by IBGE (Brazilian Institute for Geography and Statistics). Another database, owned by INMET (Brazilian Institute for Meteorology), records the time series distribution of rainfall and temperature variations. These databanks have been continuously updated using satellite images and rural surveys. Additionally, there is software that compares pedological and climate information with the agronomic characteristics of different crops, considering available varieties. The software generates maps indicating the favourable conditions for each crop. Furthermore, this zoning indicates the appropriate sowing dates, crop variety and the technological package to be used. It must be periodically updated, mainly because of the variations of climate conditions and the technology in agriculture, which is in a process of continuous improvement through new varieties, new fertilizers and pesticides.

In Brazil farmers must follow technical information from the zoning, or else they will not get loans from banks or access to insurance. This rule, associated with the use of new agricultural technologies, has induced significant gains in the crop yields. Figure 3.1 illustrates the increase in productivity in Brazil, and suggests opportunities for African countries to incorporate better tropical-based agricultural technologies. For example, from the crop-season year of 1990/1991 to 2010/2011 the grain production in Brazil rose 171.9 per cent or 8.2 per cent per year, while the planted area increased only 30.0 per cent or 1.4 per cent per year. This represents a productivity gain over the past 21 years of 109.2 per cent or 5.2 per cent per year, as shown in Figure 3.1, thanks to advances in technology and agricultural management in association with zonings.

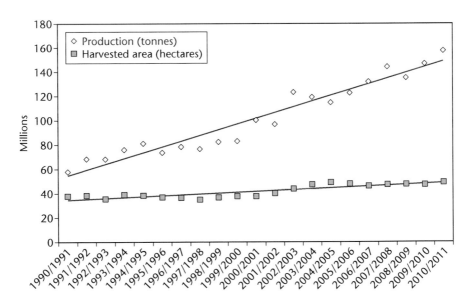

FIGURE 3.1 Evolution of Brazilian grain production and harvested area – 1990 to 2011 (source: MAPA (CONAB, 2010); crop-season 2010/2011 forecasted by MAPA (CONAB, 2011))

Agro-ecological zoning

Agro-ecological zoning (AEZ) aims to provide supporting information to decision makers regarding land use options in development projects. It is a prerequisite for optimal land allocation according to different purposes and circumstances. AEZ requires collection of biophysical and socio-economic data, systematic data analysis and making information available for all people involved in the process of zoning. An agro-ecological zoning must be understood as a complement to economic-ecological zoning (EEZ), which in addition to the physical-biotic information, also includes socio-economic, strategic and political aspects. In a more comprehensive sense, AEZ is a well-described and planned spatial framework for assessing sustainable land use potential and providing a set of conditions and recommendations for land use, considering the region's available resources, environmental requirements and development objectives.

As a reference for African countries, Brazil has already published AEZ for two important crops: sugar cane and oil palm, in 2009 and 2010 respectively. Regarding the technical methodologies, the oil palm AEZ methodology had already been published in 2008 (Ramalho-Filho, 2008), before the sugar cane zoning, which basically used the same principles.

Agro-ecological zoning offers a powerful tool to supply a technical basis for supporting public policies as well as private decisions for sustainable land use planning. AEZ considers additional variables beyond those in climate risk zoning, including current land use, topography, water resources and environmental protection areas. In a stricter sense, agro-ecological zoning involves spatializing the land potential of a region for a given crop or product as a base for planning sustainable land use in harmony with biodiversity and conservation objectives. Hence, some important benefits of this zoning are to:

*   Have well-planned management and improved control of agricultural activities, aimed at wise use of scarce and fragile resources, through a participatory approach.
*   Show the potential, vulnerability and availability of land at regional and local levels, as a first step towards achieving sustainable use of natural resources.
*   Improve land use economics, using land according to its suitability to increase productivity and lower costs, thereby increasing economic competitiveness.
*   Support social, economic and environmental planning on a sustainable basis.
*   Identify alternative uses of land and water resources for different crops or products and regions, by using yield criteria for different production technologies to optimize output.
*   Ground research in the biophysical data and support extension and technical assistance to farmers and entrepreneurs.
*   Promote institutional strengthening and cross-disciplinary analysis.

AEZ can] help both the public and private sectors to anticipate, manage and steer expansion in less developed areas so as to achieve optimal but sustainable land use.

Special attention should be given to areas where people are already settled and/or where there is evidence of sustainable resource use and opportunities for well-planned rural development. Different types of environmental attributes must be considered in agro-ecological zoning and used to perform land and climate evaluations based on crop requirements. Thus, the agro-ecological zoning supports policy decisions on land use, with consideration for distinct technological levels, so as to break the downward spiral of extensive agro-industrial production and the accompanying environmental degradation.

Sugar cane agro-ecological zoning

The sugar cane agro-ecological zoning was coordinated by the Department of Sugar Cane and Agroenergy at the Brazilian Ministry of Agriculture, Livestock and Food Supply (MAPA) through an official working group established by a Ministerial Decree,[3] which involved an interdisciplinary team, including representatives from the Ministry of the Environment. The results of this zoning were published by the Brazilian government as a legal framework,[4] and also as GIS (Global Information System) maps edited by Embrapa (Embrapa Soils Geoportal, 2010; Manzatto *et al.*, 2009).

The objective of the sugar cane AEZ was to stimulate further expansion of this crop for ethanol and sugar production, where the environmental impacts would be lower and the investments welcome. Therefore, in order to avoid random sugar cane expansion under a laissez-faire policy, the following areas were excluded, as a precautionary measure:

- Amazonia and Pantanal biomes;
- the hydrographical basin of the Paraguay River;
- areas with any type of native vegetation, in any Brazilian biome;
- areas without soil and climate-favourable conditions;
- areas that require full irrigation system;
- areas with declivity of more than 12 per cent;
- protected areas;
- indigenous reserves;
- areas with a high conservation value for biodiversity.

The sugar cane AEZ was a pioneering initiative for biofuels worldwide. It represents a high-quality technical tool for designing innovative public policies; construction of new mills and financial support from banks is only permitted for suitable areas under AEZ. The favourable areas are predominantly pasturelands, where the livestock production could be intensified, thus releasing areas for other crops. Table 3.1 shows the current Brazilian land use.

Pastureland in Brazil has been used inefficiently in many areas. These areas will likely be reduced in the coming decades, by incorporating new livestock technologies. Therefore, it is possible to convert some pasturelands for other crops without

TABLE 3.1 The Brazilian land use distribution

Land use	Area	
	million hectares	%
Pasture	172.3	20.2
Crops (annual and perennial), except sugar cane	68.9	8.1
Sugar cane★	8.1	0.9
Woods and agricultural forests	99.8	11.7
Amazon forest, protected areas, cities and others	502.2	59.0
Total	851.4	100.0

Source: IBGE (2010), based on Census 2006.

Note
★ MAPA (CONAB, 2010).

reducing livestock production. For example, in the State of São Paulo sugar cane has been expanding mainly over pastureland, which is decreasing in area; however, contrary to expectations, both beef and milk production have been increasing in this state. Such productivity gains are reflected to a lesser extent in Brazil as a whole, which is now the world's largest beef exporter, whereas 15 years ago it was a beef importer.

Regarding the environmental protection of the Amazon region, it must also be noted that currently it is not economically viable to produce sugar cane in this region, because of the long rainy season of equatorial climate, distance from the consumer market and precarious infrastructure conditions. Even so, AEZ is an important cautious instrument that could avoid potential future risks to both the biofuels sustainability and Amazon conservation. The same concern is valid for the Pantanal region. The observance of such zoning for financing and official authorization is legally mandatory in Brazil, in addition to fulfilling the environmental laws. As a result, there is no new sugar cane industrial project foreseen or being implemented in these two regions to date.

Figure 3.2 and Plate 3 show a summary of the potential areas for sugar cane expansion according to AEZ, excluding the Amazon and Pantanal regions and the Paraguay River basins. The potential suitable areas are 64.7 million hectares, amounting to 7.5 per cent of Brazilian territory and almost eight times the current sugar cane area (8.1 million hectares), although only some fraction of that potential is expected to be used for sugar cane.

Another important measure adopted by the sugar cane AEZ was to exclude areas with high declivity in order to stimulate the adoption of mechanical harvesting, since manual harvesting of sugar cane is undesirable for environmental and socio-economic reasons. The aim is to guide crop expansion only into areas suitable for mechanization, i.e. areas with declivity lower than 12 per cent, which is the highest declivity under which mechanical harvesters can operate viably at present. However, despite the fact that manual harvesting is arduous work, around 500,000 people

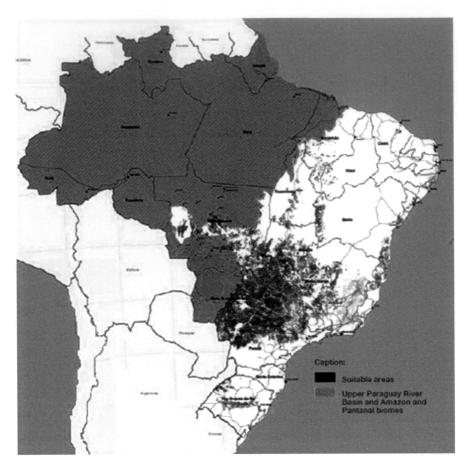

FIGURE 3.2 Sugar cane agro-ecological zoning in Brazil

work as sugar cane cutters out of approximately 1.3 million people formally working in the sugar cane sector in Brazil (Sousa and Macedo, 2009). Hence the conversion from manual to mechanical harvesting must be gradual in order to allow the incorporation of these workers into other activities. Some of these employees could work at the sugar and ethanol industries, for example. Even with mechanization, ethanol generates many more jobs than gasoline, per unit of energy delivered.

Therefore, the current challenge is to stimulate the best labour practices while also promoting mechanical harvesting. This social concern must also be addressed by African countries interested in the production of sustainable biofuel, since sustainability under the United Nations concept rests not only upon the environmental pillar, but upon the economic and social pillars too. In this sense, another important and useful example from Brazil is the "Social Agreement on Sugarcane", launched in 2009, which was voluntarily signed by workers' and millers' representatives; it establishes best working conditions in addition to the already strict labour legislation. More than 75 per cent of the Brazilian sugar cane millers have already signed that agreement, which was coordinated by the federal government.

Despite the social concerns related to job losses, the mechanical harvesting presents an important environmental gain in that it avoids the cane burning, which is a traditional agronomic technique used to facilitate the manual harvesting. The burning system is very common worldwide, including in most of the African sugar cane farms that supply the sugar industries. In this regard, another reference from Brazil is the Federal Law Project no. 6.077/2009 that establishes a phase-out period for cane burning. Moreover, avoiding cane burning benefits soil conservation: keeping a certain amount of sugar cane straw (trash) on the soil increases its carbon content, limits erosion and reduces the demand for herbicides. Cane burning also damages air quality, due to emissions of NOx, SOx, CO and aromatic compounds, particulate materials and hydrocarbon emissions (Strapasson and Job, 2007). In the future, it will likely be profitable to use part of the sugar cane straw for energy generation, but transportation costs are still a barrier (see Chapter 4). Another possibility would be to produce ethanol from both the straw and the bagasse, by ligno-cellulosic hydrolysis technology (still under research and development) or by the Fischer–Tropsch process. In this new scenario, other crops (e.g. sweet sorghum, switchgrass, *Pennisetum* sp. and eucalyptus) may become relevant for biofuel production, and the AEZ method can be applied to guide the expansion of these new crops as well.

Agro-ecological zoning for other crops

By using the AEZ approach for several different biofuels crops, it will be possible to obtain a better comparison as to the optimal location for different biofuels crops that might be considered in particular African countries or regions. In Brazil, the AEZ has also been completed for oil palm; the technical publications on the oil palm AEZ were published by Embrapa (Embrapa Soils Geoportal, 2010; Ramalho-Filho et al., 2008, 2010) and made official by the Brazilian government as a legal framework.[5] The method used for evaluating land was the "land suitability assessment system", developed by Ramalho-Filho and Beek (1995). Oil palm might be appropriate in some African equatorial regions with considerable annual rainfall (2,500–3,000 mm). The Brazilian experience could thus be adopted as another reference point for agro-ecological zoning, and similar approaches could be applied in the humid tropical areas of Africa (middle-west and middle-east coasts), observing their different environmental conditions and economic development strategies.

Oil palm production can generate income and jobs and in some cases can thereby reduce the pressure on native forests. However, to promote the production of palm and other crops in sensitive rainforest regions, basic environmental requirements must be strictly respected so as to achieve a symbiosis of forest conservation and agricultural activities (Araújo and Strapasson, 2009). There are already some successful cases of oil palm production in the Amazon region or in Brazil that might serve as useful examples in Africa, based on integrated production systems shared between companies and small and medium-scale farmers. However, there are significant opportunities for increasing cultivation of this crop, since Brazil is still a net importer of palm oil. The demand for vegetable oils in Africa for food and fuel is also likely to grow in the future. There are opportunities for intercropping (although not with sugar cane) that can be assessed through the AEZ methods, which can thereby support improvements in agro-biodiversity.

Partnerships and agro-ecological zonings for biofuels in Africa

Many African countries present high agronomical potentials for biofuels production, especially for sugar cane ethanol. In 2009 the Brazilian Ministries of Agriculture (MAPA) and Foreign Affairs (MRE), Embrapa and the Brazilian Centre for Management and Strategic Studies (CGEE) started a capacity-building programme on AEZ and biofuels in seven African countries, including Angola, Botswana, South Africa, Mozambique, Tanzania, Zambia and Zimbabwe. The aim of the programme was to share experiences with those countries on mapping tools in order to increase agricultural productivity and to reach a desirable equilibrium between food and biofuel production, since in some cases biofuels can even increase food production by promoting improvements in the infrastructure and the dissemination of agricultural technologies and markets.

African countries with existing sugar cane production generally have favourable conditions to start ethanol production by integrating new distilleries with the traditional sugar mill. This kind of combined sugar and ethanol mill can reduce economic risks and increase industrial efficiency as a whole, once it is possible to have an improved use of both molasses and sugar cane bagasse, for example (see Chapters 5, 6 and 10). Sudan offers a successful example of such integration. There are also many sugar cane-based ethanol projects (small and large scales) under analysis in Mozambique, Tanzania, Angola and Ghana, among other countries. In some cases, sweet sorghum and cassava would be other interesting alternatives for ethanol production.

Besides the high potential for sugar cane ethanol, there are many opportunities for biodiesel production in Africa as well. The oil palm, which is a native African crop, is an example of a profitable crop for both food and fuel production and could be an alternative for humid equatorial African countries. However, sustainable public policies (e.g. AEZ) must be put in place so as to avoid or reduce the environmental impacts of its expansion. Other potential biodiesel crops are *Jatropha curcas*, sunflower, soybean and other oil crops, but they still depend on public support. It is a continuous learning process and requires new agricultural partnerships between Brazil and African countries in biofuels as well as other areas. Embrapa has an office in Accra, Ghana, which works as a hub for technical cooperation in Africa.

Figure 3.3 and Plate 4 present overviews of the land cover in Africa, from which a small proportion could be used for biofuels production. AEZ and related analyses can help to optimize the energy yields – and thus minimize the land required – by demonstrating the best regions for particular crops and strategies. For some African

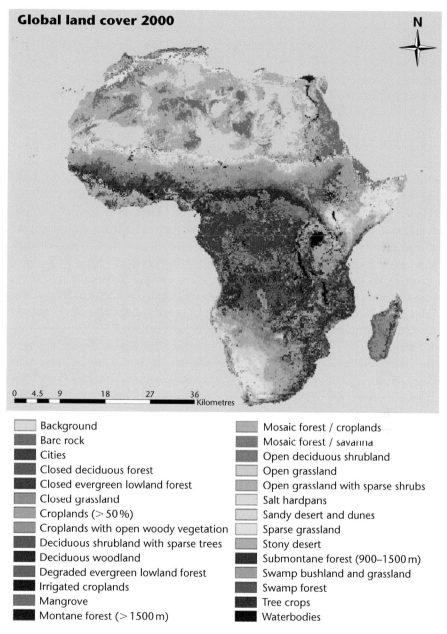

Global land cover 2000

N

Background		Mosaic forest / croplands
Bare rock		Mosaic forest / savanna
Cities		Open deciduous shrubland
Closed deciduous forest		Open grassland
Closed evergreen lowland forest		Open grassland with sparse shrubs
Closed grassland		Salt hardpans
Croplands (> 50%)		Sandy desert and dunes
Croplands with open woody vegetation		Sparse grassland
Deciduous shrubland with sparse trees		Stony desert
Deciduous woodland		Submontane forest (900–1500 m)
Degraded evergreen lowland forest		Swamp bushland and grassland
Irrigated croplands		Swamp forest
Mangrove		Tree crops
Montane forest (> 1500 m)		Waterbodies

0   4.5   9   18   27   36
                              Kilometres

FIGURE 3.3 Spatial extent on global land covers in Africa, 2000 (source: Compete (2010))

countries a more accurate scale and updated information are available, and these would allow more detailed technical analysis.

The scale of biofuels industries for African countries will depend on the characteristics of the specific project and region in which the plant will be built. For example, there are many African countries with small-scale bioenergy initiatives – including: Mali (jatropha electrification); Senegal (chardust briquettes and typha charcoal); Tanzania (sisal biogas and palm oil); Kenya (afforestation charcoal); and Ethiopia (ethanol stoves) – that are being monitored by the Bioenergy and Food Security Project (BEFS) of the FAO (2010a). In Tanzania there is also a mapping analysis on bioenergy and food security through the BEFS (FAO, 2010b). However, it is important to highlight that biofuels projects usually have a decreasing marginal cost, which imposes a minimum size on a project's economic viability.

There are many biofuels partnerships between Africa and Brazil, including those relating to AEZ. For example, the UEMOA (West African and Monetary Union) countries – Benin, Burkina Faso, Côte d'Ivoire, Guinea-Bissau, Mali, Niger, Senegal and Togo – have a memorandum of understanding (MOU) with Brazil on biofuels. The UEMOA is promoting common biofuels strategies for its member countries to improve access to modern biomass and to develop a domestic biofuels market (UEMOA and Le Hub, 2008). There is also a triangular partnership among Brazil, the European Union and the African Union, signed in October 2009 at the Third Brazil–EU Summit, which pledges to promote the sustainable development of biofuels in Africa. Another similar partnership is the Brazil–USA MOU on biofuels, in which both countries aim to promote a global biofuels market and to help some other countries to develop biofuels projects, such as Guinea-Bissau and Senegal. In the case of Senegal, the FGV (Fundação Getúlio Vargas, Brazil) has published a prospective study on biofuels, including zoning models (FGV, 2010). Brazil also has many bilateral MOUs with African countries on bioenergy, zoning and agricultural development.

In addition, another important partnership is the Bioenergy Contribution of Latin America, the Caribbean and Africa (LACAf-Cane) to the Global Sustainable Bioenergy Project (GSB). The LACAf-Cane project is supported by São Paulo Research Foundation (FAPESP) and hosted by the Brazilian Bioethanol Science and Technology Laboratory (CTBE). The main institutions involved in this project are: University of Campinas (Unicamp); Imperial College London; CTBE; Stellenbosch University (SU, South Africa); Oak Ridge National Laboratory (ONRL, USA); Stockholm Environment Institute (SEI, Sweden); eRcane (Réunion); National University of Colombia (UNAL) and Industrial University of Santander (UIS, Colombia); International Society of Sugar Cane Technologists (ISSCT, Mauritius); Centre for Energy, Environment and Engineering Zambia (CEEEZ, Zambia); University of São Paulo (USP); Brazilian Agricultural Research Corporation (Embrapa); Federal University of Rio de Janeiro (UFRJ); Federal University of Itajuba (Unifei); Management Centre for Strategic Studies (CGEE); Institute for International Trade Negotiations (Icone); and São Paulo Agency of Agribusiness Technology (APTA). Moreover, FAPESP is also sponsoring a research programme on bioenergy, the BIOEN programme, assembling Brazilian universities and industries towards advancing sugar cane-based ethanol in Brazil with potential contributions by other tropical countries from Latin America, Africa and Asia. This programme also involves international partners; e.g. Imperial College London. In spite of many important biofuels initiatives that are being implemented in Africa, many challenges are still ahead.

Biofuels policies and the zoning approaches

Zoning policies could promote biomass in a balanced way as a source of a number of "bio-products"; not only food and fuel, but also fibre, feed, fodder, wood, resins, chemical compounds and carbon stock. Bio-products are a new way to understand the agricultural potential as a whole. The concept is to explore all biochemical structures obtained from the conversion of solar radiation through photosynthesis. For instance, there are technologies under development to use sugar cane to produce also diesel, gasoline, kerosene, ethylene and many other bioproducts, besides sugar, electricity and ethanol. In addition, to attain the increasing demand for bioproducts, best practices of low carbon agriculture must be addressed, such as: no-tillage farming; crop-livestock integration; agro-forestry and silvo-pastoral systems; biochar; and crop rotation, especially with leguminous species (Strapasson, 2010).

There is a global concern on energy supply because of the need to strengthen energy security and reduce environmental impacts related to its production and use, especially regarding greenhouse gas emissions. Within this context, biofuels represent new and modern uses of agricultural biomass for energy purposes. Investments in both biomass power plants and liquid biofuels have grown rapidly in several parts of the world. Thus, biofuels represent a growing market for agriculture, which creates socio-economic benefits by increasing the number of jobs and income in rural areas. However, it also presents concerns, especially regarding two dimensions: the environmental risks associated with uncontrolled expansion of arable land, and the negative impact of food prices upon low-income populations (Vieira, 2004).

According to Vieira (2002), the relative increase in food prices could be positive and desirable as a way to increase food production worldwide. A continuous reduction in the relative price of food has occurred since the 1970s; as a consequence, the governments of developed countries increased their subsidies to agriculture in order to protect their farmers. With this subsidy policy it is possible to maintain or increase food production, even if market prices are below production costs, strongly penalizing unprotected farmers in the poorest countries. AEZ can be combined with spatial information on food security and food prices to the extent that data is available, so as better to assess the relation to biofuels expansion.

Regarding environmental impacts, the expansion of agriculture based on deforestation can endanger sensitive ecosystems and biomes. The intensive use of fertilizers, pesticides and water resources pose further environmental risks; these risks can be reduced by using analytical instruments such as AEZ. According to the Brazilian experience, the identification of the most appropriate areas to grow different crops is an essential way to increase productivity and avoid inappropriate use of lands, even without any additional investment in technology or new inputs. This statement seems clear when the productivity of major crops in different countries are compared. Taking maize as an example: while in the US the average yield is approximately ten tonnes per hectare, the Brazilian yield is less than four tonnes per hectare and the yield of maize in Mozambique is less than one tonne per hectare; the low yield in Mozambique is particularly noteworthy as it has similar climate and soil conditions to Brazil. Differences of this magnitude can also be observed when comparing the production of wheat, rice and beans. The productivity gap between agriculture in developed countries and developing countries can thus be quite large. The main reason for this productivity gap is the difficulty faced by developing countries in accessing technology and information. Tools such as AEZ can be very useful to reduce this gap. The identification of the best areas and appropriate sowing dates is the first step to increasing efficiency in agriculture.

Increasing efficiency in production is a necessary prerequisite for creating new demands, such as biofuels. In this regard, it is important to bear in mind the size of the world market for oil and oil products. It is easy to conclude that any programme to produce biofuels and substitute fossil fuels, implemented on a global scale, would impact on traditional agricultural markets. In some countries, for example, the replacement of 10 per cent of fossil fuels by biofuels would require the occupation of the whole agricultural area for feedstock production. However, this is not the case in Brazil and many other developing countries, including several in Africa. In many countries there is a paradoxical situation in which, despite the great suitability for agriculture potential, a large portion of the population is still deprived of access to food. In such cases, biofuels could make important contributions to strengthening energy security, job creation and income generation, thus enabling access to food. The AEZ and EEZ can be used together with geographical data on socioeconomic conditions to target biofuels where they are most needed to support development.

Looking more optimistically, biofuels can be an important instrument for promoting economic development. Ethanol and biodiesel are relevant alternatives to provide access to energy in rural areas; where spatial data on energy access is available, such information can be usefully compared to the AEZ and EEZ analyses. The availability of energy is essential for the development of agribusiness. As a result, it is possible to create jobs and avoid exodus to the urban areas. Besides that, for several developing countries that are oil importers, biofuels can be an important way of strengthening energy security. In this case, the positive impacts extend throughout the economy.

In the near future, the world will face the challenge of fulfilling the increasing demand for food and also the even greater increasing demand for energy. Technology will play an essential role by allowing maximization of production and use of biomass. The possible conflict between food and fuel may diminish in this context. In addition, biofuels should be seen as a complementary energy source, focused on the transport sector but able to provide a significant contribution to the renewable energies share in the global energy mix. Energy efficiency and the rational use of other energy sources must also be considered as part of the progression towards a low-carbon society (Strapasson and Fagá, 2007), including bioenergy carbon capture and storage technologies (Strapasson and Dean, 2011).

Conclusions

Biomass is the most important energy source for many developing countries. In the current context, in which energy security is strictly connected to economic and social development, the challenge of moving towards a modern conception of bioenergy involves reducing the risks associated with food security and environmental conservation. Bioenergy programmes must be designed to promote an increase in agricultural efficiency as a whole, which involves identifying potentials and limitations for different regions and giving appropriate technical support for farmers to explore these potentials. The most efficient use of natural resources, including the land, is a key issue for harmonizing different agricultural demands, such as food, feed, fibres and fuels.

Agro-ecological zoning for biofuels production can be an important tool for sustainable territorial planning, allowing the harmonious production of food and fuel and maintaining protected areas. But to do so, governmental rules and incentives are fundamental requirements. The Brazilian experience can be "adopted and adapted" for countries with available areas to produce biofuels, especially some developing countries in the tropical regions of Latin America, Africa and Asia, which generally have the best natural conditions for biomass production. Sharing agro-ecological zoning lessons with other countries is one of the pillars of the Brazilian strategy for international cooperation, especially with Africa.

The development of bioenergy programmes in Africa could be enhanced by the careful consideration of previous experiences from other tropical countries. It is likely that several of these countries could experience a success in bioenergy similar to that of Brazil, in different orders of magnitude. Nowadays there are already many African countries developing bioenergy programmes; thus to share experiences among themselves is also an important way to gain time and efficiency. In this sense agro-ecological zoning is an innovative approach to promote both food and fuel production in synergy with the environment.

Notes

1 In Brazil a comprehensive inventory of technical information is held by different institutions, such as: The National Institute for Meteorology (INMET); the National Institute for Geography and Economy (IBGE); Geological Survey of Brazil (CPRM); The Amazon Protection System (SIPAM); and the Brazilian Agricultural Research Corp (Embrapa).

2 The States of Acre, Rondônia and Roraima, all located in the Amazon region, have concluded their EEZs. Some other states are very advanced in the process of zoning.
3 MAPA Decree 333/2007 (Portaria).
4 Presidential Decree 6.961/2009; MAPA Normative Instruction 57/2009; Bacen (Central Bank of Brazil) Resolutions 3.813/2009 and 3.814/2009; Federal Law Project 6.077/2009.
5 Presidential Decree 7.172/2010, Federal Law Project 7.326/2010, Bacen (Central Bank of Brazil) Resolution 3.852/2010.

References

Araújo, G.F. and Strapasson, A.B. (2009) "Comparative Analysis of the Climate Change and Environmental Policies in Brazil: Challenges, Risks and Opportunities for the Amazon Region", United Kingdom Environmental Law Association (UKELA) and presented at a side event of the COP15/UNFCCC, Copenhagen, Denmark.

COMPETE: Competence Platform on Energy Crop and Agroforestry Systems for Arid and Semi-arid Ecosystems, Africa (2010) Land Use and Environmental Maps of Africa, at: www.compete-bioafrica.net.

CONAB (2010) Companhia Nacional de Abastecimento, Relatórios Técnicos, MAPA/ Conab, Brasília.

CONAB (2011) Acompanhamento da Safra Brasileira: Grãos, safra 2010/2011, sétimo levantamento, April, MAPA/CONAB, Brasília.

Embrapa Soils Geoportal (2010) Maps of the Sugar Cane and Oil Palm Agro-ecological Zonings, MAPA/Embrapa, at http://mapoteca.cnps.embrapa.br.

EPE: Empresa de Pesquisa Energética (2010) "Balanço Energético Nacional 2010: Resultados Preliminares, ano base 2009", Ministério de Minas e Energia, Brazil.

FAO (2010a) "Bioenergy and Food Security: The BEFS Analysis for Tanzania", Environment and Natural Resources Management Working Paper, vol. 35.

FAO (2010b) "Small Scale Bioenergy Initiatives: Brief Description and Preliminary Lessons on Livelihood Impact from Case Studies in Asia, Latin-America and Africa", Environment and Natural Resources Management Working Paper, vol. 31.

FGV (Fundação Getúlio Vargas) (2008) "Food Price Determining Factors: The Impact of Biofuels", technical report, Rio de Janeiro.

FGV (2010) "Biofuel Production in the Republic of Senegal Stage 1: Feasibility Study. Summary of the Final Report of Stage 1", November, FGV Projetos, Rio de Janeiro.

IBGE (Brazilian Institute for Geography and Statistics) (2010) Statistical Database of Census 2006, Brazilian Ministry of Planning, Budget and Management, Brasília, http://www.ibge. gov.br.

Manzatto, C.V., Assad, E.D., Bacca, J.F.M., Zaroni, M.J. and Periera, S.E.M. (2009) "Zoneamento Agroecológico da Cana-de-açúcar", Embrapa Soils' technical report, MAPA/Embrapa, Brasília.

Ramalho-Filho, A. (2008) "Agroecological Zoning for Oil Palm in Deforested Areas of the Brazilian Amazon Region", Presentation at "Bioenergy and Biodiversity: Joint International Workshop on High Nature Value Criteria and Potential for Sustainable Use of Degraded Lands", 30 June–1 July, UNEP, Paris, http://www.bioenergywiki.net/ images/6/6d/EMBRAPA_%282008%29_Ramalho-Filho_Degraded_land_Brasil_ Paris_01_Juli_08.pdf.

Ramalho-Filho, A. and Beek, K.J. (1995) *Sistema de Avaliação da Aptidão Agrícola das Terras*, 3rd edn, Embrapa-CNPS, Rio de Janeiro.

Ramalho-Filho, A., da Motta, P.E.F., de Freitas, P.L. and Teixeira, W.G. (2010) *Zoneamento Agroecológico, Produção e Manejo da Cultura da Palma de Óleo na Amazônia*, MAPA/Embrapa, Rio de Janeiro.

Ramalho-Filho, A., Naime, U.J., Motta, P.E.F., Bacca, J.M., Claessen, M.E.C., Meirelles, M.S., Teixeira, W.G., Gonçalves, A.O., Bastos, T.X., Ferraz, R.P.D. and Marin, F.R. (2008) "Zoneamento Agroecológico do Dendê para as Áreas Desmatadas da Amazônia Legal", in *Anais do Congresso Brasileiro de Plantas Oleaginosas, Óleos, Gorduras e Biodiesel*, 5, UFLA, Lavras, MG (CD-ROM), http://oleo.ufla.br/anais_05/artigos/a5__565.pdf.

SIPAM: Sistema de Proteção da Amazônia (2004) "Base Pedológica da Amazônia Legal – Base Digital em escala compatível com a escala 1:250,000", Convênio SIVAM-IBGE, SIPAM, Brasília.

Sousa, E.L. and Macedo, I.C. (eds) (2009) *Etanol e Bioeletricidade: A Cana-de-Açúcar no Futuro da Matriz Energética*, UNICA, São Paulo.

Strapasson, A.B. (2010) "Building a low carbon society: lessons from agriculture", in G.F. de Araújo, *The Challenge of International and Comparative Law in the Context of Climate Justice and Climate Change Law, Post Copenhagen Scenario*, Author House, Bloomington, IN.

Strapasson, A.B. and Dean, C. (2011) "Negative CO_2 Emissions Technology: Database and Recommendation", edited in collaboration with N. Shah and M. Workman, Technical Report submitted to the UK DECC, Grantham Institute for Climate Change, Imperial College London.

Strapasson, A.B. and Fagá, M.T.W. (2007) "Energy efficiency and heat generation: an integrated analysis of the Brazilian energy mix", *International Energy Journal*, 8, pp. 50–59.

Strapasson, A.B. and Job, L.C.M.A. (2007) "Ethanol, environment and technology: reflections on the Brazilian experience", *Journal of Agricultural Policy (Revista de Política Agrícola – English version)*, vol. 16, no. 3 (July/August/September), pp. 50–62.

UEMOA (West African and Monetary Union) and Le Hub (The Hub for Rural Development in West and Central Africa) (2008) *Sustainable Bioenergy Development in UEMOA Member Countries*, UN Foundation, in partnership with UEMOA, Le Hub, ICTSD (International Centre for Trade) and ESG (Energy and Security Group), Washington, DC.

Vieira, J.N.S. (2002) "Biofuels and their Role in the Strengthening of Non-Food Grain Demand: A New Horizon for the Expansion of Agricultural Frontiers", XIV International Symposium on Alcohol Fuel, November, Phuket Island, Thailand.

Vieira, J.N.S. (2004) "O Biodiesel e o desafio da inclusão social", in *O Biodiesel e a Inclusão Social – Videoconferencing 8.10. Cadernos de Altos Estudos I/2004*, Câmara dos Deputados, Brasília.

4

LAND SUITABILITY AND CROP HANDLING

Helen K. Watson and Brian S. Purchase

Introduction

This chapter is structured into two sections; namely land suitability and crop handling. The land suitability section commences with a description of the land and resource requirements for the economically viable production of sugar cane. It then reviews studies into where these requirements are met, carried out at different scales. The most recent of these consider sugar cane as a bioenergy feedstock and therefore additionally assess where it can be grown without detrimental impacts on biodiversity, food security and rural livelihoods. The studies clearly reveal that Africa has enormous potential to grow rain-fed sugar cane as a bioenergy feedstock. However, estimates of suitable and available land vary considerably because the different models use different key sugar cane growth determinants, integrate them differently, and exclude different land types as inappropriate for bioenergy feedstock production. It concludes that land suitability and land availability mapping should be the leading tool in sugar cane bioenergy investment planning in Africa, and suggests the path that needs to be followed to achieve this. The second section describes the practices as well as the challenges of harvesting and transporting the crop to factories in Africa. The need for efficient coordination of the logistics is highlighted, together with the need to rethink the widespread practice of burning cane before harvest. Costs and opportunities for using crop residues as a source of energy, instead of burning them in-field, are considered, leading to the conclusion that this is attractive under some conditions but that the agronomic value of cane residues has tended to be underestimated and must be considered carefully when designing systems for the collection of residues.

Section 1: land suitability

Land and resource requirements

This section provides a brief overview of the land resources and conditions needed to support growth of sugar cane, including growth habitat, climate, soils and terrain. There is some emphasis on conditions and experience from Southern Africa, although most of the principles are applicable to anywhere that sugar cane can grow.

Growth habit

The sugar cane physiology has been thoroughly described in Chapter 2. As mentioned, the crop is a perennial grass that is usually established by planting setts into the soil. Depending on where it is grown and the conditions under which it is grown, the crop may take up to two years or more to complete growth, flower and die. Sugar cane is photoperiod sensitive, with both ripening and flowering being triggered by a continuous reduction in day length. Sucrose levels peak for a relatively short period once the crop has attained full maturity and are substantially increased if this period is dry and moderately cool. After harvesting, new shoots, known as ratoons – and roots – develop from the underground part of the stalk. Sucrose content and yield generally decline with successive ratoons, mainly because soil preparation for ratoons is less intensive than for plantings. Where harvesting is mechanized, soil compaction by heavy equipment accelerates this diminishing trend (Blume, 1985).

Climatic requirements

Sugar cane thrives in wet-dry tropical and subtropical climates where the wet season rainfall is sufficient to provide a soil-water surplus capable of sustaining withdrawal by the crop during the shorter dry season, and where there is freedom from frost and cyclones. Even light frosts can damage cane (Griffee, 2000). A summary of climatic requirements for optimal growth of sugar cane is given in Table 4.1. Temperature is the main climatic determinant of sugar cane growth. While cane agriculture clearly requires high temperatures, the highest sucrose contents occur where there is a marked difference between day and nighttime temperatures (Blume, 1985) and sucrose accumulation with ripening is optimal with cooler temperatures (Griffee, 2000). To achieve optimal growth, sugar cane requires more than 1,200 hours of sunshine per annum. It requires high levels of solar radiation during its initial rapid growth phase and during ripening. The higher the intensity of insolation: (1) the higher the biomass production during the growth period, and (2) the higher the sucrose production during ripening. Optimal growth under rain-fed conditions occurs where the mean annual rainfall ranges between 1,200 mm and 1,500 mm. The crop requires a moist soil to germinate or to produce a ratoon. Its water requirement increases until it covers the ground fully, and then stabilizes through the rest of the rapid-growth phase. Drought during this period reduces biomass production. However, water stress during the crop's ripening phase arrests vegetative growth and encourages sucrose accumulation in the stalks. According to Blume (1985), irrigation is essential where the mean annual rainfall (MAR) is less than 800 mm. Schulze (1997) set this threshold at 850 mm. Although the crop can tolerate precipitation in excess of 1,500 mm per annum, with occasional flooding and high water tables, both cane and sugar yields are reduced. While winds during the growth period reduce biomass production, during ripening they increase the sugar content. Excessive winds break the cane stalks, rendering the crop susceptible to diseases and pests (Blume, 1985).

Terrain requirements

Although sugar cane can be cultivated on various terrains, the lowest input costs and highest productivities are associated with gentle slopes. The steeper the slope, the more difficult it is to mechanize land preparation, planting and harvesting and the greater the need for soil erosion and drainage control structures. Sys and colleagues (1993) set the slope threshold at 16 per cent, above which commercial cultivation under rain-fed conditions is constrained. Nevertheless, slopes steeper than this threshold are well represented in commercial sugar cane-producing areas in South Africa. Such areas are covered by a long-established intricate pattern of conservation terraces, extraction roads and waterways; machinery was intensively employed to create this pattern. Mechanical ploughing has decreased following a recommendation by SASEX (1999) that compartments between terraces with slopes steeper than 20 per cent be re-established using minimum tillage. For erodible sandy soils the slope threshold is set at 10 per cent. Sugar cane found on steep slopes elsewhere in Africa generally belongs to peasant or small-scale farmers who use hand-held implements or animal-drawn ploughs to till the soil (Blume, 1985).

TABLE 4.1 Climatic requirements for optimal growth of sugar cane

Climatic parameter	Requirement/effect
Temperature	Frost-free areas; even light frosts cause damage (Griffee, 2000) Germination occurs >8.5 °C (Griffee, 2000) Sprouting and rooting requires >20 °C (Blume, 1985) Optimal mean annual range: 26–34 °C (Blume, 1985) Growth arrested <15 °C and >38 °C (Blume, 1985) Ripening range: 10–20 °C (Griffee, 2000)
Sunshine	>1,200 hours per annum (Blume, 1985)
Rainfall	Optimal mean annual range: 1,200 mm–1,500 mm (Blume, 1985)
Relative humidity	<70% during maturity and harvesting phases (Schulze, 1997)

Closer to the Equator, where frost is less common at high altitudes, sugar cane is cultivated up to 1,600 m above sea level. The higher the altitude, the longer the growth cycle (Griffee, 2000).

Soil requirements

Sugar cane ideally requires soils with a fair rate of drainage and with more than 150 mm of available water (Schulze, 1997). Medium-textured loams are best. They are easy to plough and their granular structure provides good aeration and drainage that encourage root growth (Sys et al., 1993). Coarse-textured soils impede root development, while fine-textured soils have poor drainage and may be waterlogged for periods during which the crop becomes susceptible to viruses, diseases and bacterial infections. Sugar cane ideally requires soils deeper than 1 m (Schulze, 1997). Shallower soils pose tillage constraints and soils with effective rooting depths of less than 400 mm are non-arable or unsuitable for sugar cane cultivation (Griffee, 2000). Sugar cane can tolerate a pH range of 4.5–8.5, and has an optimal pH range of 5.5–7.5 (Blume, 1985; Fischer et al., 2008). Sugar cane requires nitrogen, potassium, phosphorus and calcium in large quantities. Deficiencies of these essential macronutrients can reduce vegetative growth, reduce resistance to diseases, cause root rot and even plant death (Sys et al., 1993). Excessive aluminium concentrations are toxic (SASEX, 1999). Sugar cane is moderately sensitive to soil salinity. The electrical conductivity (EC) of a soil's solution should be less than 2.0 deciSiemens per metre (dS/m) for there to be no impact upon cane yields (Griffee, 2000). Sugar cane farming in Swaziland, Tanzania and Zambia has encouraged the development of saline soil conditions caused by improper drainage or the use of saline underground aquifers for irrigation purposes. A suitable depth and texture is most important as it is easier to modify the other requirements by manipulating drainage and adding organic matter, lime and fertilizers (SASEX, 1999). According to the FAO (1978), the soils that are either suitable or marginally suitable for sugar cane cultivation include: Regosols, Andosols, Vertisols, Kastanozems, Chernozems, Phaeozems, Cambisols, Luvisols, Planosols, Nitisols and Ferralsols.

Agronomy

The South African Sugarcane Research Institute (SASRI) was founded in 1925. Although the Mauritian Sugar Industry Research Institute (MSIRI) was only founded in 1953, research into the choice and improvement of varieties had been ongoing since 1891. These institutions have conducted consistent, dedicated research into a wide range of sugar cane production aspects, including: site selection; clearing; planning farm layout and field design including soil conservation and drainage structures, in-field roads, etc.; land preparation and fertilizer application; seed versus sett planting; weed management; irrigation management; monitoring field parameters and cane quality; harvesting; ratoon management; pest and disease control, and the development of varieties with higher sucrose content and yield and resistance to drought, frost, pests and diseases. Research at SASRI has focused on managing the following constraints prevalent within the South African sugar cane belt: hilly terrain, steep slopes, acidic soils, infestations of the borer *Eldana saccharina* and various diseases, particularly smut and mosaic. Breeding programmes aim at developing varieties suited to the range of different climate/environment "types" found in sugar cane cultivation regions. The main predictor of "genotype by environment interaction" (see Chapter 2) has been found to be variety harvest age, followed by soil type. The former reflects the major influence of radiation, temperature and water availability on plant growth. Breeding and selection focuses on sucrose yield, sucrose content, and resistance to *Eldana* and disease. Consequently, commercial varieties vary widely in agronomic traits such as growth habit, stalk population, stalk diameter, straightness, amount and type of trash and harvesting and milling characteristics. Breeding for resistance and the use of molecular markers associated with resistance is increasing the frequency of resistant material within the germplasm pool. It is therefore likely that in the future there will be more opportunity to focus on these other agronomic traits in breeding and selection (Butterfield and Nuss, 2002).

While Mauritius is also characterized by hilly terrain and steep slopes, clearing land of large boulders has presented a bigger challenge but is nowadays being undertaken following the restructuring of the sugar industry in the country. Good agronomic practices for sugar cane are well established and generally very well represented in both large and small-scale sugar cane farms throughout Africa. Substantial numbers of soil and leaf samples from sugar cane grown all over Africa are routinely analysed by SASRI's Fertilizer Advisory Services. In South Africa and Mauritius growers are routinely visited by and can call on the expertise of the extension services developed by the industry. Elsewhere, these services are generally provided to outgrowers by the milling company.

Review of relevant studies

This section reviews relevant studies on where cane can be suitably grown in selected African countries. The chronological progression of detail in the studies illustrates that the evaluation of the suitability of sugar cane as a bioenergy feedstock in particular areas must be accompanied by assessments to determine the potential impacts on key environmental and socio-economic factors such as biodiversity, food security and rural livelihoods.

South Africa and Swaziland

Schulze (1997) used the ACRU-Thompson model to produce a $2\,km^2$-resolution map of South Africa and Swaziland showing areas falling within nine yield (t/ha/season) categories as well as those unsuitable for sugar cane. Information extracted from this map is summarized in Table 4.2. The ACRU model is a physical-conceptual, daily time-step, multilevel, multipurpose model developed at the

University of KwaZulu-Natal (Schulze, 1995; Schulze and Smithers, 2004). It has been applied extensively in South Africa for both land use and climate change impact studies (Schulze, 2000). The Thompson equation, which has stood the test of time, relates sugar cane water use to yield as follows:

$$Y_{sc} = 9.53 \star (E_{an}/100) - 2.36$$

where Y_{sc} = annual sugar cane yield (t/ha) and E_{an} = annual total evaporation (mm).

ACRU usually separates evaporation into water from the soil surface, and from transpiration. However, because sugar cane is phenologically a relatively simple crop, Schulze (1997) calculated it as a single process from day-by-day rainfall, reference evaporation (A-pan equivalent[1] or computed from minimum and maximum temperature), stage of crop growth and multilayer soil water budget. ACRU revolves around this budget and is linked to catchment scale climate and soil databases. The sugar cane yield estimates are for a rain-fed crop over an annual period. However, on average 24 per cent of the crop in South Africa, and all of it in Swaziland, is irrigated (Agrimark, 2009). Also, the ratoon duration varies from 11–12 months to 18–24 months, dependent on the bioclimatic region in which it is grown (Schulze, 1997).

Table 4.2 indicates that half of Swaziland, comprising its central, western and northern parts, is suitable for sugar cane. South Africa has 50,799 km² suitable for sugar cane extending along its eastern margin. Most of this land is in KwaZulu-Natal. The total area of South Africa is 1,220,719 km², of which 13.7 per cent or 167,238 km² is arable (Turpie et al., 2002). Table 4.2 suggests that 30.4 per cent of the country's arable land is suitable for sugar cane. However, in the 2008/2009 season only 413,556 ha of the country was planted with sugar cane (Agrimark, 2009); which is 0.81 per cent of land suitable for sugar cane and 0.2 per cent of arable land.

This major discrepancy between area under sugar cane and area suitable for it is due in part to the fact that Schulze's (1997) assessment did not consider terrain. Most of the areas deemed suitable in Swaziland and Mpumalanga are too hilly and steep for production. Four of Schulze's (1997) categories considered suitable for sugar cane have yields of less than 60 t/ha/season. While small-scale farmers do

TABLE 4.2 Areas suitable for sugar cane

	Swaziland	South Africa	South Africa's sugar cane provinces			
			Limpopo	Mpumalanga	KwaZulu-Natal	Eastern Cape
Total area, km²	17,404	1,220,719	122,433	78,238	92,285	171,221
% of total area suitable	50.0	4.2	3.2	11.2	31.1	5.5
Area suitable, km²	8,702	50,799	3,918	8,763	28,701	9,417

Source: extracted from Schulze, 1997.

obtain yields lower than this (Makhanya, 1997), 60 t/ha/season is considered a threshold for economically viable sugar cane production. If areas with yields lower than this threshold are excluded, there is still a substantial discrepancy between potentially suitable areas and areas under sugar cane. As noted previously, temperature is the principal climatic determinant of sugar cane growth. However, in Schulze's (1997) assessment, rainfall emerged as the single most important factor influencing sugar cane production. Inter-annual rainfall variability in the area potentially suitable for sugar cane is high. In the Eastern Cape, KwaZulu-Natal, Mpumalanga and Swaziland, the coefficients of variation (CV) average 15–20 per cent. However, areas in northern KwaZulu-Natal and northern Swaziland have CVs ranging from 25–30 per cent and 25–35 per cent, respectively. Most areas in Limpopo have a CV greater than 25 per cent while two have CVs greater than 35 per cent (Schulze, 1997; Bezuidenhout and Schulze, 2006). If areas with a CV greater than 20 per cent are excluded as too risky, there is still a substantial discrepancy between potentially suitable areas and areas under sugar cane. The discrepancy clearly illustrates the need to consider not only whether the land is suitable for sugar cane but also whether it is potentially available to be converted to the crop. This can be achieved by considering the contemporary land use and ownership. The discrepancy in both Swaziland and South Africa can largely be accounted for by land that is unavailable for sugar cane because: (1) it is a protected area where agriculture is not permitted, e.g. a fenced national or provincial wildlife park, or an indigenous forest; (2) is under industrial or plantation forestry; (3) belongs to the King of Swaziland or Zululand; (4) is held in trust by the state; (5) is traditional communal tribal land; or (6) is under urban or other agricultural land use. Additionally, in South Africa it may be unavailable because it is designated for restitution or redistribution under the country's Land Reform Programme, or is in too close proximity to rapidly expanding urban/peri-urban areas. Since 1994 when South Africa became a democracy, substantial areas around Durban and along the coast north and south of it that were under sugar cane have been converted to shopping complexes and residential estates (Schensul, 2009). News that new black farmers now own 11,991 ha of sugar cane – and that by 2014, 30 per cent of South Africa's sugar cane will be owned by blacks farmers does not mean that land in the categories listed above is being planted with the crop (Agrimark, 2009). Sugar cane land has been transferred to these emergent farmers from milling companies and white commercial farmers.

According to Knox and colleagues (2010), there is general consensus that by 2050 the sugar cane belt in South Africa and Swaziland will receive 5–10 per cent less rainfall and experience temperatures 1–3 °C higher than at present. In order for future production to be equivalent to current production, a 20–22 per cent increase in irrigation will be needed. The water resources of neither country will be able to sustain such an increase. Turpie and colleagues (2002) have predicted that, as a consequence of climate change, by the mid-twenty-first century the production of sugar cane in the two countries will be 10–20 per cent lower than it was at the turn of the century.

The African continent

The Food and Agriculture Organization (FAO) developed the agro-ecological zones (AEZ) methodology on the basis of an inventory of land resources and evaluation of biophysical limitations and potentials (FAO, 1978). It provides a standardized framework for assessing whether the climate, soil and terrain conditions prevailing within $10 \, km^2$ grid cells are suitable for a specific crop under specific agronomic conditions. Using this methodology, FAO (2004) produced a map (Plate 5) showing areas suitable for rain-fed sugar cane production in Africa. The areas were categorized into very suitable; suitable; moderately suitable; marginally suitable; not suitable and prohibitive.

Table 4.3 shows the estimates of areas within the first three categories extracted from this map. It suggests that the countries that offer the best prospects for planting sugar cane as a bioenergy feedstock are the Democratic Republic of Congo (DRC), Cameroon and the Congo Republic (CR). The DRC has over 1.2 million km^2 of land suitable for the crop, and most of this land falls in the very suitable and suitable categories. Most of the land identified as suitable in these three countries, as well as Gabon and Equatorial Guinea, is under closed evergreen lowland forest. It should therefore be considered unavailable for conversion to bioenergy feedstock production on account of both biodiversity and water source considerations.

The FAO (2004) map shows areas marginally suitable for sugar cane in central Guinea-Bissau, the southern tip of Mali, southern Togo and Benin, south-west Chad, south-west and south-east Sudan, south-west Ethiopia, south-west Kenya and north-east Zambia. All other countries are shown as not suitable or prohibitive. Sugar cane is grown in a number of these countries – e.g. Swaziland, Zimbabwe, Malawi, Senegal, Morocco and Egypt – under irrigation. However, the fact that South Africa and parts of other countries where rain-fed sugar cane is produced are shown as not suitable or prohibitive suggests that the coarse resolution used misses smaller areas with adequate rainfall.

Tanzania

The Mlingano Agricultural Research Institute (MARI) (2006) used the FAO AEZ methodology to assess where rain-fed sugar cane can be grown in 4 km^2 grid cells in Tanzania. While the institute used temperature and soil characteristic thresholds similar to those described in the previous section, it set the MAR threshold below which irrigation is required at 1,500 mm, which is substantially higher than the 800–850 mm noted previously. According to MARI's (2006) findings shown in Table 4.4, Tanzania has 16,378 km^2 of land suitable for rain-fed sugar cane in the Coastal Plains, and 35,466 km^2 of such land in the Eastern Plateau and Mountain Blocks. The total area is 51,844 km^2, and is equivalent to 5.90 per cent of the country. The equivalent FAO (2004) estimates (Table 4.3) are 8,699 km^2 and 0.99 per cent. Both studies used the same methodology. Despite MARI (2006) considering a narrower rainfall range as suitable, it found almost six times more land suitable for sugar cane than FAO (2004). This difference could possibly be attributed to MARI (2006) using a grid-cell resolution 2.5 times better than FAO (2004).

TABLE 4.3 Areas very to moderately suitable for rain-fed sugar cane in Africa

Country	Total country area, km^2	% of country suitable for sugar cane	Area suitable for sugar cane, km^2	Suitability ranking
Angola	1,246,700	0.75	9,350	13
Cameroon	475,440	65.00	309,036	2
Central African Republic	622,984	5.17	32,208	8
Congo Republic	342,000	39.39	134,714	3
DR Congo	2,345,410	54.42	1,276,372	1
Equatorial Guinea	28,051	67.00	18,794	12
Gabon	267,667	26.09	69,834	4
Ghana	239,460	9.30	22,270	9
Ivory Coast	322,460	12.55	40,469	7
Liberia	111,370	17.65	19,657	11
Mozambique	784,090	5.81	45,556	6
Nigeria	923,768	2.22	20,508	10
Tanzania	878,690	0.99	8,699	14
Uganda	236 040	20.00	47,208	5

Source: extracted from FAO, 2004.

TABLE 4.4 Agro-ecological zones (AEZs) suitable for rain-fed sugar cane in Tanzania

Biophysiographic region	Agro-ecological zones	Area, km²	Altitude (metres above mean sea level)
Coastal Plains	C 4	4,450	<200
	C 5	8,569	<200
	C 6	1,227	<100
	C 7	2,132	<100
Eastern Plateau and	E 3	30,332	200–750
Mountain Blocks	E 8	1,777	1,200
	E 11	3,357	500–1,000

Source: extracted from MARI, 2006.

Angola, Malawi, Mozambique, Tanzania, Zambia and Zimbabwe

Watson (2010) synthesized research carried out successively in these six Southern African countries over the period 2003–2009. This period was characterized by increasing interest from all sectors in the bioenergy potential of sugar cane in the region. There was a clear need to identify "where best to grow the crop" more accurately. Against the background of increasing international concern about the potential detrimental effects of growing bioenergy feedstocks, it was additionally clear that the identification of these areas would have to involve both land suitability and land availability assessments. As evident in the studies described above, the former assesses whether the characteristics – climatic, edaphic, terrain, etcetera – found at a particular locality would enable economically viable production under given management and input conditions. The latter interrogates whether converting the land to sugar cane is likely to have detrimental environmental and/or socio-economic impacts. If this is likely, it is deemed unavailable. Therefore "where best to grow the crop" is land that is both suitable and available.

The research established that the Miombo Network's (IGBP-IHDP, 1995; Desanker et al., 1997) 1 km² data on sugar cane land cover (SCLC) surfaces reliably identified land suitable for rain-fed sugar cane production in Southern Africa (Figure 4.1). By overlaying the SCLC surfaces on a range of other 1 km² surfaces, it then assessed whether the SCLC surfaces were available to be planted up with the crop. The SCLC surfaces were filtered out if they fell on any of the following: protected areas; slopes steeper than 16 per cent; closed-canopy forests; wetlands; land used for food and/or cash crop production (including sugar cane).

Table 4.5 shows that these six countries have 59,360 km² of land available and suitable for conversion to sugar cane bioenergy feedstock production. With over 23,000 km² of this land, Mozambique clearly offers the best potential for expanding such production, followed by Zambia and Angola, which both have over 11,000 km². Most of the work on Angola was carried out by Ackbar (2007) and that on Zimbabwe by Sibanda (2008). Their studies attempted to rate the development potential of different clusters of suitable and available areas according to proximity to potential irrigation sources (perennial rivers and dams), labour sources (villages) and transport infrastructure (primary and secondary roads and railways). Watson (2010) notes that the accuracy of identifying available areas could be enhanced by using additional databases such as High Conservation Value Areas, and the better-resolution land cover data that has become available subsequent to this research. Table 4.5 provides yet another estimate of land suitable for sugar cane in Tanzania; i.e. 16,940 km^2 or 1.93 per cent of the country, which is twice that calculated by FAO (2004) and a third of that calculated by MARI (2006). The magnitude of difference between these estimates is obviously of concern.

The African continent

The FAO collaborated with the International Institute for Applied Systems Analysis (IIASA) to identify land suitable for the commercial, fully mechanized production of rain-fed sugar cane as a bioethanol feedstock. Fischer and colleagues (2009)

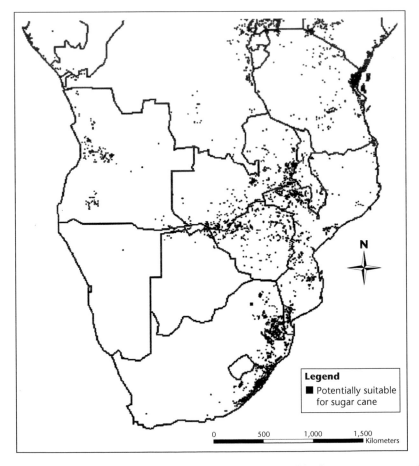

FIGURE 4.1 Areas in Southern Africa potentially suitable for sugar cane production (source: derived from IGBP/IHDP, 1995)

TABLE 4.5 Areas remaining for sugar cane cultivation after successive filtering of data

	Angola	Malawi	Mozambique	Tanzania[a]	Zambia	Zimbabwe
Country land area, km²	1,246,700	94,080	784,090	878,690	743,390	386,670
Potentially suitable, km²	16,260	7,420	49,060	16,940	35,460	29,350
Protected areas, km²	13,950	5,950	46,020	12,230	24,330	18,600
Slopes >16%, km²	13,890	5,800	45,300	12,170	24,270	18,550
Available and suitable, km²	11,270	2,060	23,380	4,670	11,780	6,200
% of country land area potentially suitable	1.30	7.89	6.26	1.93	4.77	7.59
% of country land area available and suitable for sugar cane	0.90	2.19	2.98	0.53	1.58	1.60

Source: Watson, 2010.

Note

a excluding Zanzibar and Pemba.

reported on this data. The same AEZ methodology as described for FAO (2004), above, was used. However, the inclusion of more climatic parameters, longer records and better-resolution terrain data enabled use of water balance and crop models to estimate actual evapotranspiration, length of growing period, and potential crop and biomass yields for each $10\,km^2$ grid cell. The inclusion of land use/land cover and protected areas data enabled determination of whether the grid cell falls within currently cultivated land or protected land. Because of its coarse resolution, the Africa map output of this research is virtually identical to Plate 5, with the ranking of the 14 countries with very to moderately suitable land the same as that shown in Table 4.3. Overall, Fischer and colleagues (2009) found that there is land in Africa totalling around $280,000\,km^2$ which is very suitable and suitable for rain-fed sugar cane production. Of this land, 11.2 per cent is currently cultivated, 4.9 per cent is in unprotected grass/scrub/woodland, and 24.2 per cent is in unprotected forest, giving a total availablity of 40.3 per cent, or $113,000\,km^2$. There are clearly differences of concern between the studies in terms of what land surfaces are considered available. As noted above, the studies synthesized by Watson (2010) filtered out suitable grid cells falling on cultivated land and unprotected forests.

Mozambique

Schut and colleagues (2010a, 2010b) describe the assessment of land available for large-scale rain-fed agriculture completed by the Mozambique government in 2008. Of all the country's arable land, 54 per cent was found to be available, and of this, 39 per cent is located in the country's southern and central provinces. The remainder is located in the less populated, less developed northern provinces. Schut and colleagues (2010a, 2010b) inserted the locations of existing and planned sugar cane-cultivated areas, sugar mills, ethanol distilleries, storage facilities, etcetera into the land availability map. Sugar cane and bioethanol developments and interests are clearly concentrated in the country's southern and central provinces. The authors conclude that the main factors determining where these developments and interests occur are as follows: good infrastructure (roads and ports); proximity to processing and storage facilities; availability of literate and skilled labour; and access to services and goods. Land suitability and availability, and government incentives to invest in underdeveloped provinces, are evidently less of a consideration.

Kenya

Muok and colleagues (2010) used the FAO AEZ methodology to assess where rain-fed sugar cane can be grown in $1\,km^2$ grid cells in Kenya. They set the MAR threshold below which irrigation is required at $1,000\,mm$, and considered slopes up to 45 per cent suitable. They found that suitable land was concentrated in three parts of the country: a strip along the southern part of the coast; a strip in centre of the country extending south from Meru North district; and a broad belt in the west around Lake Victoria. In total, the suitable land covered 17,332 km², which is equivalent to 3 per cent of the country's surface area. Muok and colleagues (2010) then assessed the availability of this land to be planted up with sugar cane by overlying other non-mutually exclusive surfaces on it. They found that 8.2 per cent of this land is unavailable, as it falls within protected areas: 74.5 per cent and 8.9 per cent of the land is already under food and cash crops, respectively. While the 8.4 per cent remainder is not cultivated, it is utilized for grazing livestock and harvesting veld products. They also found that 26.9 per cent of the suitable land outside the protected areas was classified as "wildlife conflict land", and 29.8 per cent of it is used as corridors in the movement of wildlife.

Senegal

Mapping areas for bioenergy feedstock production is currently taking place in Senegal using the Brazilian agro-ecological zoning for biofuels methodology (BAEZB) (Ndour, 2010; see also Chapter 3). Sugar cane is being considered with a slope threshold of 20 per cent. The methodology essentially comprises two steps: a land suitability assessment and a land availability assessment. In addition to assessing the climatic, edaphic and terrain parameters in terms of the optimal requirements of a particular crop, the land suitability assessment uses a water balance model. Protected areas are then filtered out from the land identified as suitable. The remainder is then classified as high or low biodiversity. High biodiversity areas include natural forests, woodlands, grasslands, wetlands, etcetera. Low biodiversity areas include land that is currently intensively used for cultivation or grazing, or is abandoned or degraded.

Discussion on land suitability in Africa

The review of research to date into where rain-fed sugar cane can be grown in Africa, and where best to grow it as a bioenergy feedstock, clearly shows the continent has enormous potential. However, there are discrepancies in the order of magnitude between estimates of suitable land and it is not clear where priority should be given in terms of investing in planting up land for sugar cane. These discrepancies and uncertainties are due to: (1) different methodologies being used to ascertain if the land is suitable; (2) different thresholds for MAR and slope gradient being used in ascertaining parameters for a given grid cell of land fall within the optimal range required by the crop; and (3) different studies using grid cells corresponding to different land areas.

The review has provided a number of valuable insights. First, in applying the FAO (1978) AEZ methodology to identify land suitable for rain-fed sugar cane, a wider range of conditions within a cell of land should be considered suitable, given that economically viable yields are possible under sub-optimal conditions provided good management practices are used. For example, although the optimal MAR range for sugar cane is 1,200–1,500 mm, economically viable yields can be attained in areas with an MAR as low as 800 mm, provided good management practices are used, such as ensuring the soil is well covered by mulch. Also, although Sys and colleagues (1993) assert that slopes steeper than 16 per cent should not be used for commercial sugar cane production, slopes much steeper than this are routinely used and do not present a problem provided good soil conservation measures are in place. Furthermore, it makes good sense to extend the methodology to analyse climatic parameters that directly affect growth performance. In sugar cane production, a decrease in day length and soil moisture along with an increase in the difference between day and night temperatures will increase sucrose accumulation. An attempt to capture the influence of climatic parameters on the soil's water balance was pioneered by Schulze (1997). Recognition of the importance of these parameters on the water balance has been central to the extension of the FAO (1978) methodology by Fischer and colleagues (2008), as well as to the development of the BAEZB methodology. Future assessments of land suitable for rain-fed sugar cane should analyse long-term climatic data using water balance and crop models, so that the actual evapotranspiration, length of growing period and potential crop and biomass yields of each grid cell can be estimated. Ideally, grid cells assessed should represent land areas no larger than 1 km^2. Many small commercially viable areas are evidently missed when coarser resolution data is used.

The review revealed that when considering sugar cane as a bioenergy feedstock, in addition to evaluating whether the land is suitable for the crop, it is important to assess impacts on biodiversity, food security and/or rural livelihoods. There appears to be agreement on the need to exclude protected areas. However, high conservation-value areas – natural forests, woodlands, grasslands, wetlands, wildlife movement corridors, etcetera – falling outside protected areas, as well as areas currently used for food and/or cash crops and grazing, are being treated differently in different studies.

In order for agro-ecological zone, land suitability and land availability mapping to reach its full potential role in sugar cane bioenergy planning in Africa, a common methodological approach is needed. The United Nations Environment Programme (UNEP) and the FAO, working through UN-Energy, developed a Bioenergy Decision Support Tool for national strategy planning and investment (UN-Energy, 2010). Together with the African Centre for Technology Studies (ACTS) and Policy Innovation Systems for Clean Energy Security (PISCES) project, UNEP organized an international workshop in Nairobi in 2010 to harmonize methodology for agro-environmental zoning for biofuel production. Participants concurred that strictly protected areas where agricultural use is not permitted have to be excluded. Ideally the following areas falling outside protected areas should also be excluded: (1) wildlife movement corridors; (2) routes used by nomadic pastoralists; (3) areas where there is conflict between humans and wildlife, or between humans; (4) natural forests, woodlands, grasslands, riparian areas and wetlands that are important for ecosystem services and/or providing natural resources for rural communities; and (5) areas that are of archaeological, historical and/or cultural significance. Converting land that is currently being used for food and/or cash crop production, or grazing, to biofuel feedstocks was a concern to participants, resulting in a call to develop incentives to instead encourage use of abandoned or degraded land. Participants recognized that once the land areas are identified as both suitable and available for biofuel feedstock production, they need to be analysed in relation to their proximity to: (1) infrastructure (roads, railways, electricity grid); (2) population density, literacy and poverty levels; and (3) existing biofuel processing and storage facilities. Such an analysis is necessary to enable prioritization of the areas for conversion. The Bioenergy Decision Support Tool, the subsequent report focusing on using mapping as a tool (UNEP, 2011a) and the report on the methodology developed at the Nairobi workshop (UNEP, 2011b) are anticipated to play a meaningful role in guiding sugar cane bioenergy developments in Africa in the future, as well as other types of bioenergy, given that Africa is well endowed with unexploited biomass resources.

Concluding remarks on land suitability

Land suitable for rain-fed production of sugar cane is abundant in Africa. A number of models have been used to estimate the magnitude of this land at different spatial scales. The estimates vary greatly because the models use different key sugar cane growth determinants, integrate them differently, and exclude different land types as inappropriate for bioenergy feedstock production. Land suitability mapping and land availability mapping play a critical role as leading scientific tools in guiding cost-effective and environmentally compatible investment in sugar cane bioenergy in Africa. To achieve this role, future land suitability assessments should analyse long-term climatic data using water balance and crop models, so that the major influence of radiation, temperature and water availability on plant growth in each grid cell can be incorporated. These grid cells should be equivalent to areas on the ground of no larger than 1 km^2. Suitable cells falling on any of the land types considered inappropriate for bioenergy feedstocks production in the methodology currently being developed by UNEP, FAO, ACTS and PISCES should be filtered out as unavailable.

Section 2: crop handling

This section describes supply chains; factors affecting harvesting and loading of cane, delivery arrangements and transport type and efficiency; and cross-cutting issues of harvesting additional fibre. The description of each of these aspects pertains to using the crop for sugar, against which changes necessary in using it for sugar and/or bioenergy are explored. It concludes with highlighting the research and technological developments needed to ensure such changes are sustainable. The discussion here emphasizes the experience in Southern Africa, and especially in South Africa where the most extensive experimentation and research on crop handling has been conducted; however, many of the principles are general and can be applied anywhere where sugar cane is grown at commercial scale.

Supply chains

The harvesting and delivery of sugar cane in South Africa constitute approximately 40 per cent of the operating cost of cane production, with transport amounting to 14 per cent (South African Canegrowers Association, 2010). The processes involved usually include burning, cutting, loading, in-field transport, transloading, road transport, offloading and feeding to the mill. There has been a tendency for each process to be managed separately, but for efficiency improvement there is a trend towards integration and central coordination. It is emphasized, however, that the whole chain can be affected by a change to any one condition. For instance, if cane is not burnt before harvest then the cutting rate changes with the characteristics of the harvested material, which in turn results in higher loading and transport requirements (due to the additional amount of cane residues, also known as cane trash) and additional milling work for unit sucrose recovery.

In South Africa there are two basic types of supply chains, influenced mainly by topography (slope, layout and contours of the land). Where the topography is unsuitable for large vehicles to enter the fields, cane is loaded onto small in-field trailers and subsequently transferred to road vehicles at loading zones. This "indirect delivery" adds costs as compared to "direct delivery", where road vehicles are loaded in-field. The topography must therefore be considered when benchmarking supply chains in South Africa. The major part of the industry has topography unsuitable for direct delivery. Even where the topography is suitable for direct deliveries, soil type and conditions may preclude heavy vehicles from entering the field, such as in Zimbabwe, which explains why indirect deliveries predominate there.

A small amount of cane is harvested mechanically in South Africa, but at least 90 per cent is manually harvested and delivered in whole-stick form. Approximately 88 per cent is burnt before harvest. Cut canes are mostly delivered loose with grab loading and spiller offloading, and some are delivered in chained bundles. In 2005, 6.2 per cent was delivered by train or tram, with the remainder being delivered by road (Davis and Archary, 2006).

The following description of an individual mill (Guilleman et al., 2003; Giles et al., 2005) gives the perspectives on key issues involved in a typical supply chain at a Southern African factory. Around 2.3 million tonnes of cane are processed in the sugar factory, of which 22 per cent is produced by the miller-cum-planter estate, 65 per cent by 180 large-scale farmers and 13 per cent by 5,000 small-scale farmers. The supply management involves four types of operators: growers, harvest contractors, hauliers and the mill itself. The Mill Group Board is responsible for planning and coordination between these operators.

Around 120 independently operated vehicles serve the mill area, which has a tonne-weighted-average distance of 30 km. Harvesting is done manually by the growers themselves or by harvest contractors, after which they transport and load the cane onto large road vehicles at specific loading zones. The cane stalks, which are in bundles, are weighed for payment purposes as they are transferred to road vehicles. Different modes of transport are used for hauling cane from fields to the loading zones. Growers are responsible for arranging haulage from loading zones to the mill via contracts that extend throughout the season with hauliers that may be individual contractors, large companies or the growers themselves. At the mill, cane is weighed for payment purposes and offloaded either onto a feeder table or stockpile in the factory cane yard. Sampling for quality analysis and its associated payment takes place only after the cane has been fed to the mill and shredded. When cane is visually unacceptable to the mill due to excess extraneous matter or deterioration it can be rejected without offloading but this requires concurrence of a member of the Mill Group Board.

The length of the milling season is decided by the Mill Group Board, based on estimates of the crop size provided by the Local Grower Council. Daily deliveries based on a quota system are then assigned to individual growers. The length of the milling season ranges between 210 and 290 days (Davis and Archary, 2006), which is longer than in most cane-producing countries. Many growers have insufficient cane for supply throughout the season. To prevent pressure for oversupply at mid-season (when sucrose content is highest) a relative payment system is applied that adjusts the measured quality (expressed as recoverable value (RV)) to a "relative quality" for each week according to the following formula:

Grower's relative per cent RV = grower's per cent RV − mill week average per cent RV + mill season average per cent RV.

Harvesting and loading

The harvesting process generally starts with burning. In South Africa, "Codes of Burning Practice" dictate that burning should take place only under specified wind conditions. In sensitive areas (e.g. near holiday resorts), wind predictions are conveyed to farmers' mobile phones. The restriction of burning to periods when weather conditions are favourable causes farmers to tend to burn excessively large areas when the opportunity for burning arises, which leads to excessive delays between burning and harvesting, and consequently results in sucrose deterioration.

Two types of knives are used for manual cane cutting; long-handled knives are used for cutting burnt cane, and the shorter-handled ones are used when trash has to be removed from unburnt cane. Cut cane is placed in large stacks, small stacks or windrows. The large stacks are chained into bundles and loaded by winch onto in-field trailers, which then transport the cane to road trailers. Small stacks and windrows are loaded directly into road trailers using grab loaders; alternatively, continuous conveyer-type loaders with billeters have been used, for example in Swaziland.

A variety of whole-stalk mechanical harvesters have been developed for South African conditions (e.g. Boast, 1985; de Beer and Adey, 1985) or imported for trials, but none enjoy widespread use. The interest in whole-stalk machines arises from the need to operate on slopes and irregular-shaped fields unsuitable for large chopper harvesters, with their accompanying cane transport vehicles. A small number of chopper harvesters operate where field conditions are appropriate but the economics are generally not attractive, especially because of comparatively high losses of cane (Meyer *et al.*, 2002). The availability of labour for manual harvesting and the need to preserve such employment opportunities is another reason for not converting to mechanical harvesting. Meyer and Fenwick (2003) surveyed the performance of cane cutters in South Africa and Swaziland and found considerable variation (Table 4.6).

Transport and delivery

The transport fleet used in the cane industry in South Africa consists of a variety of vehicles (Table 4.7). Where the delivery distance is short, tractor-drawn field trailers carrying bundled cane may go directly to the mill with payloads of 15–30 tonnes. For longer distances a variety of trucks with 15–20 tonne payloads are commonly used, whereas for very long distances high-sided trailers carrying loose cane are used with payloads of around 30 tonnes. The maximum size of vehicles is limited by law to 22 m in length and 56 tonnes gross mass. Software developed in Australia for efficient scheduling of deliveries has recently been adapted to South African conditions (Giles *et al.*, 2009) and applied at some factories, resulting in considerable improvements in vehicle utilization, and it is anticipated that the number of vehicles required for cane transport will steadily decline. In surrounding countries the fleets are generally less diverse and more centrally coordinated because most of the cane is produced by the miller-cum-planter.

Cross-cutting issues: cane burning practices

Decisions regarding best practices for cane supply chains are affected by numerous cross-cutting issues, among which the burning of cane prior to harvesting is quite significant. Burning improves the efficiency of harvesting, loading, transport and juice extraction, but it may lead to serious sucrose deterioration and loss of beneficial uses for the burnt material. From an environmental point of view, besides CO_2 emissions during burning, it also causes smoke and ash particulate

TABLE 4.6 Summary of cutter performances for various harvesting systems

Harvesting system	Cane yield (t/ha)	Cutter output (t/day)	Cutters per 1,000t
Cut & stack (green)	72	3.4	1.8
Cut & stack (burnt)	70	4.2	1.4
Cut & bundle (green)	74	5.6	1.1
Cut & bundle (burnt)	70	6.6	1.1
Cut & windrow (burnt)	93	8.0	1.0

Source: Meyer and Fenwick, 2003.

TABLE 4.7 Modes of cane delivery used in
South Africa, 2005 to 2006

Transport mode	%
Rail	2.3
Tram	3.9
Road	93.8
• Articulated trucks	52.0
• Rigid chassis	21.2
• Tractor drawn	20.6

Source: Davis and Archary, 2006.

pollution. Burning is thus prohibited in sensitive areas where the smoke and ash create significant problems. In Africa, however, the sensitive areas cover only a small portion of the industry, unlike on island countries such as Mauritius and Hawaii, where sensitivity to smoke and ash particulates may be the major reason for widespread prohibition of burning, or the adoption of appropriate codes of burning practices such as cool burning. The significance of these effects depends on secondary cross-cutting issues. For example, deterioration is faster with chopped cane than with whole-stick cane; however, if cane juice or molasses is used for ethanol manufacture, the impact of sucrose degradation is relatively small because the deterioration products (mainly glucose and fructose) are suitable for fermentation. Similarly, glucose and fructose that exist in relatively high concentrations in cane tops improve ethanol yield but decrease sugar recovery. The tops can therefore have either positive or negative value when delivered to a mill, depending on whether sugar or ethanol is the primary product. Also, if there is no demand for additional bagasse, as in conventional sugar factories where there is no electricity export, then unburnt cane increases the amount of fibre available at the mills. On the other hand, if soils are highly erodible and infertile then burning may result in long-term decline in productivity. These issues of burning and collection of extraneous matter require further research and analysis. There is currently negligible electricity export to the grid by most sugar factories in Africa, but this is likely to be a valuable option in the future. The existence of plentiful low-cost coal in South Africa implies that price incentives for renewable energy are needed for cane-residue collection for electricity production to become economically viable. However, in other countries such as Mauritius, which is deprived of fossil fuels, such resources are of utmost significance in sustaining power generation (see Chapter 5). In Brazil, the largest sugar cane producer, power generation from cane bagasse and trash have been thoroughly investigated through support from the UNDP and EU, and a demonstration project is currently being developed with the support of the GEF (see Chapter 7). Exploitation of such resources would enable the transformation of the cane sugar industry into one that supplies multiple products from cane biomass, especially energy products.

Collection of sugar cane agricultural residues

In South Africa there were strong perceptions that cane tops and leaves constituted the major factor affecting potential for sugar recovery from cane (Scott, 1977), and therefore burning prior to harvest was seen as critical for reducing the extraneous matter (Lamusse and Munsamy, 1979), thereby precluding the harvesting of trash for energy production and industrial uses. Subsequent developments have changed these perceptions, however, by illustrating (Plate 6) that burning tends to be a major cause of delay (Barnes *et al.*, 1998), and burnt cane deteriorates faster than unburnt cane (Lionnet, 1996). New information has been generated to assist in decisions regarding trashing (i.e. manual trash removal) instead of burning, and new incentives for trash harvesting seem to be on the horizon. Nevertheless, more than 80 per cent of cane is burnt before harvest in South Africa, and the share is even higher in other Southern African countries (Plate 8).

Comparisons across harvesting methods

Besides electricity generation from cane biomass, production of ethanol from hydrolysed bagasse was intensely researched in South Africa in the 1980s (Paterson-Jones, 1989; Purchase, 1983). This prompted some research on trash harvesting and processing (de Beer *et al.*, 1989; Reid and Lionnet, 1989; Purchase *et al.*, 1990), but commercial experience under Southern African conditions is very limited. In the research trials, four harvesting treatments were compared (all harvested manually):

1. green cane, topped but not trashed (GT);
2. burnt and topped (BT);
3. green cane, not topped nor trashed – whole plant (WP);
4. burnt, not topped (BNT).

TABLE 4.8 Productivity of men and equipment working with different harvesting techniques

	Treatment			
	GT	*BT*	*WP*	*BNT*
Total mass basis				
t/man h (cut & bundle)	1.03	1.06	1.51	1.33
t/trailer (in-field)	4.16	5.17	3.60	4.99
t/hi-lo (trailer)	20.8	26.0	18.2	24.9
Loading rate (t/h)	35.8	43.2	33.6	45.7
On clean-stalk basis				
t/man h (cut and bundle)	0.93	1.03	1.17	1.21
t/trailer (in-field)	3.76	5.00	2.80	4.55
t/hi-lo (trailer)	18.8	25.2	14.1	22.7
Loading rate (t/h)	32.3	41.8	26.1	41.7

Sources: Purchase *et al.*, 1990; de Beer *et al.*, 1989.

Extensive measurements of cutting, loading and transporting were made and enough cane was harvested for each treatment to run a factory for three hours so that the impacts on the factory could be measured. The results are summarized in Tables 4.8–4.12.

On a cost basis, the influence of harvesting method on cutting was small compared with the influence on loading and transport. On a clean-stalk basis, payloads with whole cane (WP) were 44 per cent less than those with burnt topped (BT) cane, and loading rates were 49 per cent lower. The high transport cost for trashy cane was also encountered in Brazilian studies (Hassuani *et al.*, 2005).

The fibre content of the WP treatment was 50 per cent higher than the BT (Table 4.9), but when the different tonnages harvested were taken into account, the fibre yield (t/ha) was almost 100 per cent higher, illustrating that bagasse supplies can be almost doubled by harvesting the entire plant.

TABLE 4.9 Composition of cane harvested by different methods

	Treatment			
	GT	BT	WP	BNT
Component	Composition (% wet basis)			
Stalk	90.3	96.8	77.8	91.2
Tops	2.8	2.7	14.3	8.3
Trash	6.9	0.5	7.9	0.5
	Composition (% dry basis)			
Stalk	80.5	96.1	66.2	88.8
Tops	1.8	2.4	13.0	9.8
Trash	17.6	1.4	20.8	1.4
	Cane analysis (% wet basis)			
Pol	11.5	13.6	10.6	13.4
Purity	84.6	88.4	80.4	86.4
Fibre	21.2	14.4	21.6	14./
Ash	2.5	1.1	2.8	2.1

Source: Purchase *et al.*, 1990 – based on de Beer *et al.*, 1989.

TABLE 4.10 Factory throughputs associated with different harvesting techniques

	Treatment			
	GT	BT	WP	BNT
Component	Throughput (t/h)			
Cane	140	181	127	190
Fibre	30	26	27	28
Pol (sucrose content)	16	24	13	25

Source: Purchase *et al.*, 1990 – based on Reid and Lionnet, 1989.

Whereas fibre throughputs were roughly similar across all treatments, the pol (sucrose) throughput was almost halved when the whole plant (WP) was processed (Table 4.10), which represents a considerable increase in the cost of sugar (or ethanol) production. In addition, the colour and turbidity of juice increased substantially, which has negative impacts on the recovery of sugar.

Cost implications with the supply of cane agricultural residues

The approximate breakdown of the operating costs for sugar cane production in South Africa is given in Table 4.11. A major portion of the labour, mechanical maintenance and fuel costs relate to harvesting and loading of the cane. It is therefore evident that about 40 per cent of the operating costs relate to moving the crop from field to factory, 14 per cent being cane transport. These high costs are often not appreciated by biofuels enthusiasts. Sugar cane is unique in that a large amount of fibre is moved to the factory in the course of sugar production. The cost structure dictates, however, that a minimal amount of extraneous matter should be moved if its "stand-alone" value is unable to compete with other sources of fuel delivered to the factory. In countries such as Mauritius, where coal is imported for power generation, cane fibre tends to have a higher value compared with countries that have abundant domestic coal, such as South Africa.

The cost of trash collection is therefore a barrier to extensive trash utilization. One option is probably that of leaving a limited amount of trash on cane, such that it does not impact significantly on the bulk density (low impact on transport costs) and does not completely remove agronomic benefits from cane fields. To avoid the negative effects of feeding trash to the mill, it could be separated from cane using a dry-cleaning process (Hassuani et al., 2005), which is, however, still an emerging technology. Such partial collection of extraneous matter is easiest with mechanical harvesting where the fan speed can be set to control the amount of trash left on the billeted cane. Various factors, however, including cane sampling issues, dictate that for Southern Africa the early development of trash harvesting is likely to involve trash separation in the field followed by baling of some of the partially dried trash. The costs associated with such a system are given in Table 4.12.

TABLE 4.11 Indicative operating costs of sugar cane production in South Africa

Item	%
Labour	34
Fertilizer	14
Chemicals	5
Sundries	19
Cane transport	14
Mechanical maintenance	7
Fuel and lubricants	7

Source: South African Canegrowers Association, 2010.

TABLE 4.12 Indicative direct costs and quantities for trash harvesting

	Cost of bales (US$/t dry basis)			
	Northern irrigated cane			Coastal cane
Area covered (ha)	250	500	800	800
Activity				
Raking	2.5	2.0	1.7	2.5
Baling	10.5	7.1	6.0	8.9
In-field loading	0.8	0.9	0.8	0.9
In-field haulage	1.9	1.6	1.7	1.6
Transloading	1.2	1.2	1.2	1.3
Road haulage	5.2	5.2	4.9	5.2
Twine	2.4	2.4	2.4	2.4
Total	**24.5**	**20.4**	**18.7**	**22.8**
Quantities				
Item	t/ha(dry basis)			
Residue left in field	5.2			3.6
Baled trash	8.2			5.6
Bales (dry mass) % cane (wet mass)				
Baled trash % cane	7.5			7.5

Source: Purchase *et al.*, 2008.

Note
US$1 = R7.5 (ZAR, South African rands).

Estimates of costs of trash collection are misleading if they do not cover indirect costs such as the cost of soil fertility decline, soil erosion and weed control. Fortunately, the state of knowledge of sugar cane agronomy has reached a point where such costs can begin to be quantified, based on long-term measurements of soil fertility trends under different burning and trashing regimes. Models of the impact of trashing (green cane harvesting) compared to burning have been developed (Wynne and van Antwerpen, 2004). The Wynne and van Antwerpen model takes into account 249 factors that affect long-term economics. Purchase and colleagues (2008) extended the model to assess the economics of collecting trash as a replacement for coal. The results (Table 4.13) show that, without trash sales, trashing was a better option than burning (US$149 benefit on coastal sands and US$94 benefit for irrigated cane per hectare). If trash was collected and sold, the price realized needed to be at least US$53.3 per tonne (dry basis) on coastal sands and US$26.6 per tonne for irrigated cane, otherwise the gross margin would be less than if no trash was sold.

The indication (Table 4.13) that trashing (i.e. green cane manual harvesting) is more profitable than burning raises the question of why more than 80 per cent of cane in Africa is burnt prior to harvest. The model used to simulate the data in Table 4.13 is based largely on a burning-versus-trashing experiment that is now more than 70 years old – i.e. it assesses long-term effects. In such long-term

TABLE 4.13 Simulated differences in grower margins for burnt versus trashed cane; (a) without trash sales, and (b) with partial trash sales from the trashed cane, at different values for the trash and in different regions

| | Difference in grower margins** (trashed margin − burnt margin) | |
| | US$/ha under cane | |
	Coastal sands	Northern irrigated
(a) No trash sold	149	94
(b) Trash sold @ (US$/t dry matter)		
13.3 *(R100)*	2	28
26.6	61	125
40.0	121	221
53.3	180	318
66.6 *(R500)*	239	414

Note
** Grower margins reflect only those costs that are relevant to the comparison between burning and trashing; US$1 = R7.5 (ZAR, South African rands).

assessments, yield decline (9 tonnes cane/ha) is an important cost of burning compared to trashing. In short-term assessments used by most cane growers the yield decline tends to be ignored or underestimated, while the added costs of harvesting and processing green cane are highlighted. It should be noted, however, that the positive effect of trash on long-term yield is most obvious in rain-fed cane because the trash assists with moisture conservation. Cane grown with high rainfall or irrigation shows less benefit of trashing over burning (e.g. Northern irrigated scenario in Table 4.13).

Table 4.14 shows the relationship between coal price and trash value. In computing the trash value it was assumed that the trash would be left in the field for a few days so as to dry to 30 per cent moisture. This is a reasonable assumption for areas where harvesting takes place mainly in dry winter conditions – i.e. most of Africa. The energy equivalent value of trash is simply calculated as the value of coal that could be replaced by the trash, taking into account the moisture, ash and Brix (total soluble matter) content of the trash. This is the basis of some longstanding agreements in South Africa, where bagasse is sold to paper factories on the basis of the coal required to replace it. The coal price (delivered) has recently oscillated between US$70 and US$110 per tonne, suggesting that despite high coal prices the consistent collection of trash could only be justified from high-yielding, irrigated cane on level, fertile soils (Table 4.14 shows that even at US$67 per tonne for coal the trash value would be above the required profitability level of US$26 per tonne derived from Table 4.13). For coastal, rain-fed, sloping fields, the trash generally has more value for soil protection and fertility than as a coal replacement, the latter being profitable only at the higher range of coal prices.

TABLE 4.14 Coal equivalent value of trash, assuming energy content for trash equivalent to that containing 30 per cent moisture but expressing this as value per unit dry matter

Price of coal (US$/t delivered)	Coal equivalent value of trash (30% moisture) (US$/t dry matter)
53	30.8
67	38.5
80	46.3
93	53.9
107	61.6

Note
US$1 = R7.5 (ZAR, South African rands).

Currently, the opportunity to sell trash depends on demand for additional boiler fuel at factories. Only those factories with back-end refineries or by-product activities need additional fuel. Such demand, in South Africa, is about 270,000 tonnes of coal equivalent (Purchase *et al.*, 2008). Generation of electricity for irrigation creates some demand but, in South Africa, very little electricity is sold to the national grid. This apparent anomaly is due to surplus capacity and low generation costs of the national supplier in the past decades. Competing with the huge economies of scale, low coal costs and high boiler pressures available to the national supplier has not been possible. Changing circumstances are offering some opportunities but if electricity sales from a sugar factory would require investment in additional boiler capacity, then the economics are generally not attractive. Use of existing boilers during off-crop (using stored bagasse) may be attractive in the future if a market for seasonal electricity can be found. Some incentives may arise through national or regional commitments to use more renewable energy, as is already common in Europe.

It is emphasized that such scenarios relate mainly to South Africa, and even within South Africa the conclusions differ for different areas (Table 4.13), although the general principles may well apply in the future to other African countries. The incentives will differ in different countries and regions, depending mainly on the price offered for energy and on the local value of leaving residues in-field. This field value of the residues is in turn affected by agricultural conditions, particularly the availability of water. Comparisons between assessments made in different countries are complicated by the assumption of different moisture levels. A comprehensive analysis of the potential for electricity generation from cane agricultural residues in Mauritius is given by Seebaluck and Seeruttun (2009), who conclude that burning a mixture of 70 per cent bagasse with 30 per cent residues (trash) is optimum and would involve collecting 35 per cent of the total residues.

Productivity improvement

In Southern Africa or even in sub-Saharan Africa more generally, the greatest improvement in productivity is likely to arise from moving cane production to more appropriate land in tropical regions. This shift has already begun, with recent investments being mainly in tropical countries rather than in the marginal growing conditions that exist in the continent's major cane producer (South Africa). Within the production system, there is scope for improvement regarding the high-cost areas of harvesting and transport; the introduction of a logistics benchmarking system (Perry and Wynne, 2004) and vehicle scheduling software (Giles *et al.*, 2009) have already contributed to improved productivity. Emphasis on the design of haulage vehicles (Roberts *et al.*, 2009) is also likely to increase productivity. The training and support of competent farmers constitutes another major factor that will affect future productivity.

Research and technological developments

Technological developments likely to impact on the sustainable development of biofuels from sugar cane in Southern Africa are those relating to:

- Efficient use of irrigation water (new tropical schemes may depend on irrigation).
- Biorefineries and improved use of all components of the cane plant using appropriate technologies, including gasification, torrefaction, cellulose hydrolysis and anaerobic digestion (see also Chapter 7). Systems with a relatively small economy of scale are particularly pertinent, for instance partial pyrolysis (torrefaction) of cane residues enables reduction in transport costs, which may open new markets, such as biomass-to-liquid fuel production, or charcoal replacement.
- Breeding of varieties suited to regional conditions and to the changing values of plant components (see Chapter 2).
- Improved machinery for residue collection, including baling machines and dry-cleaning machines.
- Data collection enabling accurate assessment of environmental impact of biofuels made from sugar cane in the region (existing land use change protocols and assumptions are sometimes not relevant to African conditions).
- In sub-Saharan Africa, charcoal is widely used for cooking and leads to some detrimental ecological impacts; substitution of bagasse-based charcoal could offer a solution, if further research and testing were provided.

Concluding remarks on crop handling

The harvesting and delivery of cane is a costly operation for sugar production. Rationalization of transport equipment and coordination has potential for reducing production costs, and is a current focus area in Southern Africa. Most of the cane grown in Africa is burnt prior to harvest and is then harvested in whole-stick form. Changing values of the different components of cane plants, especially the anticipated increase in the value of fibre, suggest that less burning and increased gathering of crop residues may be a future trend in Africa. The aversion to pollution caused by cane burning has contributed to a focus on alternative uses for fibre. Countries such as Mauritius and India are providing insights to systems that use fibre for creating revenue from electricity sales. Gathering of additional fibre usually requires cessation of burning as well as major changes to harvesting and transport systems. It calls for careful analysis of the costs of various harvesting systems and of the value of crop residues in different roles. The optimum sustainable system will be different in different countries and even in different parts of a single country. Assessments based on short-term economics may give different results from those that take long-term crop-yield effects into account, but there is usually scope for harvesting a portion of the non-stalk components of cane without adverse long-term effects. In Africa, sustainable economics for electricity generation from harvested cane trash inevitably depends on price incentives for electricity derived from renewable fuels.

Note

1 A class A-pan is a standardized method for measuring evaporation, developed by the United States of America's National Weather Service. The pan is a cylinder, 120.7 cm in diameter and 25 cm deep. It is placed on a level surface and fenced so that animals are unable to drink from it. At a set time daily, evaporation is measured as the amount of water needed to refill the pan to exactly 5 cm from the top. Variations through the day can be recorded using automated water-level sensors.

References

Ackbar, L.A. (2007) "A Land Suitability Assessment for Sugarcane Cultivation in Angola – Bioenergy Implications", MSc Thesis, University of KwaZulu-Natal, Durban.

South African Canegrowers Association (2010) www.sacanegrowers.co.za, accessed January 2010.

Agrimark (2009) Agricultural Market Trends – Sugar Outlook, April 2009, www.agrimark. co.za, accessed 15 September 2010.

Barnes, A.J., Meyer, E., Hansey, A.C., de la Harpe, E.R. and Lyne, P.W.L. (1998) "Simulation modelling of sugarcane harvesting and transport delays", *Proceedings of the South African Sugar Technologists Association*, 72, pp. 18–23.

de Beer, A.G. and Adey, A. (1985) "The mini-rotor chopper harvester", *Proceedings of the South African Sugar Technologists Association*, 59, pp. 229–231.

de Beer, A.G., Boast, M.M.W. and Worlock, W. (1989) "The agricultural consequences of harvesting sugarcane containing various amounts of tops and trash", *Proceedings of the South African Sugar Technologists Association*, 63, pp. 107–110.

Bezuidenhout, C.N. and Schulze, R.E. (2006) "Application of seasonal climate outlooks to forecast sugarcane production in South Africa", *Climate Research*, 30, pp. 239–246.

Boast, M.M. (1985) "Progress report on the SASABY whole stalk cane harvester", *Proceedings of the South African Sugar Technologists Association*, 59, pp. 225–228.

Blume, H. (1985) *Geography of Sugarcane*, Verlag, Berlin.

Butterfield, M.K. and Nuss, K.J. (2002) "Prospects for new varieties in the medium to long term: the effects of current and future breeding strategy on variety characteristics", *Proceedings of the South African Sugar Industry Agronomy Association AGM*, pp. 41–48. Internet: www.sasa.org.za/sasex/about/agronomy/aapdfs/2002/butterfield.pdf.

Davis, S.B. and Archary, M. (2006) "Eighty-first annual review of the milling season in Southern Africa (2005–2006)", *Proceedings of the South African Sugar Technologists Association*, 80, pp. 1–27.

Desanker, P.V., Frost, P.G.H., Frost, C.O., Justice, C.O. and Scholes, R.J. (1997) *The Miombo Network: Framework for a Terrestrial Transect Study of Land-Use and Land-Cover Change in the Miombo Ecosystems of Central Africa*, International Geosphere-Biosphere Programme (IGBP) Report 41, pp. 1–109, Stockholm.

FAO (1978) "Report on the Agro-Ecological Zones Project: Methodology and Results for Africa", Rome, pp. 1–158.

FAO (2004) Crop Suitability Assessment, www.fao.org/ag/agl/agll/cropsuit.asp, accessed 17 August 2004.

Fischer, G., Teixeira, E., Tothne Hizsnyik, E. and van Velthuizen, H. (2008) "Land use dynamics and sugarcane production", in Z. Zuurbier and J. van de Vooren (eds), *Sugarcane Ethanol – Contributions to Climate Change Mitigation and the Environment*, Ch. 2, pp. 29–62, Wageningen Academic Publishers, Waginengen, The Netherlands.

Giles, R.C., Bezuidenhout, C.N. and Lyne, P.W.L. (2005) "A simulation study on cane transport system improvements in the Sezela mill area", *Proceedings of the South African Sugar Technologists Association*, 79, pp. 402–408.

Giles, R.C., Lyne, P.W.L., Venter, R., van Niekerk, J.F. and Dines, G.R. (2009) "Vehicle scheduling project success at South African and Swaziland sugar mills", *Proceedings of the South African Sugar Technologists Association*, 82, pp. 151–163.

Griffee, P. (2000) "Ecology of Sugar Cane", www.ecoport.org/EP.exe$EntFull?ID=1884, accessed 6 May 2004.

Guilleman, E., le Gal, P.Y., Meyer, E. and Schmidt, E. (2003) "Assessing the potential for improving mill area profitability by modifying cane supply and harvest scheduling: A South African study", *Proceedings of the South African Sugar Technologists Association*, 77, pp. 566–579.

Hassuani, S.J., Leal, M.R.L.V. and Macedo, I. de C. (2005) *Biomass Power Generation: Sugar Cane Bagasse and Trash*, Programa das Nacoes Unidas para o Desenvolvimento and Centro de Technologi a Canavieriva, Peracicaba, Brazil.

IGBP-IHDP (1995) "The Miombo Network: Project on Land Use and Land Cover Change in Southern Africa's Miombo Region", http://miombo.gecp.virginia.edu.

Knox, J.W., Rodríguez Díaz, J.A., Nixon, D.J. and Mkhwanazi, M. (2010) "A preliminary assessment of climate change impacts on sugarcane in Swaziland", *Agricultural Systems*, 103, pp. 63–72.

Lamusse, J.P. and Munsamy, S. (1979) "Extraneous matter in cane and its effect on the extraction plant", *Proceedings of the South African Sugar Technologists Association*, 53, pp. 84–89.

Lionnet, G.R.E. (1996) "Cane deterioration", *Proceedings of the South African Sugar Technologists Association*, 70, pp. 287–289.

Makhanya, E.M. (1997) "Factors influencing the viability and sustainability of smallholder sugarcane production in Umbumbulu", *South African Geographical Journal*, 79, pp. 19–26.

Meyer, E. and Fenwick, L.J. (2003) "Manual sugarcane cutter performances in the Southern Africa region", *Proceedings of the South African Sugar Technologists Association*, 77, pp. 150–157.

Meyer, E., Govender, N. and Clowes, M. St J. (2002) "Trials comparing semi-mechanised and chopper harvesting methods over three seasons in Swaziland", *Proceedings of the South African Sugar Technologists Association*, 76, pp. 120–134.

Muok, B.O., Nyabenge, M., Ouma, B.O., Esilaba, A.O., Nandokha, T. and Owuor, B. (2010) *Environmental Suitability and Agro-Ecological Zoning of Kenya for Biofuel Production*, ACTS, PISCES and UNEP, Nairobi.

Ndour, A. (2010) "Bioenergy feedstock mapping in Senegal", *Proceedings of the International Workshop on Methodology for Agro-Environmental Zoning for Biofuel Production*, UNEP, Nairobi, 7–9 September.

Paterson-Jones, J.C. (ed.) (1989) "The Biological Utilization of Bagasse, a Lignocelluloses Waste", South African National Scientific Programmes, Report no. 149, 111 pages.

Perry, I.W. and Wynne, A.T. (2004) "The sugar logistics improvement plan: An initiative to improve supply chain efficiencies in the South African sugar industry", *Proceedings of the South African Sugar Technologists Association*, 78, pp. 69–80.

Purchase, B.S. (1983) "Perspectives in the production of ethanol from bagasse", *Proceedings of the South African Sugar Technologists Association*, 57, pp. 75–78.

Purchase, B.S., Lionnet, G.R.E., Reid, M.J., Wienese, A. and de Beer, A.G. (1990) "Options for, and implications of, increasing the supply of bagasse by including tops and trash with cane", *Proceedings of the 1990 Sugar Processing Research Conference, San Francisco*, pp. 229–243.

Purchase, B.S., Wynne, A.T., Meyer, E. and van Antwerpen, R. (2008) "Is there profit in cane trash? Another dimension to the assessment of trashing versus burning", *Proceedings of the South African Sugar Technologists Association*, 81, pp. 86–99.

Reid, M.J. and Lionnet, G.R.E. (1989) "The effects of tops and trash on cane milling based on trials at Maidstone", *Proceedings of the South African Sugar Technologists Association*, 63, pp. 3–6.

Roberts, J.A., Giles, R.C., Lyne, P.W.L. and Hellberg, F.W.J. (2009) "A study of sugar industry vehicle configurations and the impact of risks and opportunities on haulage costs", *Proceedings of the South African Sugar Technologists Association*, 82, pp. 118–131.

SASEX (1999) *Identification and Management of the Soils of the South African Sugar Industry*, 3rd edn, South African Sugar Association Experimental Station, Mount Edgecombe.

Seebaluck, V. and Seeruttun, D. (2009) "Utilisation of sugarcane agricultural residues: electricity production and climate mitigation", *Progress in Industrial Ecology – An International Journal*, 6, pp. 168–184.

Schensul, D. (2009) "Remaking an Apartheid City State-Led Spatial Transformation in Post-Apartheid Durban, South Africa", PhD thesis, Brown University, Rhode Island.

Schulze, R.E. (1995) "Hydrology and Agrohydrology: A Text to Accompany the ACRU 3.00 Agrohydrological Modelling System", Water Research Commission, Pretoria.

Schulze, R.E. (1997) "South African Atlas of Agrohydrology and Climatology", Report TT82/96, Water Research Commission, Pretoria.

Schulze, R.E. (2000) "Modelling hydrological responses to land use and climate change: A southern African Perspective", *Ambio*, 29, pp. 12–22.

Schulze, R.E. and Smithers, J.C. (2004) "The ACRU modelling system as of 2002: background, concepts, structure, output, typical applications and operations", in R.E. Schulze (ed.) "Modelling as a Tool in Integrated Water Resources Management: Conceptual Issues and Case Study Applications", Report 749/2/04, Water Research Commission, Pretoria, pp. 47–83.

Schut, M., Bos, S., Machuma, L. and Slingerland, M. (2010a) *Working Towards Sustainability. Learning Experiences for Sustainable Biofuel Strategies in Mozambique*, Wageningen University and Research Centre, Wageningen, The Netherlands in collaboration with CEPAGRI, Maputo, Mozambique.

Schut, M., Slingerland, M. and Locke, A. (2010b) "Biofuel developments in Mozambique. Update and analysis of policy, potential and reality", *Energy Policy*, 38, pp. 5151–5165.

Scott, R.P. (1977) "The limitations imposed on crushing rate by tops and trash", *Proceedings of the South African Sugar Technologists Association*, 51, pp. 164–166.

Sibanda, D.P. (2008) "A Land Suitability Assessment of Zimbabwe for Sugarcane and Sweet Sorghum as Bioenergy Crops", MSc Thesis, University of KwaZulu-Natal, Durban.

Sys, I.C., van Ranst, E., Debaveye, I.J. and Beernaert, F. (1993) *Land Evaluation*, Parts I–III, Agricultural Publications, General Publications for Development Cooperation, Brussels.

Turpie, J., Winkler, H., Spalding-Fecher, R. and Midgley, G. (2002) *Economic Impacts of Climate Change in South Africa: A Preliminary Analysis of Unmitigated Damage Costs*, Southern Waters Ecological Research & Consulting & Energy & Development Research Centre, University of Cape Town.

UN-Energy (2010) "A Decision Support Tool for Sustainable Bioenergy", www.fao.org/docrep/013/am237e/am237e00.pdf.

UNEP (2011a) Protecting Biodiversity When Planning and Implementing Bioenergy Policy and Projects: The Role of Agro-Environmental Zoning, http://www.cbd.int/agriculture/2011-121/UNEP-WCMC-sep11-en.pdf.

UNEP (2011b) Agro-environmental mapping and zoning for biofuel production: Methodology review and guidance, unpublished report, Policy Unit, UNEP's Energy Branch, Paris.

Watson, H.K. (2010) "Potential to expand sustainable bioenergy from sugarcane in Southern Africa", *Energy Policy*, doi:10.1016/j.enpol.2010.07.035.

Wynne, A.T. and van Antwerpen, R. (2004) "Factors affecting the economics of trashing", *Proceedings of the South African Sugar Technologists Association*, 78, pp. 207–214.

PART II
Industry

5

SUGAR CANE PROCESSING AND ENERGY GENERATION FROM FIBRE RESOURCES

Vikram Seebaluck and P.R.K. Sobhanbabu

Introduction

Sugar cane is a renewable resource that can sustain highly diverse multifunctional roles, going far beyond its main economic use for sugar production. The biomass potential of the crop can transform the sugar industry into a more versatile and competitive one, with a broader product portfolio that can generate valuable energy products and services. In this chapter, we present a review of cane utilization focused on the structure of sugar production processes and power generation from fibre resources. The resource requirements, technology use, performance indicators, benchmarks, best practices and opportunities for higher productivity are given with emphasis on operations in Southern Africa but also draw comparisons to the largest cane producers, namely Brazil and India.

Sugar cane utilization

Sugar cane is mainly cultivated for the production of food (sugar). However, during the processing of cane into sugar, a series of "by-products" or "waste products" are obtained – namely bagasse, molasses (final), filter cake and ash – which are or can be used for the production of other commercial products. Bagasse, the fibrous residue left after juice extraction from cane, accounts for around 30 per cent of the weight of cane processed and is used for energy (steam and electricity) generation for the sugar and/or ethanol-making processes with any surplus electricity generated being exported to the grid. Final molasses is the highly viscous liquid obtained after maximum sucrose exhaustion in the boiling house of a sugar factory. It accounts for around 3 per cent of the weight of cane processed and is mainly used for fuel ethanol production, but can be used for the production of other value-added products. Filter cake is the residue obtained from the juice clarification process and contains a major part of the precipitated impurities from cane. It is produced at a rate of around 5 per cent of the weight of cane processed and is generally used as fertilizer in cane fields. Ash, consisting of fly ash and bottom ash, is the residue remaining after bagasse combustion in the boiler house; it is sometimes mixed with filter cake for application in cane fields but can also be used in the construction industry. In addition to these factory by-products, sugar cane agricultural residues (SAR), consisting of cane tops and leaves and trash, is obtained as another field waste product during cane harvesting. This can be transformed into other value-added products, but is usually left in cane fields.

In some cane-producing countries, such as Brazil, the juice obtained after juice–fibre separation is directly used for the production of bioethanol instead of sugar recovery, thus shifting the use of cane to energy production. However, both sugar and ethanol are produced in appropriate ratio from cane juice in "flexi" factories, which dominate the cane industry in Brazil. Sugar itself can be used as feedstock for other products, such as industrial chemicals. Hence, sugar cane resources support a variety of uses and products, based on different resource streams derived from the plant (Figure 5.1).

However, bagasse-based cogenerated electricity and bioethanol production from molasses/cane juice remain the most important cane co-products that are commonly and commercially exploited on a large scale in many cane-producing countries. Sugar production and fibre-resource utilization for energy production is reviewed in this chapter, while sucrose utilization for fuel ethanol production is presented in Chapter 6. The many other commercial co-products that can or are produced from cane as elaborated by Paturau (1989), who identified more than 100 alternative applications, and by Rao (1997), who synthesized the key developments for by-products and co-products from sugar cane biomass, presented in Chapter 7.

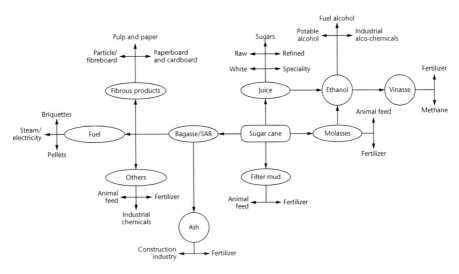

FIGURE 5.1   Sugar cane resources and key by-products

The combination of Chapters 5, 6 and 7 in this volume enables a complete understanding of the industry phase of cane resources and the multitude of products supported by the crop.

Sugar production, types of sugars and sugar factory setup

Sugar is a basic food commodity that is consumed worldwide. It can be produced from a range of sucrose-containing plants. However, around 78 per cent of global commercial sugar is produced from sugar cane (Illovo, 2007), which is cultivated in more than 100 countries. Sugar beet, grown in countries with temperate or cold climates, is second only to sugar cane as a major source of the world's sugar, while other sweeteners are produced in insignificant amounts when compared to the first two major sources of sugar. World sugar production has grown by an average of 2.8 per cent annually since the year 2000, reaching 162.5 million tonnes in 2008 (see Chapter 8 for trends in world sugar production and the share from cane).

A variety of sugars are produced from cane; namely raw, white, refined and special sugars. Raw sugars of yellowish-brown colour having a pol[1] of around 98–99 per cent are produced with minimal chemical processes and are intermediates to white and refined sugar production. White sugar, also called plantation white sugar, is raw sugar whose coloured impurities have been bleached, usually with sulphur dioxide. This is a common form of sugar in many sugar cane-growing areas, but it is difficult to store or ship because during long storage, its impurities cause discoloration and clumping. Refined sugar is the most common form of sugar, which is made by dissolving raw sugar and purifying it with various treatments such as carbonatation, phosphotation, sulphitation, ion exchange and activated carbon to produce a high-quality sugar at a pol of 99.9 per cent. Special sugars, with different sizes, shapes, colours and tastes, are usually produced by modifying the boiling-house configuration of a sugar factory. Some of the different types of special sugars produced are demerara, muscovado, golden caster, molasses sugar and turbinado. Mauritius is one of the most significant exporters of such speciality sugars. Raw sugars can also be prepared as sugar cakes or loaves, known as *jaggery* or *gur* in India, *pingbian tong* in China, and *panela*, *panocha*, *pile* and *piloncillo* in various parts of Latin America. Such types of sugars are produced with minimal processing and are of low quality.

Cane is generally processed into raw sugar, which is then refined in either a back-end or a stand-alone refinery. In back-end refineries, cane is processed into refined sugar in the same factory; the refining process being undertaken as part of the integrated sugar-making process. There are two options for producing refined sugar in back-end refineries. The first option consists of producing raw sugar in a conventional raw sugar factory and subsequently using it as the raw material in an annexed/integrated back-end refinery for refined sugar production. The second option involves the initial production of plantation white sugar by using a revamped initial clarification process (including a carbonatation and/or sulphitation process), and using that as the input for the production of refined sugar in the back-end refinery. The main benefits of back-end refineries are that energy required for the refining process is derived from bagasse and refinery by-products are recycled back to the sugar-making process. In addition, this option eliminates the inconvenience of raw sugar transportation from raw sugar factories to outlying stand-alone refineries. Rein (2007) reports that back-end refineries are associated with lower production costs by around 45 per cent when compared to stand-alone refineries. Hence, back-end refineries have become more popular in recent years given that they provide many advantages in terms of energy utilization, cost of production and the new sugar-production environment. Stand-alone refineries, which have traditionally been the most common type of refinery, use raw sugar as the raw material for the production of refined sugar, and are usually located close to substantial sugar markets. The refined sugar-making process is basically of the same type as the first option for a back-end refinery, except that the refinery is not annexed to a raw sugar factory. With the increasing size of sugar factories it becomes more profitable to refine sugar on site, and refined sugar also fetches a higher price of around 20 per cent more than raw sugar. However, the recovery rate of refined sugar from raw sugar, which is 93.45 per cent according to USEPA (1997) and generally depends on the quality of raw sugar processed and the efficiency of the equipment used, as well as the additional steam and electricity consumption of around 100 kg/tc and 4 kWh/tc respectively (Seebaluck, 2009), should be considered.

The standards for refined sugar are gradually being improved and are already of considerably high quality. The specifications for high-grade sugar, for instance in the EU market where a very low coloration is required, urge the production of high-quality refined sugar. Many attempts for the direct production of refined sugar from cane have been made through the use of emerging and innovative separation techniques, such as membrane ultra-filtration and re-crystallization, with a view to eliminating the need for a refined sugar-production unit attached to the raw sugar factory. Although not yet commercial or fully developed, these processes are of some interest and potential. However, due to the complexity of the equipment involved, these processes are economically less attractive and generally more difficult to operate. Indeed, a simpler option for achieving the production of high-quality refined sugar is through cleaner production, leading to the extraction of fewer impurities from the cane in the juice-extraction process.

Cane composition, quality and influence on cane processing

The efficiency of sugar recovery from cane depends first on the inherent properties of the cane crop and associated extraneous matter derived during harvesting, and second on the technology used in the sugar-making process.

Sugar cane biomass consists of three major parts: namely the underground rhizomes and roots, above-ground stalk with trash, and cane tops and leaves. The yield of above-ground sugar cane biomass is around 100 tonnes per hectare, out of which around half of its mass is dry matter in the form of fibres, sugars and a minor amount of chemical constituents. Sugar cane stalks, the long internodes with soft centres and an external hard rind with nodes in between, which are of concern for sugar recovery, consist essentially of a combination of around 12–19 per cent insoluble fibre and around more than 80 per cent juice, which is an aqueous solution of sucrose and a minor amount of impurities in the form of inorganic and organic substances. Sucrose content in cane is generally around 11–15 per cent.

Ideally, cane stalks free of any cane tops and leaves, trash or roots are desirable for processing in a sugar factory, but this is seldom realized; cane is known to be hetero-geneous in composition even within a given cane stalk. However, variation in the amounts of impurities or non-sugars present in each of the different parts of the cane plant should be avoided as far as possible, as this would otherwise interfere with the sugar-recovery process and reduce sugar productivity from cane. Depending on cane-harvesting practices (full mechanization, semi-mechanization, manual harvesting or burnt cane), cane of different quality is sent to sugar factories whereby the juice extracted in the milling department is not of the highest purity (Plates 9–12).

Indeed, the impurity content of juice obtained in the milling department does not depend only on the extraneous matter content of cane sent to the sugar factory, but also on the inherent impurity content of the cane stalk itself. In the sugar cane stalk, sugars are stored in large, soft-walled parenchyma cells, which occupy roughly 70 per cent of the volume of cane (Rein, 2007). These cells contain a high concentration of sucrose as well as a small amount of non-sucrose components. In addition, the rind of the cane stalk contains relatively little juice and it has a high impurity content. Gener-ally, the nature and impurity content of cane stalks depend on a number of factors, such as cane variety, fertilization, maturity and soil conditions, amongst others. However, it has been clearly established that for obtaining a higher yield of cane per hectare of land in a sustainable manner, the application of adequate amounts of nutri-ents – mainly in the form of nitrogen, phosphorus and potassium that are absorbed by the plant – is essential on the agricultural side. On the cane processing side, absolute or raw cane juice obtained from the juice-extraction process expresses to a great extent the quality of cane processed with regard to assimilation of these nutrients during plant growth, and as such gives an indication of the level of impurities in the plant. Hence during the juice-extraction process both sucrose and impurities are extracted, and the latter must be removed in subsequent processes for sucrose recovery.

It is important to emphasize that the main interest of practical sugar manufactur-ing with respect to the impurity content of delivered millable cane to sugar factories is the extent to which the impurities are extracted in cane juice. It is equally impor-tant to know the level to which these non-sugars present in raw juice are removed during the clarification process, and how those that are not removed at this stage affect subsequent processes in sugar factories, ending up mostly in final molasses and, in minor amounts, in commercial sugar.

Review of the sugar manufacturing process with emphasis on key factors affecting sugar recovery

Sugar recovery from sugar cane is among the most complex processes, comparable to that of the petrochemical industry. It is undertaken on a large industrial scale and under strict norms, given that it deals with the production of a foodstuff. Together with the boiler for steam and electricity production and the distillery for ethanol production, a sugar cane complex comprises almost all the unit operations that could be found in the chemical industry. The following gives only a review of the basics of sugar recovery from cane and is focused on raw sugar production, which is intermediate to other direct-consumption sugars.

A sugar factory consists of three distinct departments; namely the milling department for the extraction of juice from cane, the boiling house for the recovery of crystalline sugar from extracted juice, and the boiler for the generation of process steam and electricity. A schematic diagram of a typical raw sugar manufacturing process, together with the corresponding energy utilization, is shown in Figure 5.2.

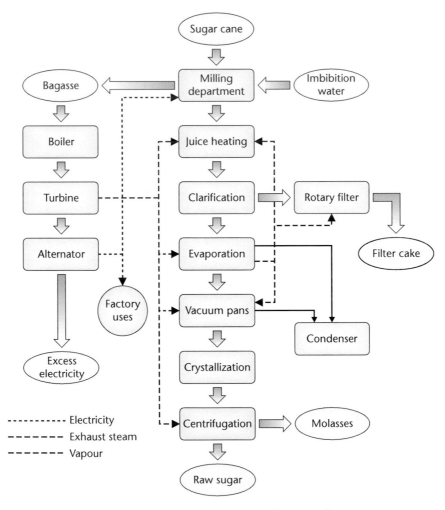

FIGURE 5.2 Schematic diagram of a raw sugar manufacturing plant

Milling department

The milling department is comprised of two major workstations – namely cane preparation and juice extraction – and the latter is undertaken by two long-lasting industrial techniques: cane milling using milling tandems or diffusion using cane diffusers.

Prior to juice extraction, the cane is unloaded, placed in large piles and sometimes cleaned before being conveyed for cane preparation. Cane preparation consists of breaking the hard structure of the stalks with a combination of knives and shredders, during which most of the juice-containing cells are ruptured. The aim of this workstation is to present to the milling tandem an evenly prepared mat of cane with a high level of opened juice-containing cells at a relatively constant rate. Concurrently, a more appropriate material (bagasse) that burns more efficiently in the boiler is produced. Thorough cane preparation is essential to obtain maximum sucrose extraction in the milling tandems and to reduce sucrose loss in bagasse. It has been reported from a number of studies that the overall mill extraction (sucrose extracted in raw juice as a percentage of sucrose in cane) increases with improvement in cane preparation (Moor, 1974; Payne, 1982; Rein, 2007).

After cane preparation, juice is extracted in milling tandems consisting of multiple sets (between four and seven) of the three-roller milling units commonly used in sugar factories, although in other countries some milling units consist of four, five, or six rollers. During the squeezing action in the mills, most of the juice is drained out of the ruptured cells, while the bagasse leaving the mills carries the residual juice. Sucrose present in the bagasse mat goes on reducing in subsequent mills put in series, and ultimately whatever sucrose is carried after the last mill is lost in the final bagasse. To enhance juice extraction, residual sucrose retained in the bagasse mat is reduced by diluting the sugar-bearing juice with water (known as imbibition water) or recirculated thin juice. This process has no effect on the unbroken juice cells, unless the liquid used is so hot that the cells are broken. Juice obtained in the milling department is roughly equivalent to the weight of cane processed and has a Brix[2] of around 12–15 (MSIRI, 2009). The average mill or sucrose extraction in cane-producing countries such as Brazil, India and Southern Africa varies between 96 and 97 (STAI, 2003; SMRI, 1993–2002; Seebaluck *et al.*, 2008).

The energy consumption of the milling department (cane handling, preparation and juice extraction) is the highest in a sugar factory, accounting for around 60 per cent of the total electricity requirements in a sugar plant (Seebaluck, 2009). It is thus evident that energy-reduction efforts should be focused on this section of the plant, to enable higher export to the grid in case surplus electricity is produced. There have been a number of recent developments with respect to cane milling. A recently developed type of two-roller mill in a sugar factory in China has achieved a juice extraction efficiency of 62 per cent and consumes 35 per cent less energy than the three-roller mills (Xu Si-Xin and Gu Yu-Keng, 1996). Low-pressure extraction, which employs a series of light mills each comprising of two pairs of rollers, has been reported to consume less energy (Bhagat and Jadhav, 1999); however, these mills are designed to separate the "easy" juice from the bagasse, and hence final dewatering by a conventional high-compression mill is required to reduce the bagasse moisture to an acceptable level for use as boiler fuel. Such equipment has been installed in India, Brazil and Mexico, but little performance data has yet been published. The Bundaberg's high-extraction mill, a large-diameter two-roll mill, has been reported to achieve an extraction rate comparable with multi-roll mills with 25–40 per cent less power consumption than a four-roll mill, and 30–50 per cent less than a six-roll mill of comparable performance (Batstone *et al.*, 2001; Rein, 2007).

An alternative method for juice extraction from sugar cane is diffusion, which is common in South African sugar factories. Diffusion, a combination of leaching and dialysis processes for sucrose recovery, originated in the beet sugar industry, but has rapidly been adopted in the manufacture of cane sugar, particularly in Southern Africa, where more than 90 per cent of all cane is processed in diffusers (Rein, 2007). However, the installation of diffusers in other cane-producing areas of the world has been much slower. A well-operated diffuser can achieve an extraction efficiency of 98 per cent or above and produces a very clean juice; however, the resulting bagasse has higher ash content, and the boiler design, operation and maintenance must take this into consideration. With respect to energy consumption, the diffuser uses more steam than milling; however, it is low-pressure steam, thereby enabling more of the high-pressure steam to be used for electricity generation. Thus, where electricity generation is to receive high priority, a diffuser may be preferred to milling.

Boiling house

Juice extracted in the milling department is a turbid, greyish-green liquid consisting of a solution of sucrose mixed with soluble and insoluble impurities. The impurities are removed by treating the juice with heat and lime, and after settling in a clarifier, clear juice containing some impurities is obtained together with an underflow called mud. Mud is a thick fluid containing the precipitated impurities and about 90 per cent clear juice, which is recovered in a modern and efficient factory by filtration in rotary vacuum filters. This unit recovers the sucrose in the mud as far as possible and separates the solid impurity fraction from the juice to produce filter cake, the second by-product in a sugar factory.

After removal of most of the impurities from raw juice that would otherwise interfere with subsequent processing, water amounting to roughly 80 per cent of the juice weight is evaporated in a multiple-effect evaporator station to yield syrup. Sugar boiling, also known as crystallization, is then undertaken. This starts just before the point where crystals begin to appear in the syrup and continues up to the maximum concentration, yielding a semi-solid fluid called massecuite. Crystallization is then continued in open vessels of semi-cylindrical shape, known as crystallizers. The massecuite is stirred and cooled for a certain time in order to crystallize the recoverable sucrose remaining in the solutions, and at the same time forcing further the exhaustion of molasses. The product from the crystallizers consists of solid sucrose particles suspended in a mother liquor, which is removed by processing in centrifugals where water/steam is used to wash the sugar crystals to the required polarization (sucrose content). Final molasses is obtained as another key by-product at this stage. The sugar obtained is moist and hot, and is then dried to produce raw sugar, the main product in the factory.

Sugar refining, undertaken in refineries, aims as far as possible at the removal of colour as well as non-sucrose components from raw sugar and/or plantation white sugar. The bulk of refined sugar generally produced has a pol of 99.9 per cent and a colour of less than 45 ICUMSA (Rein, 2007).[3] A number of unit operations similar to those in the boiling house of a raw sugar factory are used in sugar refineries, but with slight changes in design and operational configurations.

Impurities not removed from juice have major negative impacts on sugar processing. They tend to become more concentrated during further sugar processing and are mainly eliminated in final molasses, but a small amount remains in commercial sugar. In general, higher impurity content in cane juice substantially affects the sugar-boiling process, making sugar crystallization less efficient and more difficult, energy intensive and expensive in terms of equipment. Most of the impurities interfere with the crystallization and purging processes, thereby causing high sucrose losses into final molasses.

The boiling house requires electricity and almost all of the low-pressure steam used in the plant, which amounts to around 400 kg/tc, depending on the efficiency of steam use in the factory (Seebaluck, 2009). In this section of the plant, efforts are concentrated on the reduction of steam used. This is achieved using emerging energy-efficient equipment, revamping and retrofitting certain workstations, and applying energy-efficiency tools.

Comparative analysis of factory-process controls in Southern Africa with other large sugar producers

The main objective of comparing mill performances is to identify areas of the mill that deserve more attention with respect to performance and efficiency improvement. The performance will depend mainly on the characteristics of the material processed, the technique and equipment used, as well as process monitoring. Comparisons should be made with caution, since differences in aspects like sugar cane quality and variety, mill equipment type, size, age and efficiency and other specific conditions can distort the results. This is even more important in cases where one tries to compare the performance of mills in different countries.

The main performance indicators for sugar-mill operation include the crushing season, capacity utilization, sugar content (or pol) in cane, fibre content in cane, mill extraction, boiling-house recovery, overall sucrose recovery, molasses recovery, sucrose losses, sugar yield and steam and power consumption. These performance indicators reflect the overall sugar-mill efficiency, right from the quality of cane received to the production of sugar. Such factors vary within and across different regions of the world, including the African region, so that the performance of sugar mills also varies across countries and regions. Table 5.1 gives the physical performance indicators for sugar mills in three continents in the world – Mauritius (Africa); India (Asia); and Brazil (South America) and Table 5.2 gives selected physical performance indicators for sugar mills in four Southern African countries: Malawi; South Africa; Swaziland; and Zimbabwe.

The sugar industry worldwide is a very mature industry that has existed for decades, and based on this long-standing experience significant improvements have been brought to the sugar-recovery process. Dedicated research institutes such as the SMRI and MSIRI (in Africa) and many others worldwide have contributed significantly to improving sugar recovery from cane. On the cane processing side there is less room for major improvement unless there is a major breakthrough in the use of emerging equipment. This is illustrated by the rather similar factory performances recorded in different cane processing regions in the world (Tables 5.1 and 5.2), whereby the differences in overall sugar recovery from cane or even sucrose losses in bagasse, filter cake or molasses can be considered negligible. However, with the size of sugar mills increasing to accommodate more cane, which in some countries is due to centralization of cane-milling activities and to benefit from economies of scale, the industry is more likely to experience higher sucrose losses, but with the trade-off of higher production.

Benchmarks and best practices in sugar factories

Benchmarks are very important comparative performance measurement parameters that indicate a factory's level of performance against industry standards. In the sugar cane sector, benchmarks are established for both the agricultural and industrial areas. When establishing the benchmarks for a specific mill, attention is paid to the configuration of the factory with respect to quality of the raw material, technology status/condition, maximum capacity of each processing stage and level of automation. For example, a four-mill tandem cannot match the performance of a six-mill tandem, either in terms of extraction efficiency or throughput. However, high mill performance standards can be established as long-term targets, and mills can then assess how far they are from best practice, and which areas deserve close attention.

Given the maturity of the sugar-making process, opportunities for new technology development are limited, although prospects still exist for process innovation and optimization, especially in relation to improved productivity. Many African sugar producers face financial constraints around investing in new technologies, but opportunities exist for process optimization and improved productivity through efficient use of resources, including energy, and developments in equipment design. The potential of some new technologies and best practices for each key stage in sugar processing are outlined below.

Milling department

Two-roller milling units are one of the more recent innovations, and have been used in Australia, India and Mauritius. Comparable extraction efficiency to three-roller mills can be obtained, but the main benefit is lower power consumption. The use of electrically driven hydrostatic drives simplifies mill operations and allows easier mill maintenance than a traditional steam turbine; such equipment is generally

TABLE 5.1 Typical physical performance indicators of selected sugar mills

Indicator	Mauritius[a]	India[b]	Brazil[c]
Sugar cane yield (tc/ha/yr*)	71.9	64.5	77.0
Crushing season (days)	116	100 to 180	210
Sucrose % cane	12.1	11.5–15.0	14.0
Fibre % cane	15.3	12.5–15.0	12.9
No. of mills	11	453	320
Crushing capacity (tc/h)	70–275	100–150	100–1,500
Sugar recovered % cane (average)	10.7	10.4	11.0
Sucrose % bagasse (average)	1.26	0.50–0.75	2.50
Moisture % bagasse (average.)	48.6	45.0	50.0
Fibre % bagasse (average)	49.7	48.0	47.0
Bagasse weight % cane (average)	30.8	30.0	27.4
Mixed juice weight % cane (average)	102	90–100	100
Filter cake weight % cane (average)	5.1	3.0–3.1	0–3.5
Final molasses weight % cane @ 85 Brix (average)	3.0	4.5	–
Total sugar recovered % cane (average)	10.7	10.4	12.6[d]
Tonnes cane per tonne sugar at 96 pol (average)	9.12	9.65[e]	8.75
Milling work imbibition water % cane (average)	33.3	25.0	28
Milling work imbibition water % fibre (average)	218	167–200	223
Reduced mill extraction (average)	97.5	96.0	96.5
Pol in open cells	87–89	80–85	88
Sucrose lost in bagasse % cane (average)	0.39	0.75	0.7
Sucrose lost in filter cake % cane (average)	0.07	0.06	1.0
Sucrose lost in molasses % cane (average)	0.91	1.10	–
Total losses % cane (average)	1.62	2.00	–
Reduced boiling-house recovery	88.2	90.0	–
Reduced overall recovery (average)	86.0	86.5	–
Boiling-house efficiency (average)	98.9	88.0	–
Steam to bagasse ratio	1.9–2.5	2.0–2.5	2.0
Steam consumption (kg/tc)	325–550	420–480	450–550
Power consumption (kWh/tc)	28–32	25	28
Excess power generated (kWh/tc)	30–126	30–130	0–60

Notes

a Average technical results of raw sugar mills over the years 1993–2002 (Source: MSIRI, 2003).

b Average technical results of 453 sugar mills in India during 2002–2003 season (Source: STAI, 2003).

c Technical results of raw sugar mills in Brazil found in Centre-South and North-North East.

d Total sugars will include sugars other than sucrose where ethanol production is involved, as in Brazil.

e Indian sugar mills produce direct-consumption plantation white sugar of 99.8 polarization.

* Actual yields at harvest are adjusted according to the duration of the growth cycle.

a good investment where cogeneration is practised. It is equally important to determine a cut-off point wherein sugar recovery and energy utilization are optimized in the extraction process of the milling tandem of a sugar factory producing raw/refined sugar, given that the milling department is the highest energy consumer and that any energy savings could be exported to the grid (Seebaluck, 2009).

Clarification station

Rotary vacuum filters for underflow treatment is the preferred technology in raw sugar factories. Some sugar technologists suggest eliminating the filters completely and using instead a mat of shredded cane in a diffuser to recycle the underflow. The results of trials in South Africa have shown that this recycling technique has no adverse effect on extraction and on bed percolation, and does not result in additional sucrose loss in the diffuser (Clarke, 1999).

The most intensely discussed technology is membrane filtration, expected to replace the standard clarification techniques and lead to improved sugar quality, as well as increased recovery. Other claimed benefits include reduced evaporator scaling, juice sterilization, and increased utilization of downstream equipment.

TABLE 5.2 Typical physical performance indicators of selected sugar factories in Southern Africa

Indicator[a]	Malawi	South Africa	Swaziland	Zimbabwe
Sugar cane yield (t/ha/yr)[b]	105	59.4	93.8	97.7
Sucrose % cane	13.8	13.0	14.0	13.9
Fibre % cane	15.3	14.9	13.8	14.6
No. of mills	2	15	3	2
Average crushing capacity (tc/h)	215	304	313	455
Crushing capacity (range – tc/h)	160–300	90–550	270–360	400–500
Sucrose % bagasse	1.23	1.00	1.35	1.39
Moisture % bagasse	48.4	51.0	50.4	50.2
Fibre % bagasse	49.4	47.1	47.3	47.7
Bagasse weight % cane	30.5	31.1	28.7	30.1
Mixed juice weight % cane	111.5	121.5	119.0	119.4
Filter cake weight % cane	1.36	2.67	2.71	1.43
Final molasses weight % cane @ 85 Brix	4.24	4.06	3.75	3.94
Milling work imbibition water % fibre	278.8	357.5	354.8	344.3
Reduced mill extraction	96.8	97.6	96.7	96.9
Preparation index	90.2	89.1	91.1	91.7
Sucrose lost in bagasse % sucrose in cane	2.84	2.28	2.64	2.72
Sucrose lost in filter cake % sucrose in cane	0.15	0.23	0.18	0.12
Sucrose lost in molasses % sucrose in cane	11.59	9.98	7.71	8.48
Lower calorific value of bagasse (kJ/kg)	6,923	6,902	7,150	7,704

Notes
a Average technical performance of sugar mills over the years 1993–2002 (Source: SMRI, 2003).
b Source: FAOSTAT, 2004.

However, this is yet to be demonstrated on a commercial scale; there are questions still regarding its reliability and high operating costs.

Evaporation station

Most sugar factories use a forward feed, four or five effects, short-tube rising film or Roberts evaporator, with vapours bled from the first two or three effects for juice heating, and vacuum pans. Five-effects evaporation has proven to be the most efficient from Mauritian experience.

The most promising technology for steam savings in the evaporator station is the use of long-tube, falling-film (LTFF) evaporators. The primary advantage of LTFF evaporators is the higher juice-flow velocity, which both enhances heat transfer and shortens juice residence times. LTFF evaporators are quite common in the beet sugar industry and are being experimented in the cane sugar industry. Research shows that in an efficient factory, the evaporator station acts as a boiler to supply low-pressure steam for heating and concentration of juice in the boiling house, while the main boiler is used for supplying high-pressure steam for electricity generation. The use of plate and platular heaters in place of tubular heaters is also emphasized in modern and energy-efficient sugar factories.

Pan boiling and centrifugation

Continuous vacuum pans offer advantages both in the stability of steam demand and overall steam requirement. They also have increased capacity, reduced sugar losses, and are easier to control. On the other hand, batch pans need a longer time to get the equipment in operation, and it is necessary to empty the equipment during the weekly shutdown (Rao, 2001). Batch pans are nevertheless still widely used since improvement in productivity of batch pans can be achieved by improved circulation, improved feed quality and instrument control. Batch centrifugals have become much larger and more energy efficient, but the major development has been in the use of continuous centrifugals for high-grade sugar. There are problems of reliability at high throughput, lump formation and crystal breakage, which need to be resolved for continuous centrifugals to become widely accepted for high-grade sugar.

Fibre resources from cane

Fibre resources from cane can be classified into field resources and factory resources, depending on the location where these resources are generated or left. Bagasse is a factory resource, which is obtained as a main by-product during sugar cane processing and is mainly used for process steam and power generation but can also be used for the production of other value-added products (see Chapter 7). Field-based fibre resources are sugar cane agricultural residues (SAR) consisting of cane tops and leaves and trash that are left in cane fields. SAR is available in almost equal proportion to bagasse and possesses more or less similar physico-chemical characteristics for combustion in bagasse cogeneration plants; as such it represents an important part of the cane crop that can be converted to energy. However, except for a minor amount used for cattle feed, it is left in sugar cane fields where it provides certain benefits such as preservation of soil moisture and control of weed growth.

Cane bagasse

Bagasse is the fibrous residue leaving the last mill of a milling tandem or diffuser of a sugar factory. It is a mixture of hard fibre, with soft and smooth parenchymatous (pith) tissue of highly hygroscopic nature. The composition of bagasse is given in Table 5.3 and it contains mainly cellulose, hemicellulose, pentosans, lignin, sugar and minerals. The quantity of bagasse obtained varies from 22 to 36 per cent of the weight of cane processed and is affected mainly by the fibre content and the cleanliness of cane supplied, which in turn depends on the harvesting practices.

The gross calorific value of bagasse is 19,250 kJ/kg at 0 per cent moisture and 9,950 kJ/kg at 48 per cent moisture. According to Rao (1997), it has varying gross calorific values (GCVs) in the range of 19,037–19,497 kJ/kg. The net calorific value or the practical energy that can be derived from bagasse is 7,985 kJ/kg at 48 per cent moisture content. Moisture content is indeed the main determinant of the net calorific value; a good milling process will result in low moisture content whereas high moisture content is the result of poor milling efficiency. Bagasse contains 45–52 per cent moisture as it comes out of the milling plant, and is generally known as "mill-wet bagasse". Most mills produce a bagasse of 48 per cent moisture content and as such most boilers are designed to burn bagasse with around 50 per cent moisture. Generally, bagasse provides all the energy needs of sugar factories. After exiting the milling tandem or the dewatering mills of the diffuser system, bagasse is transported by conveyors and fed directly to the boilers. Tables 5.4 and 5.5 provide the proximate and ultimate analyses of sugar cane bagasse, which are the basic components used in determining combustion performance.

TABLE 5.3 Composition of bagasse

Chemical components in bagasse (dry basis) %	
Cellulose	26–43
Hemicellulose	17–23
Pentosans	20–33
Lignin	13–22
Average composition of mill-wet bagasse	
Moisture	50
Fibre-pith	47
Sugar	2.5
Minerals	0.5

Source: Rao, 1997.

Bagasse is difficult to store and is prone to fermentation and chemical reactions that can trigger slow internal combustion resulting in fire risks. There is also a loss in its sugar content, which ultimately results in a drop in its calorific value. Bagasse can be stored up to a period of one year if it is dried to a moisture content of less than 30 per cent. Bagasse drying is generally not practised in the sugar industry and is currently an area of research and development.

Bagasse has a low bulk density of around 160 kgm[4] (Paturau, 1989), which poses handling and storage problems. Hence, it has historically been common practice to burn bagasse obtained from the mill continuously in order to avoid disposing it in stacking areas, which is costly in terms of equipment and facilities. Any excess bagasse generated is stored in the open or under covered areas for use during weekend shutdowns or during off-crop season to produce steam and electricity for factory use as well as for export to the grid. Due to the inherent low bulk density of the material, compaction and briquetting can be practised to reduce handling and storage problems, which at the same time decrease the moisture content. Pelletization is another densification process that can be adopted to improve the combustion properties of bagasse as well as facilitating storage.

TABLE 5.4 Proximate analysis of bagasse

	As received	Dry	Air dry
Moisture	50.2	0.0	2.7
Ash	2.7	5.5	5.3
Volatile	40.4	81.2	79.0
Fixed carbon	6.7	13.4	13.0
Total	100.0	100.0	100.0

Source: Beeharry, 1998.

TABLE 5.5 Ultimate analysis of bagasse

	As received	Dry	Air dry
Moisture	50.2	0.0	2.7
Carbon	24.8	49.9	48.6
Hydrogen	2.9	5.9	5.7
Nitrogen	0.2	0.4	0.4
Sulphur	0.0	0.1	0.1
Ash	2.7	5.5	5.3
Oxygen[a]	19.1	38.4	37.4
Total	100.0	100.0	100.0
Chlorine[b]	<0.01	0.01	0.01

Source: Beeharry, 1998.

Notes
a By difference.
b Not usually reported as part of an ultimate analysis.

Technique of bagasse conversion into steam and electricity

The technology for electricity generation using solid fuels, through the rankine cycle route, is conventional and is adopted by the power industry worldwide. The basic alteration in the technology for generating exportable surplus electricity from bagasse in sugar mills depends on the peculiarities of the material that are addressed during the design of the furnace in the boiler house. All other technologies and equipment are similar to the conventional power plants using other solid fuels for power generation (Plate 14).

Basically, cogeneration is defined as the concurrent generation of process heat and electrical power in an industrial plant by the sequential use of energy from a common fuel source. Depending on the quality of process heat used, cogeneration may be based on the "topping" or the "bottom cycle". In the "bottom cycle", the required process temperature is high and hence power is generated through a suitable waste-heat recovery system. In the "topping cycle", the required process temperature is lower, and therefore power generation is performed first. All sugar mills employ this cycle for cogenerating power and heat. For sugar mill cogeneration, a combination of stored and fresh bagasse is usually fed to a specially designed furnace to generate steam in a boiler at typical pressures and temperatures of usually between 20–87 bars and 350–525 °C respectively. The high-pressure steam is then expanded either in a back pressure or single-extraction back pressure or single-extraction condensing or double-extraction cum condensing-type turbo generator operating at similar inlet steam conditions. Due to high pressure and temperature, as well as extraction and condensing modes of the turbine, a higher quantum of power gets generated in the turbine generator set, over and above the power required for the sugar process, other by-products and cogeneration plant auxiliaries. The excess power generated in the turbine generator set is then stepped up to an extra-high voltage, depending on the nearby substation configuration, and fed into the nearby utility grid. As the sugar industry operates seasonally, the boilers are normally designed for multi-fuel operations, so as to utilize mill bagasse, procured bagasse, other biomass, coal and fossil fuel (supplementary fuel or in exigencies), so as to ensure year-round operation of the power plant for export to the grid.

Bagasse cogeneration technologies

There are mainly three bagasse cogeneration technologies: the extraction cum back pressure route; the extraction and condensing route; and the condensing route based on a dual-fuel system as shown in Figure 5.3.

In the extraction cum back pressure route the sugar factory produces only as much steam as is needed for the sugar processing side. By upgrading the steam parameters, surplus power is produced after meeting captive requirements. It is the cheapest option from the point of view of the initial cost and efficiency of the system. However, one major drawback is that fluctuations in surplus power supply

The capital costs of the bagasse cogeneration plant are among the lowest of all renewable forms of power generation. The cost of electricity from bagasse is comparable to small hydro, but is much lower than solar photovoltaic and wind. The technologies used for bagasse cogeneration are proven and well established, and such projects have a short gestation period (18 to 24 months) as compared to the longer period (96 to 120 months) required for a coal-based power plant. Where domestic supplies of coal are plentiful and the environmental costs of coal are not considered it is more difficult for bagasse cogeneration to compete economically.[3] However, where the environmental costs of coal are addressed to some extent, such as through legislation or through carbon finance mechanisms, then bagasse cogeneration can be cheaper than coal.

The use of bagasse as a local fuel source also guarantees security of energy supply, reducing the dependency on imported fuels like coal, and therefore achieves more balanced trade and saves foreign exchange. Use of other biomass fuels during off-season enables the cogeneration plants to operate beyond the crushing season for up to between 300 and 330 days. Sugar mills that produce and export electricity also increase grid stability and reliability, as well as decreasing the need for high-capital investments that would otherwise be required to upgrade transmission equipment and maintain reliable power supply.

Bagasse cogeneration, being a decentralized mode of generating electricity, reduces transmission and distribution losses significantly by supplying electricity near its generation point whilst reducing load on the grid. In most developing countries like India and Brazil, transmission and distribution losses are extremely high (over 20 per cent), due to long distances between power generation and consumption points. The transmission and distribution losses in Mauritius are much lower at around 10 per cent.

The economics of cogeneration depend on the capital cost, mode of implementation and cost of electricity generation. Capital costs are dependent on the technology adopted (back pressure or condensing cum extraction), the pressure/temperature level of boilers/turbines and automation and controls of the cogeneration plant. The capital cost of bagasse cogeneration is in the range of US$700 to US$1,500 per kW. The cost of generation of electricity from bagasse cogeneration plants again depends on a number of parameters and is site specific.

High-efficiency bagasse cogeneration systems also result in a number of additional technical and organizational benefits. First, an efficient cogeneration system protects the sugar mill from vagaries of the power supply and resultant interruptions of the manufacturing process. Likewise, it provides an initiative to sugar mills to concentrate on the conservation of energy and the reduction of steam consumption (in sugar factories and other annexed plants), thereby improving the profitability of the operation. The cogeneration plant usually places no financial or administrative burdens on the sugar factory as it can be executed and managed by a separate entity. It is also notable that the technical staffs of sugar mills are quite conversant with power generation.

Benchmarks and opportunities for bagasse energy generation

The assessment of bagasse electricity-production potential available to the public grid must be based on statistical data available over a long period of time so that the evaluation attains a good degree of confidence. The amount of surplus electricity is highly dependent on the fibre content of the raw material, the efficiency of steam generation and conversion of thermal power to electrical energy, and on the efficiency of energy (steam, vapour and electricity) use on the processing side of the sugar cane complex. Improved and emerging technologies, which are less energy intensive, are determining factors in enhancing the bagasse energy potential.

Prior to erection of a bagasse cogeneration plant, it is of utmost significance to improve the efficiency of steam and electricity use in the sugar-making process and/or any annexed plant such as a distillery, since this influences surplus electricity production. In a sugar factory the material properties of the combustibles can be improved together with the adoption of energy conservation measures in the plant. On the cogeneration side, optimum boiler configuration, optimized boiler draft system and use of programmable logic controller system or distributed control system as instrumentation and control systems are some key areas that can be considered to enhance surplus electricity export to the national grid. In an annexed distillery, steam use can be optimized through the proper selection of ethanol dehydration technologies.

A set of benchmarks has been developed through actual practices of bagasse cogeneration in three parts of the world: Mauritius, India and Brazil. The surplus electricity that can be exported to the grid under specific operating conditions is given in Table 5.6.

The surplus exportable electricity in Mauritian power plants has been based on a cane fibre content of 13–16 per cent, a bagasse moisture content of around 48 per cent, process steam consumption of 350–450 kg steam/tc and power consumption of 27–32 kWh/tc in the sugar factory. For Brazilian plants, bagasse production

TABLE 5.6 Benchmarks for surplus electricity production

Country	Power mode	Operating configuration	Surplus exportable electricity
Mauritius	Continuous	20 bars & 325 °C	25 kWh/tc
	Continuous	31 bars & 440 °C	45 kWh/tc
	Continuous	45 bars & 475 °C	53 kWh/tc
	Firm (CEST)	45 bars & 440 °C	75 kWh/tc
	Firm (CEST)	82 bars & 525 °C	130–140 kWh/tc
India	CEST	67 bars & 495 °C	90–120 kWh/tc
	CEST	87 bars & 515 °C	130–140 kWh/tc
Brazil	Continuous	22 bars & 300 °C	0–10 kWh/tc
	Continuous	42 bars & 440 °C	20 kWh/tc
	Continuous	67 bars & 480 °C	40–60 kWh/tc

Source: Seebaluck *et al.*, 2007.

averages 280 kg/tc, with a moisture content of 50 per cent, and with a boiler steam condition of 22 bars at 300 °C, which are gradually being raised to 42/67 bars. At low boiler pressure, very little surplus power can be produced. In Brazil, the mills traditionally focused on self-sufficiency in energy rather than optimizing for cogeneration; the price offered by the utility was not sufficient from the private sector perspective to justify investments in factory energy and process efficiency that could, otherwise, facilitate the profitable export of surplus electricity. But this trend is gradually being reviewed to install higher-pressure boilers. In India, high-pressure and temperature configurations in CEST systems have been widely implemented, starting in the 1990s, in part through international technical cooperation programmes (USAID, 2002).

Options for improving the energy potential of sugar cane fibres

There are a number of options for increasing the potential of cane fibres for steam and electricity generation, which include: conservative options like improving the physico-chemical properties of the combustible or process efficiency optimization; the use of additional materials like SAR; and the use of upcoming technologies such as gasification. Many research and development projects are currently being undertaken in a number of countries for improving the bioenergy potential of cane biomass. The following sections review the main options for enhancing the cane-fibre energy potential.

Use of high-pressure boilers

Thermodynamically, the use of higher pressure enables the generation of more electricity from the same quantity of bagasse or biomass fuel, thereby increasing productivity. Thus, higher-pressure and temperature configurations are key factors for increasing exportable surplus electricity. Modern power plants use high pressures up to 82–87 bars; for instance, the state-of-the-art bagasse cum coal cogeneration plants in Mauritius use boiler configurations of 82 bars and 525 °C. Nowadays, the boiler pressure can be increased beyond 82–87 bars up to 110 bars, which, however, necessitates proper water-treatment technologies to ensure that a high-quality fluid is sent to the turbine. This development is highly relevant with respect to the construction of new bagasse power plants.

Optimization of the physico-chemical properties of the combustible

Optimization of the physico-chemical properties of bagasse, such as the moisture, sucrose and fibre content for subsequent combustion, can enhance the electricity generation potential of the material. Of the 20 per cent condensation loss in boiler efficiency, 14 per cent is due to the 50 per cent moisture content of bagasse (Narendranath and Rao, 2005). Bagasse drying by making use of flue gas or steam can bring significant reduction in the moisture content, thereby improving the calorific value of the material. Typical projects, mostly pilot plants, have been undertaken in a number of sugar factories and research is still ongoing. Seebaluck (2009) evaluated that an increase of 1 per cent moisture in bagasse leads to a drop in 1.07 kWh/tc of surplus electricity, while an increase of 1 per cent sucrose in bagasse leads to an increase in 4.58 kWh/tc of surplus electricity. The most significant impact is seen with an increase of 1 per cent fibre in cane, which leads to an increase of 13.34 kWh/tc in surplus electricity.

Combustion or co-firing of SAR with bagasse

Sugar cane agricultural residues (SAR) appear to be a very attractive option for enhancing electricity production, particularly in places where cogeneration of bagasse energy is already being carried out, such as in Mauritius, India and Brazil. Its availability, equivalent to roughly the same amount as bagasse, provides the opportunity to increase the volume of combustibles on a single site so that the economic return is more attractive.

Deepchand (1984) reported that combustion of SAR for the production of electricity is technically feasible and creates opportunities for increasing the renewable energy share in sugar cane-producing countries. Seebaluck et al. (2008) reported the net calorific value or energy content of SAR to be 5,980 kJ/kg at a moisture content of 54.5 per cent, which was slightly lower than that of bagasse. The combustion of a mixture of 70 per cent bagasse and 30 per cent SAR amongst different potential options for SAR conversion to electricity was found to be the most appropriate for the short term; it entails the lowest risk with respect to fouling or slagging of the furnace, given that SAR can be treated in a dewatering process to reduce the moisture content and wash out alkali species. This option required the collection of 35 per cent of SAR from cane fields that would increase the share of electricity exported to the grid per tonne of cane by about 37 per cent. The cost of collection, handling and transportation of SAR was found to be around US$7–8 per tonne, or around US$1 per GJ, while the additional equipment required for the processing of SAR prior to its combustion in the power plant was about US$16 per tonne. The cost of electricity generation from this mix of bagasse and SAR was estimated to be 6 US cents per kWh, which was found to be very competitive compared to other renewable energy resources.

Besides being a viable renewable energy business, SAR conversion into electricity favours green harvesting that avoids significant CO_2 emission during cane burning in fields and its utilization for energy generation displaces around 230 kg of sub-bituminous coal and 560 kg of CO_2 per tonne of cane (Seebaluck et al., 2008). Such a project could even qualify for CERs (Certified Emission Reductions). It has also the potential for the creation of livelihoods, given that one job is likely to be created for each 2 GWh of electricity produced from SAR. This project could be ideal for increasing the revenue and bioenergy potential of the cane industry.

Combustion of SAR has indeed been tried in a number of sugar cane-producing countries on the international front. In São Paulo, Brazil, there are five sugar mills that are known to produce and sell surplus electricity to the state by burning mixtures of 20:30 per cent cane trash and bagasse. It was estimated that the full utilization of bagasse and SAR for the production of electricity could serve 5.5–7 million people respectively per year in Brazil (Ripoli *et al.*, 2000). In 1994 the Cruangi Mill in Brazil used 100 per cent cane trash for combustion and operated continuously for more than one month (Leal, 1995). In 1993, two sugar factories in Hawaii, namely Ka'u Agn' Business Co. and Waialua Sugar Co., burned around 27,000 tonnes and 70,976 tonnes of cane trash respectively, in addition to bagasse utilization for cogeneration purposes (Rao, 1997).

In Brazil the Sugarcane Renewable Electricity (SUCRE) project, supported by UNDP and the EU, will build on an earlier project, BRA/96/G31 – Biomass Power Generation: Sugar Cane Bagasse and Trash (CTC, 2005) to catalyse the transformation of the sugar cane industry in Brazil. This transformation will involve supplying renewable electricity from sugar cane biomass to the grid, making this a significant and core aspect of the industry alongside sugar and ethanol production. To maximize the potential of electricity generation from sugar cane, the project will facilitate the expanded use of bagasse and launch the widespread use of SAR that has historically been burned in the cane fields (UNDP, 2009).

Considering SAR and bagasse, the biomass resource from sugar cane is effectively doubled with the additional SAR harvested solely used for surplus electricity generation and which can be burnt in existing cogeneration plants. Research is still being carried out in many parts of the world such as India, Thailand, Jamaica, Cuba, the Philippines and Mexico on pilot scales to exploit SAR fully.

Bagasse gasification

The electricity generation potential of bagasse using the conventional rankine steam cycle could be more than doubled by the adoption of a bagasse integrated gasification-combine cycle (BIG-CC), which is the process of converting biomass energy into an energy-rich gas. Gasification is a thermo-chemical process occurring at high temperature that converts carbon-containing fuels, such as SAR and bagasse, into a synthesis or producer gas consisting primarily of carbon monoxide (CO), carbon dioxide (CO_2) and hydrogen (H_2), through incomplete combustion and reduction. The product of gasification is electricity or heat.

Research on bagasse gasification has been carried out on pilot scales and this technology is yet to emerge on a commercial scale. In Maharashtra, India, the Nimbkar Agricultural Research Institute (NARI) developed a commercial-scale (0.3 MW) model of a gasifier in 1997, which can handle both bagasse and cane residues, for generation of both heat and electricity (Rajvanshi and Jorapur, 1997). According to BRET (2010), the BRET COMPACT biomass power plant, having a capacity of 1.0 MW, gasifies about 20–25 tonnes of cane biomass per day to produce electrical and thermal power. Currently the TPS's ACFBG plant in Brazil, having a nominal capacity of 2 MW, converts approximately 500 kg dry bagasse into syngas per hour for the generation of heat and power (Hassuani *et al.*, 2005).

The BRA/96/G31 project had the specific aim to develop and evaluate the technology required in the complete fuel-to-electricity chain; starting from sugar cane by-product fuels right through to electric power generation with gasification systems integrated in a sugar mill. The estimated investment cost of a BIG-CC plant of capacity 30 MW was found to be too high in Brazil; the resulting electricity production cost of around US$74 per MWh was considered too high for the Brazilian situation. Since this would have been the first plant of its kind, these high costs could have been considered justified given that it would have enabled a breakthrough in technology with bagasse as raw material. However, no investor was willing to provide support to the plant, despite the promised assistance from the World Bank, which was interested in providing sufficient funds to make the project economically viable. Therefore, the project was abandoned (CTC, 2005).

Research undertaken in this area is very scattered between different regions where sugar cane is grown, and is also conducted in countries with cheap alternative energy sources where further research investments in this area could hardly be economically feasible. In addition, research on bagasse gasification is very costly when undertaken by individual countries, so a collegial undertaking involving potential sugar cane-producing countries as well as funding organizations would probably yield better outcomes. The creation of a common platform, gathering expertise from different countries, to work on this common project will be beneficial to the larger or global sugar cane community.

High-fibre cane

Genetic engineering of the cane plant can result in varied composition of fibres from around 10 per cent to more that 30 per cent (see Chapter 2), with the resulting plants classified as normal cane, high-quality cane or fuel cane, depending on their utilization for sugar, electricity or ethanol production or a combination thereof. This concept departs from the traditional production of sugar cane for the manufacture of sugar to that of very high-fibre cane for the primary production of consumer electricity or fuel ethanol. These cane varieties, if successfully and commercially developed, could give a big boost to the sugar and energy potential of the cane crop. As mentioned earlier, an increase in 1 per cent fibre in cane leads to an increase in 13.34 kWh/tc of surplus electricity (Seebaluck, 2009). Research on energy cane from Barbados indicated that the fibre content of cane can be increased up to around 25 per cent (Albert-Thenet et al., 2006). However, research conducted in South Africa showed that the transfer of research findings to commercial-scale application can be very different from those obtained under experimental conditions (ISSCT, 2007). In light of this, further research is required to replicate the experimental outcomes in the fields, and any transfer and adaptation of these new varieties to other regions would take several years. Research in the development of energy and fuel canes is already being undertaken in Mauritius.

Energy conservation and efficiency improvement

Energy optimization and conservation in sugar cane plants provides a practical approach for enhancing electricity export to the grid. Reduction in the steam and electricity consumption on the processing side can be achieved through the adoption of energy conservation devices and automated and continuous processes. The use of energy management tools, such as pinch technology and exergy analysis, re-engineering of the plant or retrofitting of emerging equipment amongst a number of other options, could bring substantial reduction in the internal energy requirements of the sugar cane plant, thereby allowing additional export to the grid. Seebaluck (2009) evaluated that a decrease in consumption of 100 kg steam per tonne of cane leads to an increase of 14.42 kWh/tc of surplus electricity. This area of research is indeed among the promising options for improving energy recovery from cane biomass.

As regards the options for enhancing the energy potential of cane biomass, it has been shown that energy optimization and conservation opportunities, use of high-pressure boilers and use of SAR, listed in order of priority, are among the most appropriate conservative measures that could be adopted by the sugar industry, unless there is a major breakthrough of other projects such as gasification.

Concluding remarks

The process of sugar manufacture from sugar cane is fairly standardized worldwide and is a mature process where opportunities for improved sugar processing technology are limited. Prospects still exist, though, for process innovation and optimization, especially related to improved productivity. The viability of the industry can be increased by the production of value-added products, among which production of electricity from bagasse in cogeneration systems is an attractive and well-established undertaking and would improve the profitability and competitiveness of the sugar industry. These systems have already been successfully demonstrated and implemented in a number of cane-producing countries and the challenge is now for other sugar cane-producing countries to replicate, expand or adapt similar systems. A wealth of experience exists in those countries that have successfully developed these systems, and the potential of this resource for Africa is significant with regard to the current state-of-the-art technologies that can be adopted. Additional potential exists by further optimizing of the system performance, use of cane residues and by the use of emerging technologies. The initiation and expansion of biomass cogeneration in African countries will nevertheless require that proper infrastructure and policy measures are in place to facilitate the markets for independent power production.

Notes

1 Polarity is the apparent sucrose content of a sugar product (determined by polarimetric method).
2 Brix is the apparent percentage of soluble solid matter (sucrose and soluble non-sucrose), determined densitometrically.

3 The ICUMSA (International Commission for Uniform Methods of Sugar Analysis) is an international unit for expressing the purity of sugar in solution and is directly related to the colour of the sugar.

4 In South Africa, which has plentiful domestic coal supplies, bagasse is sometimes sold by sugar factories to paper producers, with the price being roughly equivalent to the cost of coal for electricity generation; i.e. the value of bagasse in this case is roughly equivalent to the coal that is replacing it (Purchase, 2007).

References

Albert-Thenet, J.R., Simpson, C.O. and Rao, P.S. (2006) "The BAMC Fuel Cane Project", Ministry of Agriculture and Rural Development, Barbados, www.agriculture.gov.bb/.

Batstone, D.B., Hatt, R.J., Evans, B.D. and Mitchell, G.E. (2001) "The new Bundaberg two roll mill", *Proceedings of the International Society of Sugar Cane Technologists*, 24, pp. 209–214.

Beehary, R.P. (1998) "Biofuel Characterization and Life Cycle Assessment of Sugarcane Bioenergy Systems", PhD Thesis, University of Mauritius.

Bhagat, J.J. and Jadhav, J.T. (1999) "Low pressure extraction technology for juice extraction – an Indian experience", *Proceedings of the International Society of Sugar Cane Technologists*, 1, pp. 246–254.

BRET (2010) "Utility-Grade Biomass Power", Bridge Renewable Energy Technologies, www.bridgeret.com.

CTC (2005) "Biomass Power Generation: Sugar Cane Bagasse and Trash", Copersucar Technology Center, Project BRA/96/G31, Report to UNDP/MCT/GEF.

Deepchand, K. (1984) "Systems Study for the Utilisation of Cane Tops and Leaves", PhD Thesis, University of Mauritius.

FAOSTAT (2004) "Statistical Database of the UN Food and Agriculture Organization", www.faostat.fao.org.

Hassuani, S.J., Leal, M.R.V.L. and Macedo, I.C. (2005) "Biomass Power Generation – Sugar Cane Bagasse and Trash", Centro de Tecnologia, Canavieira.

Illovo (2007) "World of Sugar – International Sugar Statistics", www.illovo.co.za/worldof-sugar/internationalSugarStats.htm/.

ISSCT (2007) "Workshop of the International Society of Sugar Cane Technologists", 26th Congress, Durban, South Africa.

Leal, M.R.L.V. (1995) "Brazilian mill burns cane trash", *International Cane Energy News*, August, Winrock International (US), www.winrock.org/clean_energy/files/icen1995.pdf.

Moor, B.S.C. (1974) "An evaluation of very fine shredding", *Proceedings of the International Society of Sugar Cane Technologists*, 15, pp. 1,590–1,603.

MSIRI (2003) "Annual Report 2003", Mauritius Sugar Industry Research Institute.

MSIRI (2009) "Annual Report 2009", Mauritius Sugar Industry Research Institute.

Narendranath, M. and Rao, G.V.S.P. (2005) "Improvement of the Calorific Value of Bagasse using Flue Gas Drying", International Society of Sugar Cane Technologists Workshop, 7–11 October, Berlin.

Paturau, J.M. (1989) *By-products of the Cane Sugar Industry*, Elsevier Scientific Publishing Company, Amsterdam.

Payne, J.H. (1982) *Unit Operations in Cane Sugar Production*, Elsevier Scientific Publishing Company, Amsterdam.

Purchase, B. (2007) Personal communication.

Rajvanshi, A.K. and Jorapur, R. (1997) "Sugar Cane Leaf-Bagasse Gasifiers for Industrial Heating Applications", www.nariphaltan.virtualave.net/Gasifier.pdf.

Rao, P.J.M. (1997) *Industrial Utilisation of Sugar Cane and its Co-Products*, ISPCK Publishers and Distributors, New Delhi.

Rao, P.J.M. (2001) *Energy Conservation and Alternative Sources of Energy in Sugar Factories and Distilleries*, ISPCK Publishers and Distributors, New Delhi.

Rein, P. (2007) *Cane Sugar Engineering*, Albert Bartens, Berlin.

Ripoli, T.C.C., Molina, W.F. and Ripoli, M.L.C. (2000) "Energy Potential of sugarcane biomass in Brazil", *Scientia Agricola*, 57, pp. 677–681, www.scielo.br/pdf/sa/v57n4/a13v57n4.pdf.

Seebaluck, V. (2009) "Study of Sugar and Energy Recovery from Sugarcane with Emphasis on the Milling Department", PhD Thesis, University of Mauritius.

Seebaluck, V., Leal, M.R.L.V., Rosillo-Calle, F., Sobhanbabu, P.R.K. and Johnson, F.X. (2007) "Sugar cane bagasse cogeneration as a renewable energy resource for southern Africa", Proceedings of the 3rd International Green Energy Conference, 17–21 June, Malardalen University, Vasteras, Sweden, pp. 658–670.

Seebaluck, V., Mohee, R., Sobhanbabu, P.R.K., Rosillo-Calle, F., Leal, M.R.L.V. and Johnson, F.X. (2008) "Bioenergy for Sustainable Development and Global Competitiveness: The Case of Sugarcane in Southern Africa", Thematic Report 2: Industry, Cane Resources Network for Southern Africa, Stockholm Environment Institute.

Seebaluck, V. and Seeruttun, D. (2009) "Utilisation of sugarcane agricultural residues: electricity production and climate mitigation", *Progress in Industrial Ecology – An International Journal*, 6, pp. 168–184.

SMRI (1993–2002) "Chemical Control Data from Sugar Factories", Durban.

STAI (2003) "Technical Data Directory", Sugar Technologists Association of India Sugar Industry.

UNDP (2009) "Sugarcane Renewable Electricity ('SUCRE')", UNDP Project document.

USAID (2002) "USAID (United States Agency for International Development) International Conference and Exhibition on Bagasse Cogeneration", 24–26 October, New Delhi.

USEPA (1997) "Emission Factor Documentation for AP-42 – Sugar Cane Processing", Midwest Research Institute, US Environmental Protection Agency.

Xu Si-Xin and Gu Yu-Keng (1996) "The extraction mechanism of the new type two-roller mill", *International Sugar Journal*, vol. 98, no. 1165E, pp. 4–8.

6

ETHANOL PRODUCTION FROM CANE RESOURCES

Manoel Regis Lima Verde Leal

Introduction

Ethanol from sugar cane is certainly the most promising first-generation biofuel (IEA, 2008) from economic, energy and environmental perspectives: it has a highly positive energy balance, high yield per hectare, low production costs and excellent greenhouse gas (GHG) abatement potential. Nevertheless, more than half of world ethanol production is based on grains as feedstock (mainly US), which fares poorly in comparison to sugar cane. For countries already producing sugar from sugar cane, rising market demand would make it economical to use some of the intermediate products (juice, syrup and molasses) to expand the production of ethanol. The technology for such undertaking is mature and widely available and the experience from Brazil and elsewhere could be quickly disseminated to other cane-producing countries. Optimal distillery configurations and related modifications in sugar factories can be adapted to the country specificities in terms of size, level of automation, equipment design and operating strategies. This chapter focuses on the techno-economic and environmental aspects of ethanol production from cane resources, with special reference to the experiences from Brazil and the lessons that might be learned in expanding ethanol production in African countries. The chapter also demonstrates the versatility of cane in producing fuel ethanol to diversify global energy supply in the transport sector.

Sugar cane as feedstock for bioethanol production

Sugar cane is the most commercially important feedstock for ethanol production and will remain so for quite some time, for a number of reasons. First, it is widely produced in over 100 countries. Second, it is among the most efficient synthesizers of solar energy into biomass, especially in the form of sugars and cellulose; its high productivity and widespread cultivation translates into excellent potential in many tropical and subtropical regions. Third, there are very few economical alternative feedstocks for ethanol production; the long historical experience due to its cultivation for sugar gives a significant advantage compared to other crops. Last, it has additional benefits through energy self-sufficiency and the possibility to generate many commercial by-products that provide additional revenue and economic flexibility (see Chapter 7).

There are several feedstocks available during cane processing that can be used for ethanol production, ranging from juice to various intermediate products or by-products derived from the sugar crystallization process. These feedstocks are discussed in the following sections. The sugar factory configuration and related industrial processes determine to some extent the quantity and quality of intermediate products and by-products (see Chapter 5).

Cane juice

The components of sugar cane can be classified into three main groups (CTC, 1999): water (65–75 per cent), dissolved solids (12–23 per cent, of which sugars represent 11–18 per cent) and fibres (8–14 per cent). In the dissolved solids group, the main components are sugars (75–93 per cent, including 70–91 per cent sucrose, and glucose and fructose at 2–4 per cent each), salts (3–4 per cent), protein (0.5–0.6 per cent) and colour components (3–5 per cent). The juice, as it leaves the milling tandem or a diffuser, also contains some extraneous matter brought in with the cane, formed by vegetal (pieces of tops and leaves and trash) and mineral (sand, dirt, pieces of metal and rocks) impurities (Rein, 2007). Ash is formed by silica and salts of potassium, calcium, magnesium and sodium as well as chlorides, sulphates and phosphate, and its content in cane can vary from 0.6 per cent (baseline value) to above 2 per cent, in which case serious problems can occur in terms of equipment wear, juice cleaning system overload and boiler tube erosion. In smaller quantities, there are several other components in the juice, mostly derived from cane deterioration, such as dextran, ethanol, oligosaccharides and acids.

Since the non-sucrose materials are more difficult to extract than sucrose, the purity of the juice tends to decrease along the milling tandem and therefore it is a common practice in Brazil to send the juice extracted by the first mill for sugar recovery in the sugar factory. The secondary juice (extracted from the other mills, which is more impure and more diluted due to the imbibition water) and the filtered juice (obtained from the vacuum filters in the clarification station) are sent to the distillery; the presence of non-sucrose sugars (e.g. glucose and fructose) in the juice reduces the efficiency of sugar recovery while it has no effect on the fermentation process and the non-sucrose sugars are even converted into ethanol. However, juice treatment similar to that in a sugar factory (see Chapter 5) is required in a distillery in order to facilitate high yield in the process of fermentation into ethanol.

Syrup

Syrup is the juice concentrated in the multi-effect evaporators; it normally has a Brix (a measurement of the soluble solids) of 60–70 per cent. In distilleries annexed to sugar factories, syrup can be one of the components, used together with the molasses and juice, to prepare the fermentation mash (broth) at the correct Brix. In autonomous distilleries it is desirable, most of the time, to use part of the juice to be concentrated by evaporation to syrup, which can then be blended with the normal juice to adjust the Brix of the mash.

Molasses

Molasses, which can be produced from any sugar-containing material, is the main feedstock for ethanol production in many countries, although in Brazil it does not play such a key role due to the use of cane juice. Molasses can be classified as follows:

- Integral or unclarified molasses
  Integral high-test molasses (IHTM) is produced from unclarified sugar cane juice that has been partially inverted to prevent crystallization, and then concentrated by evaporation to about 80 per cent dry matter (DM) content.
- High-test molasses
  High-test molasses are basically the same as IHTM but the juice has already been clarified for removal of impurities prior to evaporation.
- "A" molasses
  A molasses is an intermediate product obtained upon centrifugation of A massecuite[1] in a sugar factory; approximately 77 per cent of the total available raw sugar (sucrose) is extracted at this stage. A molasses is produced simultaneously with the "first" or "A" sugar, containing 80–85 per cent of DM.
- "B" molasses
  B molasses (or second molasses) is another intermediate product obtained by boiling together "seed-sugar" and A molasses to obtain B massecuite, which is centrifuged to extract an additional 12 per cent of the sucrose. After this step about 89 per cent of the total recoverable sugar (TRS) in the processed cane has been extracted; B molasses contains 80–85 per cent of DM and usually does not crystallize spontaneously.
- "C" molasses
  C molasses (final or blackstrap molasses) is the end product obtained by the combination of virgin sugar crystals obtained from syrup crystallization as seed, syrup and B molasses to form C massecuite, which after boiling and centrifuging produces C sugar and C molasses. C molasses is the end product although it has a sugar (sucrose) content of 32–42 per cent, which cannot be economically recovered.[2]

Choice of feedstocks

There are specific advantages and disadvantages of using juice, A, B or C molasses, depending mainly on the market values of sugar and ethanol. If the aim is to produce mainly sugar, the preferred option is to use A and B molasses because of their higher sugar content. This is not necessarily the case if ethanol is the major product because it can be produced from lower-quality feedstock such as B and C molasses.[3]

Deciding which feedstocks (juice, various types of molasses) and whether to incorporate flexible production in the factory set-up (i.e. to produce either sugar or ethanol) requires in-depth analysis to ascertain the main technical and economic costs and benefits (see Chapter 9). The question is whether other countries should use the same flexible operational configuration as that in Brazil, of switching between A, B and C molasses and juice for ethanol production. In India, all 297 distilleries use only C molasses for the production of ethanol. For African countries, an alternative could be to switch to B molasses or to employ an additional feedstock such as sweet sorghum to complement C molasses, which could substantially expand the production of ethanol (Johnson and Matsika, 2006). However, any new alternative feedstock involves many changes to both the agricultural and industrial operations and will generally be more expensive, at least in the short term.

Ethanol production and use as a transportation fuel

Brazil is the world's largest producer of ethanol from sugar cane. In the 2008/2009 crushing season, ethanol production from cane juice and molasses reached 27.5 billion litres.

Brazil has become a global model for the use of bioethanol as a transportation fuel, which started with the massive PROALCOOL programme initiated in the 1970s due to the oil crisis. The PROALCOOL programme has seen anhydrous ethanol blend with gasoline increase from 4.5 per cent in 1977 to the current level of 25 per cent.[4] There are basically two types of fuel ethanol used in Brazil: *anhydrous*, which is blended with gasoline, and *hydrous*, used in neat ethanol cars or in flexible-fuel vehicles (FFVs). Figure 6.1 shows the trends in the production of the two types of ethanol, as well as sugar cane and sugar, in Brazil since 1970.

Flexible-fuel vehicles

Since March 2003, the so-called flexible-fuel vehicles (FFV), which allow the use of gasoline, neat ethanol or any blend of these two fuels, have been marketed successfully in Brazil. Nearly all car models made in Brazil are now offered in both the gasoline and FFV versions, but the gasoline versions are likely to be discontinued as FFVs already represent close to 90 per cent of new cars sold in 2009. The total FFV fleet at the end of 2009 was 9.6 million vehicles (ANFAVEA, 2010), representing around 35 per cent of the light duty fleet (passenger cars and light commercial vehicles). Close to three million new FFVs were added in 2010, bringing the present FFV fleet to over 12 million units.

Ethanol in India

India is another major world producer of sugar and ethanol, with an annual alcohol production of around 1,300 million litres.[5] With a view to protect the environment, the government of India introduced a scheme to blend 5 per cent anhydrous ethanol in October 2003, and a follow-up phase was planned to increase the blend to 10 per cent. The 10 per cent target has not been met so far, due especially to two barriers. First, because of variations in sugar cane production and consequently molasses production, insufficient ethanol has been produced to meet the 1 billion litres of ethanol required for the 10 per cent blend (after meeting the ethanol demand by the chemical and beverage industries). Second, the distilleries are demanding a higher price for ethanol, which reduces the economic viability. However, the government of India is keen to make this scheme a success, even if it requires importing alcohol from other countries.

Fuel ethanol in Africa

According to REN21 (2010), by the end of 2009 there were mandates for ethanol blending in 19 countries, including several countries in Latin America, but none in Africa. The preferences concerning biofuels in Africa and Asia have tended to favour biodiesel, due mainly to the higher reliance on diesel compared to gasoline. However, there are several countries in these two regions that are considering the adoption of ethanol mandates in the near future. Africa produces 10.6 million

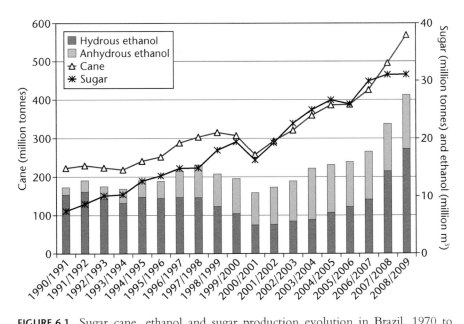

FIGURE 6.1 Sugar cane, ethanol and sugar production evolution in Brazil, 1970 to 2008

tonnes of sugar and consumes 10.1 million tonnes, which suggests that any significant increase in ethanol production will require additional sugar cane cultivation; nevertheless, there is adequate land and appropriate climate in several sub-Saharan countries to provide additional feedstock (see Chapter 4). Kenya, Madagascar, Malawi, Mozambique, Senegal, South Africa, Tanzania and Zambia are among those countries that are planning to introduce large-scale use of fuel ethanol. The main constraint to ethanol production is linked to the availability and subsequent development of suitable agricultural areas, since the installation of sugar cane distilleries does not pose any technical problem. Better dissemination of the lessons learnt from ethanol production are required to transform this biofuel into an international commodity, since 88 per cent of world ethanol production is currently concentrated in two countries (USA and Brazil), with the five main producers representing 93 per cent of global production (REN21, 2010).

Types of distilleries

There are two main types of distilleries for ethanol production from sugar cane: annexed and autonomous distilleries. Distillery sizes vary considerably from those with a capacity of a few thousand litres per day to those with a capacity of several million litres per day. In the early days of the PROALCOOL programme in Brazil, new distilleries were in the range of 30–500 kl/day, but those in the range 120–180 kl/day were preferred. Today, distilleries having capacities between 500 kl/day and 1,000 kl/day are the new standard.[6]

Annexed distillery

An annexed distillery is built alongside a sugar mill, sharing several common systems, such as cane reception, juice extraction system, boilers and effluent treatment, in a synergistic process. Furthermore, this set-up provides considerable flexibility to adjust the sugar/ethanol production to suit market demands and sugar cane quality variations. The feedstock is a blend of cane juice and molasses (B or C) that results in an excellent mash (broth) with respect to nutrients and acceptable level of inhibitors of fermentation. In 2008 in Brazil, around 85 per cent of sugar cane was milled in annexed distilleries (CONAB, 2010), which are generally larger since they are really two factories combined in one.

Autonomous distillery

This type of distillery is a facility where all sugar cane is used to produce ethanol. It was widely used at the beginning of the PROALCOOL programme in Brazil, but later most of the autonomous distilleries were converted to sugar/ethanol plants, to increase production flexibility in order to follow the ethanol/sugar market fluctuations. An autonomous distillery can also be built to operate with molasses purchased from other sugar factories, as is common in India and other countries.

Evolution of distilleries in Brazil

In the past two decades, most of the sugar cane processing capacity has been accomplished through the expansion of existing mills rather than through installation of new mills. It is estimated that from 1992 to 2002, 94 per cent of new ethanol production capacity resulted from mill expansion. Table 6.1 gives the mill set-up in 1990, 2002 and 2009 in Brazil.

The number of sugar mills with annexed distilleries increased with a concurrent drop in both new sugar mills and autonomous distilleries until 2004, when the number of autonomous distilleries started to increase again as a consequence of the faster growth of ethanol production relative to sugar. Overall, the average capacity of the plants has increased, with flexible mills accounting in 2009 for around 75 per cent of the total production.

Since the mid 1990s, sugar cane processing has been equally divided between the sugar factory and the distillery, fluctuating slightly around the 50 : 50 ratio from one season to the next. More recently, the ethanol/sugar ratio increased to 55 : 45 in 2009 (CONAB, 2010).

Technologies for ethanol production

The technology used in Brazil for cane preparation, juice extraction and juice treatment is similar to that found in most cane sugar factories in the world (see Chapter 5). The following discussion gives an overview of the technology used in Brazil, emphasizing the differences when cane is processed for ethanol production rather than sugar.

Cane reception and preparation

When working with whole cane (hand harvesting), it is first sprayed (washed) with water at the top of the feeding table; water use is significant, around 5 cubic metres per tonne of cane (UNICA, 2005), and can represent an environmental problem unless a closed circuit is used. The sucrose loss from washing is estimated to be around 0.5 per cent for hand-harvested whole cane in Brazil (CGEE, 2009), but values in the range 1.4–1.8 per cent have been measured in Mauritius (Rein, 2007).

TABLE 6.1 Evolution of mill set-up in the sugar cane sector in Brazil

	1990[a]	2002[a]	2009[b]
Sugar mills only	27	15	16
Autonomous distilleries	180	104	168
Sugar/ethanol mills	168	199	253
Total units	375	318	437
Sugar cane milled (10^6 tonne)	220	318	600

Sources: a NIPE, 2005 and b EPE, 2010.

With chopped cane (mechanical harvesting), washing with water is not used due to the resulting high sucrose losses. Dry cane cleaning with air blowing is under development in several countries, including Australia, Cuba, Colombia, Guatemala and Mauritius, as well as in Brazil. However, this technology is still in its infancy. In India, all cane is manually harvested, and hence clean cane is supplied to mills. Washing of cane is not prevalent in Indian sugar mills and is being phased out in Brazil with the increasing rate of mechanical harvesting. The efficiency of the cane preparation workstation is generally monitored through the cane preparation index,[7] which should be at least 85 per cent in the case of milling and 92 per cent for diffusion to ensure efficient sucrose extraction (see Chapter 5 for a discussion of milling and diffusion at sugar factories).

Juice extraction

The performance of both types of technologies used to extract sucrose from cane (milling and diffusion) have minor differences with respect to extraction efficiency, juice purity and energy consumption. However, milling is more flexible for operations and capacity increase, and is the most common technique used worldwide. Investment and operating costs are similar in both cases (Rein, 2007) although newer diffusion designs are reporting lower costs (Bosch, 2011).

Diffusion may produce good-quality juice that could potentially be used in the ethanol distillery without further clarification, although this approach has not yet been demonstrated.

Milling is the preferred technology for sugar cane juice extraction in Brazil, due mainly to the flexibility that it offers for capacity expansion at minimum investment costs. More recently, the modular Bosch chainless diffuser (Bosch, 2011) is having some success in penetrating the Brazilian market since it solves the flexibility problem by permitting stepwise increases in capacity by adding modules to the original installation.

In the first phase of the PROALCOOL programme, the main interest of the mills was to increase the processing capacity; very little attention was given to factory efficiency, since ambitious goals had been set for the programme (to achieve a production level of 10.6 billion litres per year in 1986). The mill technology imported from South Africa and Australia had to be modified to increase the milling capacity of the existing milling tandems (CTC, 2003). Table 6.2 shows this evolution of mill capacities in Brazil. Since milling tandems account for a significant part of the total investment cost in a sugar/ethanol mill, this improvement helped to decrease sugar cane processing costs.

Subsequently, factory efficiency has become a key issue. The evolution of mill extraction is given in Table 6.3. The improvements were brought through the following phases:

- phase 1: better cane preparation, introduction of pressure roll and mixed imbibitions;

- phase 2: Donnelly chute; automation of cane feeding; hard weld deposit on rolls;
- phase 3: Donnelly chute and automation in all mills of the tandem; and
- phase 4: improved mill setting and operation control.

The normal practice in Brazil is to send the first mill juice (which has higher sucrose content) to the sugar factory, and the secondary juice to the distillery.

Juice treatment

After the extraction process, sugar cane juice contains impurities that impair its processing into sugar or ethanol. The primary physical treatment uses screens to remove the maximum amount of insoluble impurities, such as sand, clay and bagasse fines, that are in the range of 0.1 to 1 per cent.

A second chemical treatment is then undertaken to remove the fine impurities, which could be dissolved or present in colloidal form or even as insolubles. The juice is then heated, after which it is sent to the clarifiers where the clear juice is separated from the impurities. The most common equipment used in Brazil is the trayless clarifier designed by the Sugar Research Institute (SRI), which has a retention time of less than 45 minutes. Part of the clarified juice is sent to the multi-effect evaporators to be concentrated to a Brix in the range of 60–70 per cent, which is subsequently blended with the remaining part of the clarifier juice to adjust the Brix to the 18–20 per cent range. In annexed distilleries, this concentration process is not necessary since the final mash (broth) Brix can be adjusted by blending the juice and molasses.

The impurities that are removed from the clarifier, in the form of mud with a solids concentration of around 10 per cent, are sent for filtration to recover sucrose that would otherwise be lost, and represent 15–20 per cent of the juice that enters

TABLE 6.2 Evolution of mill capacities (tonnes cane per hour)

Mill size	Evolution phases				
	Original	I	II	III	IV
54"	130	180	190	210	280
78"	270	375	400	440	480

Source: CTC, 2003.

TABLE 6.3 Evolution of improvements in mill extraction efficiencies

	Evolution phases			
	I	II	III	IV
Extraction efficiency (%)	91.0–93.0	93.5–95.0	94.5–96.0	97.0–97.5

Source: CTC, 2003.

the clarifier. The filtrated juice is returned to the process while the filter waste (called filter mud) is sent to the fields as fertilizer.

The decision on whether to pre-treat sugar cane juice prior to distillation is based on a trade-off between investment costs and fermentation efficiency. At the beginning of the PROALCOOL programme, most mills did not treat the juice sent to the distillery, especially the autonomous units. But with increasing knowledge of microbiological processes and the demonstrated correlation between juice purity and fermentation yield, pre-treatment of juice prior to fermentation (similar to the pre-treatment process in a sugar factory) is now being widely adopted, with some minor changes.

There are four possible levels of juice treatment for ethanol production, which result in correspondingly different impacts on the investment costs of the system and the main fermentation parameters. These include:

- Physical treatment only, which is the cheapest.
- Physical treatment, accompanied by heat shock and cooling treatment (this requires between 3.3 and 7.9 times more investment).
- Complete treatment, including physical operations, liming, heat shock and sedimentation (this requires 6.6 times more investment cost) (Rossell, 1988).

The impact on fermentation parameters such as productivity, fermentation time, alcohol concentration in wine, antifoaming agent consumption and sulphuric acid consumption improves with increasing number of treatment stages, and the best results are obtained with the multistage treatment. Such treatment sterilizes the mash (broth) and results in a high fermentation yield, but the increased steam consumption sometimes reduces the benefits of the increased yield. A clean juice also improves the operation and maintenance of the centrifuges, and reduces the consumption of antifoaming agent, biocides and antibiotics.

Fermentation

The raw material for the fermentation process is the mash or broth, a sugar solution with the concentration adjusted to promote yeast activity and to reach a preselected ethanol concentration in the wine at the end of the fermentation process. It is normally prepared as a mixture of sugar cane juice and molasses with a solids concentration of around 18–20° Brix. Water may eventually be used for Brix adjustment. There are basically two major categories of fermentation processes: the batch and the continuous process.

Batch fermentation: Melle–Boinot process

The most popular fermentation technology used in Brazil is the Melle–Boinot process with yeast recovery by means of wine centrifuging. The main components of batch fermentation with cell recycle (Melle–Boinot) as used in a typical Brazilian distillery are shown in Figure 6.2.

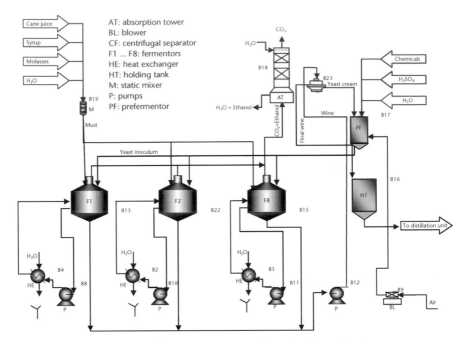

FIGURE 6.2 Melle–Boinot fermentation process for ethanol production

Fermentation takes place in carbon steel tanks, called fermentation vats, with a feeding phase that normally takes between four and ten hours. The sucrose is first hydrolysed by the yeast and then fermented to ethanol, according to the simplified Gay-Lussac reaction:

$$C_{12}H_{22}O_{11} \text{ (sucrose)} + H_2O \text{ (water)} \rightarrow C_6H_{12}O_6 \text{ (glucose)} + C_6H_{12}O_6 \text{ (fructose)}$$

$$C_6H_{12}O_6 \text{ (glucose/fructose)} \rightarrow 2CH_3CH_2OH \text{ (ethanol)} + 2CO_2 + 23.5 \text{kcal}$$

During fermentation, the release of CO_2 causes intense agitation in the vat, and heating. As fermentation is an exothermic reaction, heat is generated and external cooling is necessary to keep the temperature at optimum levels of around 32–34 °C. Besides ethanol and CO_2, higher alcohols, glycerol, aldehydes and organic acids are produced in traces and these products decrease the ethanol yield. The total fermentation time in Brazilian distilleries ranges from six to 16 hours, but stays mostly in the range of eight or nine hours, and the final ethanol concentration in the wine varies from 7 per cent to 9 per cent. The main parameters of the batch fermentation process are given in Table 6.4.

Continuous fermentation

Continuous fermentation accounts for around 20 per cent of the ethanol produced in Brazil (Rossell, 2006). Continuous fermentation evolved from the

TABLE 6.4 Typical parameters of the Melle–Boinot fermentation process

Description	Equipment and performance data		
	Quantity	Capacity	Additional information
Fermentors	8–12	300 m³	Carbon steel cylindrical fermentors with conical bottom
Centrifuges	3–6	45 m³/h	Disc centrifuges with nozzle discharge
Heat exchangers	4–6	1,200,000 kcal/h	Plate type
Productivity	4–7	kg of ethanol m³/h	Weight of ethanol (kg) per volume of fermentation tanks (m³) per unit of time (h)

Source: Rossell, 1988.

Melle–Boinot process; a typical configuration is shown in Figure 6.3.[8] The typical processing parameters of this system are given in Table 6.5. Its advantages include lower investment costs and easier automation (there are facilities producing more than 500,000 litres of ethanol per day that can be operated by one person per shift). Its main disadvantage is the difficulty to handle infections. Continuous fermentation normally uses three to five vats in series, with more than 60 per cent of the sugars being consumed in the first one. The first large-scale fermentation plant (200,000 litres ethanol/day) started operating in Brazil in 1982, in a Copersucar mill (Copersucar is a cooperative comprised of sugar and ethanol mills, located mostly in the state of São Paulo and operating independently but trading cooperatively; Rossell, 1988). In India, more distilleries have switched over

TABLE 6.5 Typical parameters of the continuous fermentation process

Equipment and performance data	
First fermentor volume	700 m³
Number of fermentors	3–5
Number of heat exchangers	3
Air flow rate in pre-fermentors and first reactor for aeration and stirring	0.01 vvm (volume per volume per minute)
Productivity	4–8 kg of ethanol m³/h
Must (broth)	Same as Melle–Boinot process
Wine:	Process data
Total reducing sugars	below 0.5%
Ethanol content	7–11 °GL (8.3 °GL on average)
Yeast concentration	7–12%
Yeast viability	40–75%
Fermentation temperature	32–36°C
Residence time	6–9 h
Fermentation time	4–7 h
Yeast innoculum	Same as Melle–Boinot process

Source: Rossell, 1988.

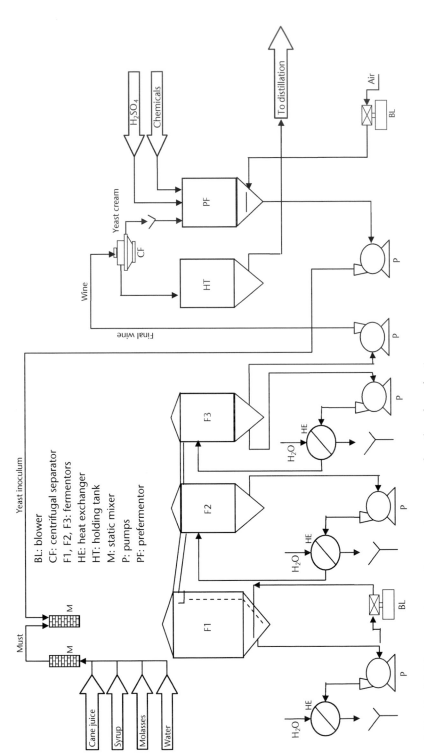

FIGURE 6.3 Multistage continuous fermentation process for ethanol production

to continuous fermentation since 1985. The equipment was originally supplied by Vogelbusch from Austria and Alfa Laval from Sweden, until local capacity was developed. Around 150 distilleries use continuous fermentation out of the 297 distilleries found in India.

The evolution of fermentation technology in Brazil can be represented by optimization of fermentation time, yield and final ethanol concentration in the wine. Fermentation process in Brazil is now fully mature having reached peak performance in the 1990s. Although there is very little potential for additional gains in yield and fermentation time reduction, there are several possible improvements that can further reduce the production cost, mainly through steam economy in the distillation plant and an increase in fermentation robustness.

Distillation

This is the highest energy-consuming process during ethanol production (in the form of process heat). The wine with an ethanol content of around 8 per cent of volume/volume concentration (v/v) is processed in successive columns, in a sequence of processes: from distillation to rectification to dehydration. Figure 6.4 shows a simplified flow diagram of the distillation/dehydration process.

The distillation phase takes place in three stacked columns (A, A1 and D in Figure 6.4), where ethanol is partially extracted from the wine as phlegm (vapour with 40 to 50°GL), which is then directed to column B for further extraction. Some undesirable contaminants are removed in this process, such as aldehydes and esters. The volatile contaminants are concentrated in column D and removed from its top.

Rectification involves concentrating the phlegm to about 96°GL, in columns B1 and B, and the removal of more impurities (higher alcohols, aldehydes, esters, acids, etc.). The main product of this phase is hydrous ethanol; the effluent is recycled or dumped as vinasse (also known as stillage).

Dehydration

Hydrous ethanol, at 96°GL, is an azeotropic mixture of ethanol and water, which by definition means that the ethanol cannot be further concentrated by mere distillation. Additional water removal is thus accomplished by dehydration. There are three main dehydration technologies used in Brazil: azeotropic distillation (Figure 6.4), extraction distillation with monoethyleneglycol (MEG), and molecular sieves technology. The choice of ethanol dehydration technologies is based on investment costs, operating costs, energy consumption and intended use of ethanol produced (fuel, beverage or industrial). Molecular sieves have the highest initial investment cost while azeotropic distillation has the lowest, and for this reason the latter continues to be the first choice in Brazil. However, the operating cost for molecular sieves is the lowest compared to azeotropic distillation, which has the highest operating cost among the three options.

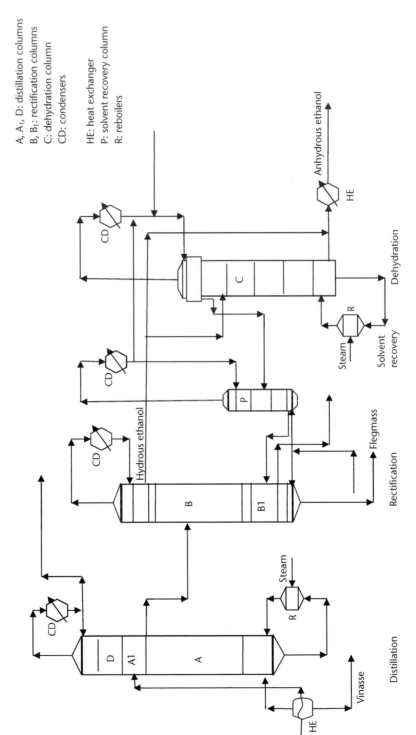

FIGURE 6.4 Flow diagram of the distillation/dehydration process

A, A₁, D: distillation columns
B, B₁: rectification columns
C: dehydration column
CD: condensers

HE: heat exchanger
P: solvent recovery column
R: reboilers

Anhydrous ethanol
Hydrous ethanol
Flegmass
Vinasse
Steam
Solvent recovery
Dehydration
Rectification
Distillation

In terms of energy economy, both molecular sieves and extraction distillation with MEG have similar performance, which is much better than azeotropic distillation. For potable ethanol, the best option is undoubtedly the molecular sieves, since it is a pure physical process with no chemicals coming into contact with the ethanol (as in the case of azeotropic distillation and MEG). Molecular sieves also allow for higher dehydration to be achieved for extra dry ethanol (99.95 per cent purity). Since 2000, Indian distilleries have adopted molecular sieve technology to produce anhydrous alcohol required for a 5 per cent blend with petrol.

Optimal distillery configuration

The configuration for ethanol plants is basically the same, with the first step involving fermentation of the substrate to ethanol, followed by separation of the ethanol from the wine. Several technological advances are important to consider in configuring a distillery. The first is continuous fermentation (through increased yeast concentration), which has become a valued alternative to batch processing, as discussed previously. The higher productivity of continuous fermentation reduces the volume capacity required for fermentation tanks, thereby reducing costs.

A second option for improving the productivity of the fermentation process is through the advanced application of flocculent yeast in the tower fermentor. Cane juice or diluted molasses is fed into the bottom of the fermentor and slowly rises up the tower through the flocculent yeast. Ethanol concentration increases as the feed flows up the tower, reaching the desired concentration at the top. Because of high yeast concentration, productivity is much higher than that of batch fermentation.

Another method is through Vacuferm technology, in which the fermentation tank is maintained at low pressure (vacuum) to enable the ethanol to evaporate off as it is produced. This allows high fermentable sugar and yeast concentrations, because at no point does the ethanol concentration approach inhibiting levels. A serious drawback of the process is the energy required to pump off the carbon dioxide to maintain the required low pressure. It has yet to be demonstrated at an industrial level.

One promising continuous fermentation method is the Biostil, produced by Alfa Laval. It consists of a fermentation tank from which the fermenting feed is continuously drawn. Yeast is centrifugally separated from the wine, which is then introduced into the distillation column. The advantage of the Biostil process is that the fermentor feed can have a much higher sugar concentration. Biostil has been demonstrated at industrial levels in Australia and Sweden. This process was used in some distilleries in Brazil at the beginning of PROALCOOL, but it resulted in stability problems and was thus discontinued. Vogelbusch of Austria and its partners in India (Praj Industries) have 150 distilleries that have successfully adopted continuous fermentation and multi-pressure distillation process.

It should be emphasized that both sugar and ethanol production technologies have reached a very mature stage in Brazil and thus major breakthroughs are unlikely. The processes are well established and technical expertise is widely available. Thus, there is very little potential for gains in efficiency and productivity on the industry side; potential efficiency improvements on the agricultural side remain significant, i.e. in yields and productivity for sugar cane. A remaining area for improvement on the industry side is in the introduction of innovative management approaches related to the use of by-products and waste streams.

There are expectations of a small gain in sugar extraction by the use of an advanced extraction system being developed at CTC – the so-called hydrodynamic extraction system, based on a patent of the late Maxime Riviere of Réunion. However, the higher gains from this innovative system are expected to result from lower investment and operating costs and improved energy economy, rather than improved extraction efficiency. In fermentation, the improvements, albeit small, should come from the use of a cleaner and sterile must (broth) that will reduce infections and, consequently, result in a more stable process, higher yields and reduced use of antifoaming agents and antibiotics.

Improvements in energy (steam) consumption are expected in distillation, motivated by the interest to generate more surplus electricity and allow operation of the power generation system on a year-round basis. Low steam utilization technologies have been introduced through heat integration using waste heat in heat exchangers, which is then re-used to increase the temperature and/or pressure of other processes. Such approaches leave more steam for electricity generation, thereby improving the economics of production (Seebaluck *et al.*, 2008).

Investment and production costs

The capacity of sugar/ethanol plants has been continuously increasing. The majority of new mills being built are designed with annual crushing capacities in the range of 1.5 to 2.5 million tonnes of cane, but capacities of around 4 million tonnes are gradually becoming popular. Sugar industry specialists consider the optimum mill size to be those with capacities of around 2 million tonnes of cane per season, with a crushing capacity of 12,000 tonnes of cane per day. Such a mill in a sugar factory with an annexed distillery consists of a six three-roller milling tandem of 78" × 37" roller size and two 150 tonnes steam per hour boilers, with a capacity for production of 1 million litres of ethanol and 1,000 tonnes of sugar per day, accounting for some spare capacity in both the distillery and sugar factory. The total cost of a 12,000 tonnes cane/day plant is in the range of US\$75 million (in 2006), including ethanol storage facilities (CGEE, 2009). Some idea of economy of scale was indicated by Oliverio (2007), when the cost of a 270,000 tc/year plant was estimated as R\$90 million (R\$333/tc/year), while that of a 2.3 million tc/year was estimated at R\$210 million (R\$93/tc/year).

Mill sizes below one million tonnes of cane per year (approximately 6,000 tc/day) are not considered cost effective. This takes into consideration both the economies of scale in the industrial plant and the logistics and management of cane fields (increasing the size will increase the fields operations complexity and cane transportation costs; see Chapter 4). In comparing this scale to typical existing sugar factories/mills in Africa, it becomes clear that there are actually few factories at the Brazilian minimum scale of two million tonnes of cane, and there are even some that are less than one million tonnes of cane. In order for ethanol production to be cost effective, many African factories would therefore have to be expanded and/or a supplementary feedstock such as sweet sorghum would have to be secured (Johnson and Matsika, 2006).

There are many studies on ethanol production costs in Brazil and elsewhere, but the basic assumptions are sometimes unclear, making comparisons difficult. In NIPE (2005) a specific study on ethanol costs revealed that for a two million tc/year distillery the total production cost was R\$570/m^3 (split into R\$390/m^3 for feedstock, R\$133/m^3 for sugar cane processing and R\$47/m^3 for administration costs such as marketing, logistics, administration salaries, etc.). The processing costs items were wages (R\$28.86/m^3), depreciation (R\$26.50), chemicals and lubricants (R\$25.06/m^3), maintenance materials (R\$20.97/m^3), outsourcing (R\$8.74/m^3) and others (R\$22.58/m^3) and the cost of cane was R\$33.16/tc. It is important to point out that these production costs have increased since 2005 due to escalating oil prices (and the accompanying impact on diesel oil and fertilizer costs) and increases in labour costs and land rental prices among other things. On the other hand, the increasing size of the mills together with higher automation enables important gains as a result of economies of scale and consequent reductions in cost. The feedstock cost accounted for more than two-thirds of the total cost, while the processing cost amounted to only 23 per cent.

Efficiencies, benchmarks and resource requirements

Key technical parameters for evaluating the efficiency of a distillery are the fermentation efficiency (the ratio of actual yield of alcohol to theoretical yield), distillation efficiency (ratio of actual quantity of alcohol recovered to alcohol contained in the wash distillate) and steam economy (the energy used in the form of steam) (Rao, 2001). Resource requirements for key inputs (water, energy, labour, chemicals, etc.) are also quite important and often need to be evaluated in detail, since there will be specific constraining factors in some regions. This section reviews briefly efficiency and resource-use aspects of the ethanol distillery, including some quantitative benchmarks as appropriate.

Industrial efficiency (IE)

For an autonomous distillery, the industrial efficiency is the total reducing sugars (TRS) equivalent of the ethanol produced divided by the TRS of the milled cane.[9] When there are different final products, such as in the case of a sugar factory with an annexed distillery, the industrial efficiency is the ratio of the sum of all products' TRS equivalent (sugar, ethanol, yeast, molasses, etc.) to the TRS of the milled cane.

In Brazil, there are several procedures to calculate the industrial efficiency (IE); the direct calculation method is as follows:

$$IE_{TRS} = 10 \times (TRS_p/TRS_c)$$

Where IE_{TRS} = industrial efficiency in TRS

TRS_p = total reducing sugars equivalent of all products (kg/tc)

$TRSc$ = total reducing sugar as a percentage of cane

It is important to emphasize that industrial efficiency is a weighted average of the efficiency of all product streams, with sugar and ethanol being the most important. Since ethanol production is generally less efficient than sugar production, due to higher fermentation losses, a mill that has a higher ethanol/sugar production ratio will have lower industrial efficiency. Therefore, it may be interesting to compare the performance under different circumstances. Benchmarks for key performance parameters for ethanol production are given in Table 6.6.

Water

Water is used in several processes in the conversion of cane into ethanol, especially for cleaning and in cooling systems, imbibition, barometric condensers and other equipment. Table 6.7 summarizes the main water uses in a sugar factory with an annexed distillery.

Most of these processes use a closed-loop (i.e. with recycling) water system and therefore the losses are mostly related to evaporation, spilling and blowdown. Typical water consumption is about 0.92 cubic metres per tonne of cane, based on 5.07 m³/tc withdrawal (collection) and discharge of 4.15 m³/tc.

The tendency, due to environmental pressures and water costs, is to reduce water collection to around 1 m³/tc and discharge to zero (i.e. a zero leakage facility). In India, the combination of serious pollution concerns and water constraints has led to significant investment in water-saving and recycling measures at sugar factories and ethanol distilleries (Tewari *et al.*, 2007). Some sugar mills in India have become self-sufficient in water (i.e. relying only on the water contained in sugar cane) so that no effluent is discharged, and a few factories are even able to supply

TABLE 6.6 Average and best values of the key performance parameters for ethanol production

Parameter	Average	Best values
Extraction efficiency (%)	96	97.5
Ethanol in wine (% w/w)	8	10
Fermentation efficiency (%)	89	91
Distillation efficiency (%)	99.5	99.9
Industrial efficiency (% TRS in cane)	85	87

Source: Fernandes, 2005.

TABLE 6.7 Water use in Brazilian mills (with annexed distillery)

Process	Use	Average requirement (m^3/tc)
Cane washing		5.33
Milling	Imbibition, bearing cooling	0.75
Juice treatment	Lime preparation, filters, sulphating	0.40
Evaporation	Condensers, molasses dilution, cooling	6.09
Power generation	Boiler make-up, turbo generator cooling	0.70
Fermentation	Must (broth) and juice cooling	4.00
Distillery	Condenser cooling	4.00
Other	Equipment and floor cleaning	0.80
Total		21.00

Source: UNICA, 2005.

drinking water to the local community from the water contained in cane (Seebaluck et al., 2007). In Brazil the zero water intake technology is being offered as a commercial option, but is not yet being adopted due to economic constraints.

Energy

Ethanol distilleries are energy-intensive installations that require thermal energy in the form of low-pressure steam (2.5 bars) at a rate of around 500 kilograms of steam per tonne of processed cane. The mechanical and electrical energy for driving the equipment, lighting, controls and auxiliary equipment results from live steam produced in bagasse-fired boilers that expands in the steam turbines to drive the larger mechanical equipment and/or turbine generators and exits at 2.5 bars as process steam in a pure cogeneration mode; electrical and mechanical power demands are around 12 kilowatt hours per tonne of cane and 16–18 kWh/tc respectively, and the tendency is to replace the steam turbine drives by electric motors with inverters or with hydraulic speed converters.

In the so-called self-sufficient mills, where all the internal energy requirements are met by bagasse cogeneration and no surplus electricity is generated, there is normally a surplus of bagasse left that is sold as fuel for other industries; the boilers for this type of mill operate at around 20 bar/300 °C. The amount of surplus bagasse generally depends on the cane fibre, boiler efficiency, process steam consumption, stability of the plant operation and the number of starts/stops/hot standby occurrences during the crushing season. The latter results in higher consumption of bagasse with the demand estimated at 5–10 per cent of the total bagasse requirement. With increasing interest in generating surplus power for sale to the grid, two changes are being made in the new mills: increase in boiler pressure to values above 60 bars with temperatures in the range of 480–530 °C, and reduction of the process steam consumption to around 400 kg/tc with smart heat recovery techniques, molecular sieves used for dehydration, full automation and optimum column tray design. For advanced distillation approaches, using multi-pressure design, Seemann (2003) and Moura (2006) have indicated that process steam consumption in distillation can be reduced from 3–5 kilograms of steam per litre of ethanol (at a pressure of 2.5 bars) to around 1.8 kg/litre of ethanol (at a pressure of 11 bars), corresponding to 260–430 kg steam/tc and 155 kg steam/tc, respectively. Other uses of low-pressure steam are for juice heating (around 0.5 kg steam/litre ethanol or 43 kg steam/tc), feedwater deaerator and steam losses (around 2 per cent of total steam consumption). According to Oliverio (2008), the technology to reduce the process steam consumption to 300 kg/tc is already commercially available in Brazil and elsewhere; however, the economics for its implementation need to be consolidated.

Chemicals

Several chemicals are used in the mills, mostly in very small quantities, such as lubricants, sulphuric acid, lime, biocides, antibiotics, polymers, sulphur, welding electrodes, antifoaming agents, sodium hydroxide, cyclohexane, etcetera. Consumption varies widely between one mill and another, within the same mill throughout the season, and from one season to another.

Labour

Labour use in sugar mills varies widely, especially across different regions in Brazil. The differences result from the size of the plant, the degree of automation, the local cost of labour (e.g. North-Northeast vs Centre-South regions) and training needs. A typical well-managed mill in the Centre-South region (C-S), crushing around two million tonnes of cane per season (12,000 tc/day) has a staff of around 350–400 industrial workers and 150–200 administrative employees (total factory and field administration). Employment at African sugar factories would be slightly higher in relation to the total production since the factories are smaller than in Brazil; a factory crushing one million tonnes of cane is estimated to require 314 employees on average (Cornland *et al.*, 2001). It should be noted that labour requirements (during regular operation) at the ethanol distillery itself are almost negligible – normally just a few persons per shift are needed – since it has few mechanical components (unlike a sugar factory) and does not require the complex monitoring, adjustment and evaluation needed at a sugar factory.

Advanced technologies for ethanol production

Various other technologies are being developed for ethanol production from alternative feedstocks such as woody biomass or other sources, as discussed further below.

Lignocellulosic pathways

Lignocellulosic materials can be processed into biofuels via two main types of pathways: thermochemical or biochemical conversion.

Thermochemical conversion uses heat to break the biomass into liquid (pyrolysis) or gaseous (gasification) components that are further processed by distillation or catalytic synthesis to produce a wide variety of liquid transport fuels, including alcohols, ethers or hydrocarbons similar to gasoline or diesel. The most popular among these is the Fischer–Tropsch synthesis, which is already used commercially to produce liquid fuels from coal.

Biochemical conversion involves the use of microorganisms to break the lignocellulosic material into simple sugars that are fermented to ethanol. Due to its synergies with conventional ethanol production this group of processes has received more attention from the sugar cane industry, considering the high availability of biomass residues at the mill (bagasse from the cane stalk) or in the fields (cane tops and leaves or cane trash when the crop is harvested without burning).

Lignocellulosic ethanol production from cane biomass

Fibres in sugar cane stalks and leaves represent approximately two-thirds of the primary energy contained in the plant. Therefore the use of this resource to produce additional ethanol can contribute significantly to improving the energy efficiency and sustainability of ethanol as a biofuel. A variety of processes have been studied for the conversion of cellulosic biomass into ethanol; enzymatic hydrolysis is a well-known approach that could improve the economic competitiveness of bioethanol compared to other liquid fuels on a large scale (Mosier *et al.*, 2004). Typical conversion processes of lignocellulosic materials into ethanol are as follows.

Separate hydrolysis and fermentation (SHF) process

This technology consists of the enzymatic breakdown of cellulose and hemicellulose into sugars, followed by their separate fermentation. A pre-treatment is generally required to remove the lignin to facilitate the reaction of the sugar polymers with the enzymes.

Simultaneous saccharification and fermentation (SSF)

The SSF process is an improvement on the SHF, and was patented by the Gulf Oil Company and the University of Arkansas in the USA. SSF is one of the most promising processes for the production of ethanol from lignocellulosics. In SSF, enzymatic hydrolysis of cellulose to glucose and fermentation of sugars to ethanol are carried out simultaneously, in the same vessel. As sugars are produced, the fermentative organisms convert them into ethanol. The rate of hydrolysis is increased by reduced glucose inhibition of the cellulases (in comparison to SHF), resulting in higher productivity, or reduced enzyme consumption for the same productivity (Bollók *et al.*, 2000). Another advantage of SSF is its lower capital cost due to the reduced number of vessels needed. One disadvantage of SSF is the difference in optimum conditions (pH and temperature) for hydrolysis and fermentation. The optimal pH is 4.8 for the cellulase enzyme activity, and between 4 and 5 for the yeast cell; if the pH is maintained between 4.5 and 5 during the SSF process, it is convenient for both the enzyme and the yeast. The optimal temperature is below 35 °C for yeasts used in ethanol production, while temperatures above 40 °C lower the viability of most yeast cultures. The rate of hydrolysis is highest at temperatures in the 40–50 °C range, as cellulase enzyme acts optimally at about 50 °C. At lower temperatures, the rate of saccharification decreases considerably.

Direct microbial conversion (DMC)

DMC is an innovation of the SSF process. The DMC method involves conversion of cellulosic biomass to ethanol in which ethanol and all the required enzymes are produced by a single microorganism. The main advantage of DMC is that a dedicated process for production of cellulase enzyme is not needed. DMC is not considered to be a leading alternative process today because there are no available organisms that produce cellulase as well as other enzymes at sufficiently high levels, and concurrently producing ethanol at the required high concentrations and yields. Most commercial applications for cellulase enzymes (for example in the textile industry) account for higher market value compared to its use in the fuel industry.

Environmental implications of cane ethanol production

The production chain of sugar cane ethanol has a series of environmental impacts on air, soil and water, due mainly to air pollution caused by pre-harvest cane burning, bagasse burning in boilers, water pollution caused by fertilizer and herbicide leaching or carryover, water use in the factory and effluent disposal in the water bodies. The soil can also be contaminated or degraded by misuse of effluents, residues and chemicals, erosion and loss of fertility. However, long-standing experience in sugar cane cultivation and processing has led to several good practices that, if adopted, can significantly reduce these environmental impacts. In Brazil cane burning is being phased out and soon it will no longer be a cause of concern. For cane-producing countries that are considering ethanol production on a large scale, the production and disposal of vinasse, the residue obtained from ethanol distillation, should be carefully evaluated. The proper use of other important residues (filter mud) also has significant economic as well as environmental impacts, as discussed below.

Vinasse/stillage production and disposal

Vinasse or stillage is produced from cane juice and/or molasses fermentation. The use of vinasse deserves particular attention due to its potentially large impacts. The sugar cane industry in Brazil has made considerable strides in controlling environmental impacts. For example, its main effluents (vinasse, filter mud, boiler ashes) are recycled and used as fertilizer, and this is thus an area where Brazil can provide considerable know-how. Production processes usually do not have significant environmental impacts (e.g. there are no extreme temperatures involved and there are no toxic chemicals or sulphur in boiler exhaust).

Chemically, vinasse composition varies according to the soil, sugar cane variety, harvesting method, and the industrial process used in the production of ethanol. Its colour, total solid contents and acidity may vary according to the type of vinasse, processes and treatments.

Vinasse contains unconverted sugars, non-fermented carbohydrates, dead yeast, and a variety of organic compounds.[10] The organic substances present in the vinasse generate a very high BOD (biological oxygen demand), ranging from 30,000–40,000 milligrams per litre, and a low pH of 4–5, and because of the presence of organic acids (which are corrosive) it requires stainless steel, plastic or fibreglass to resist it (Freire and Cortez, 2000). The large volume of vinasse and its high BOD and high COD (80,000–100,000 mg/L) pose a problem for its disposal.

Because vinasse is produced in large volumes, one possibility of reducing its polluting effects is that of recycling it in the fermentation process. Vinasse may be partly used to dilute the sugar cane juice or molasses in the fermentation step. The juice or molasses generally needs to have the Brix adjusted to allow proper yeast growth, a process that normally requires water for dilution. Alfa Laval developed a process called Biostil that uses vinasse to dilute the molasses prior to the fermentation step. Vinasse recycling to the fermentation broth has been tried in Brazil, either though the Biostil technology or directly in the conventional Melle–Boinot process, but without success, and this has apparently been dropped.

Research is also being carried out to decrease production of vinasse by developing new yeast strains capable of enduring higher alcohol concentrations. In addition, other measures such as vinasse evaporative cooling could allow significant reduction in vinasse volume and corresponding disposal costs. The best-known options for vinasse disposal may be grouped according to source or point of application as follows:

Land application of vinasse

This is the large-scale solution adopted in Brazil by all mills for vinasse disposal, due to the high value of this effluent as a fertilizer. The application area corresponds roughly to 30 per cent of total cane fields in mills producing both ethanol and sugar, essentially in the ratoon areas. In the 1960s the vinasse was dumped into water bodies, but in 1982 a federal law prohibited this practice, which led to the use of the so-called "sacrifice area"; an area close to the mill where the vinasse was dumped for slow oxidization and evaporation, with very high doses. The problems arising from this option (nauseating odours, insect proliferation, water table contamination, public complaints, etc.) has made it practically non-existent today. In some states, like São Paulo, this practice has been completely prohibited.

Detailed and extensive studies and field testing have shown that vinasse is an excellent fertilizer and improves the physical, chemical and biological properties of the soil; it increases the pH and the capacity of cation exchange, enhances the nutrient availability, improves the soil structure as a result of the addition of organic matter, increases the water retention capacity and improves the microorganisms population (Donzelli et al., 2003). This is known as "ferti–irrigation" in Brazil.[11]

To avoid environmental problems and the inefficient use of vinasse, the application dosage has to be established based on the chemical analysis of the vinasse, soil analysis, cane fields productivity and environmental conditions (water table depth, proximity to water bodies, etc.); in São Paulo state there is an environmental law defining the procedures to calculate the maximum allowable vinasse dose and to monitor the impacts on ground water and soil (CETESB, 2006). Vinasse application methods have evolved from the use of the simple tank truck, which simply spreads the stillage by means of a perforated pipe at the back of the tank, to modern sprinkler systems.[12]

Recycling

Recycling is seen as a solution to avoid high transportation costs for the large volumes of vinasse generated. At the beginning of the PROALCOOL programme in Brazil, the Biostil process was used in several mills, but the technology had to be abandoned due to difficulties in maintaining continuous operation.[13] However, Chematur Engineering has recently developed this technology further with success. The most important gain with this technology is the low production of vinasse (25 per cent of the volume from the conventional process) with solids concentration of 30–35 per cent, by weight, that reduces the costs of field application (Chematur, 2011). Chematur also claims the following advantages of the fermentation process: higher ethanol yields, lower water consumption, easier process control, compact layout and energy-saving process. There is so far no commercial application of this technology for evaluation of the real benefits.

Direct use as animal feed

This method is used around some distilleries in the USA (e.g. the Shepherd Oil Distillery at Mermentau, Louisiana), adopting the same practice used with vinasse from spirits distilleries, where high-grade, high-protein vinasse is obtained. Vinasse obtained from molasses or HTM (high-test molasses) fermentation is fed directly to animals, usually beef cattle. Although good results have been obtained, more studies on nutrition and other effects are still needed. In some countries such as Cuba, animal feed from the waste stream is an important industrial by-product. Several other alternative uses of stillage have been subjected to research and development and different degrees of success have been achieved. These include fungus production, construction material development, biodigestion and direct disposal by incineration.[14]

Filter mud

Filter mud is another by-product that has proved to have an economic value as fertilizer, due to mainly its phosphorus and organic matter content (see Chapter 7). It is produced at a rate of around 35 kg/tc and is generally used during cane plantation. Fertilizer economy is improved by using stillage in the ratoon and filter mud for plantation. A typical impact of the application of mill effluents on the fertilization rate is given in Table 6.8.

Composting and dewatering is often used to reduce the volume of this material and to improve the mineralization of nitrogen. In this process, vinasse can be added to the composting process, thus increasing the fertilizer potential of the filter mud and reducing the vinasse disposal costs. In India, sugar factories in the state of Tamil Nadu have plants for the production of biogas from filter mud. The biogas as a fuel is used in the laboratory, canteen, etcetera.

Biogas production from vinasse

Vinasse biodigestion has the advantage of on-site disposal while ensuring the production of a good fertilizer, together with energy recovery through biogas production. The heating value combined with its potential for pollution reduction merits further investigation of vinasse biodigestion. This is particularly interesting because while the sugar cane industry produces a large amount of vinasse, it still depends on diesel oil for sugar cane transportation. In Brazil, for example, many distilleries intended to produce biogas. However, due to high costs, difficulties with special spare parts for trucks, etcetera, there are hardly any distilleries currently producing biogas.

The organic matter content of vinasse qualifies it as a feedstock for biogas production, using the widely used biodigestion process. Biodigestion is a complex process involving dozens of types of microorganisms and several intermediate compounds. It can be divided into three phases: polymer hydrolysis (fibres, fats, etc.) producing basic compounds (sugars, amino acids, organic acids, etc.); the acidogenic phase where volatile acids, alcohols, CO_2, molecular hydrogen and ammonium are formed; and finally the methanogenic phase where microorganisms convert these compounds into (basically) methane and CO_2. The last phase is the slowest process and determines the retention time of the reactor.[15]

TABLE 6.8 Fertilizer application with and without use of mill effluents

Macronutrient	Application rate (kg/ha)			
	Plant cane		Ratoon	
	Case 1	Case 2	Case 1	Case 2
Nitrogen – N	30	–	80	90
Phosphorus – P_2O_5	120	50	25	–
Potassium – K_2O	120	80	120	–

Source: SMA, 2004.

Notes
Case 1: without stillage or filter mud application.
Case 2: with stillage application on ratoon and filter mud in plant cane.

For diluted effluents with high COD, such as stillage, the upflow anaerobic sludge blanket reactor (UASB) has proved to be the best option. Recently this technology has been upgraded to what is known as the expanded granular sludge bed reactor (EGSB), which includes the option of internal circulation (IC). Table 6.9 gives some typical values of main stillage characteristics before and after the biodigestion.

The following observations can be made.

- The COD is significantly reduced (70 per cent) but still remains at a high level, maintaining the polluting potential of stillage.
- The quantity of the main nutrient, potassium, is unchanged, ensuring the fertilizer value of biodigested vinasse.
- The increase in pH reduces the corrosive property of vinasse.
- The reduction of more than 90 per cent of sulphates is probably the main benefit of the biodigestion process, because it eliminates noxious odours and the threat of sulphur migrating deep down into the soil.

Biogas from stillage biodigestion has a chemical composition of approximately 60 per cent methane and 40 per cent CO_2, together with small amounts of H_2S. The lower heating value (LHV) is around $20–22 MJ/Nm^3$; it can be upgraded by the removal of CO_2 and H_2S. At the mills, biogas can be used to generate electricity or to substitute diesel oil as truck fuel. The different options for power generation are:

- Burning in the bagasse boiler, as a supplementary fuel, to extend power generation beyond the crushing season; this option is unlikely to be economically justifiable due to the high production cost of biogas and the low thermal efficiency of the mill steam cycle.
- Use in gas turbines or microturbines: this option is also economically unattractive due to the scale, the gas purification requirements and the need to compress the biogas to the gas turbine (GT) pressure. However, the possibility of employing a combined cycle is very interesting from a thermodynamic point of view.

TABLE 6.9 Variation of the main characteristics of vinasse due to biodigestion

Parameter	Stillage "in nature"	Biodigested stillage
pH	4.0	6.9
COD (mg/L)	29,000	9,000
N total (mg/L)	550	600
N ammonia (mg/L)	40	220
P total (mg/L)	17	32
Sulphate (mg/L)	450	32
K (mg/L)	1,400	1,400

Source: Feire and Cortez, 2000.

- Use as fuel in motor generator groups (compression ignition or spark ignition engines): this is the widely used alternative in biogas production facilities in Europe. Purification of the gas to remove sulphur compounds, moisture and particulates is necessary, however; the removal of CO_2 is optional, and may be interesting when gas compression is required.

Considering a distillery that produces 400,000 litres of ethanol per day, with a vinasse production of 4.8 million litres per day (COD of 50 kg/L and BOD of 20 kg/L), the estimated biogas production is 70,000 Nm^3/day (LHV of 21 MJ/Nm^3), corresponding to 17 MWt. By burning this biogas in an Otto cycle motor generator, the resulting net power would be approximately 5 MW, which corresponds to 25 kWh/tc,[16] a considerable amount of surplus power. Unfortunately, under the Brazilian conditions surplus power generation with biogas from vinasse is not economically viable due to the high investment costs required, and the lack of laws or regulations requiring vinasse treatment.

Concluding remarks

The production of ethanol from sugar cane is a mature commercial process that is widely used in several countries. It has some key synergies with sugar production given that several workstations, such as mills, boilers and effluent treatment, could be shared in the production of both sugar and ethanol. In most distilleries there is little that could be gained in efficiency and yields improvement, but production costs could be reduced by producing clean microorganisms for fermentation and improving the steam economy to increase surplus power generation. Advanced technologies to produce ethanol from lignocellulosic materials are being developed and upon commercialization they can be combined with the conventional technology by using cane fibres for higher ethanol production. The effluents from a distillery, despite having a high polluting potential, can be recycled to cane fields, thereby replacing part of the chemical fertilizer requirement.

The use of ethanol as a transportation fuel, although concentrated today in the USA and Brazil, is gradually spreading worldwide in developed and developing countries, driven by concerns about the global warming effects of fossil fuels and escalating oil prices. Blends up to 10 per cent of ethanol in gasoline have proven to be feasible by its adoption in many countries. Sugar cane-producing countries are potential candidates to become ethanol producers, the decision depending more on economic considerations than on technical ones. Since two-thirds of the ethanol production cost comes from the feedstocks, efficient and competitive sugar cane producers will stand a better chance of becoming competitive ethanol producers. Given the long experience with sugar cane in Southern Africa and the highly competitive position of many African sugar producers more generally, there are excellent opportunities in Africa to translate this experience with sugar cane into an efficient and competitive ethanol industry.

Notes

1 Massecuite is the mixture of crystals and sugar liquor discharged from a vacuum pan.
2 In some countries it is not economical to produce ethanol from C molasses; for example, in some South African mills, extraction efficiency is high and thus makes C molasses use uneconomical. On the contrary, in Australia ethanol production can only be justified if C molasses is used. The cost of a litre of ethanol from C molasses was estimated for 2001 at 56 Australian cents and 72 Australian cents per litre for B molasses. Brazil is unique in the sense that it uses A, B and C molasses, and also cane juice to produce ethanol directly, but this has not always been the case. The Brazilian National Alcohol Programme (*Programa Nacional do Alcool*) – PROALCOOL – was launched in November 1975 by the federal government in response to the first oil shock in 1973, which had caused serious problems for the country's balance of payments, since at that time 80 per cent of the oil consumed there was imported. The production of fuel ethanol was stimulated, to displace gasoline, by creating a parity price between ethanol and sugar and ethanol and gasoline (making it economically equivalent to produce either sugar or ethanol and to favour the use of ethanol instead of gasoline). The initial target was to produce enough ethanol to reach the 20 per cent blend of anhydrous ethanol in gasoline. The first phase, between 1975 and 1979, favoured the installation of distilleries annexed to existing sugar mills and produced anhydrous ethanol from C molasses. With the second oil shock in 1979 the government began a second phase of the programme (1979–1985), deciding to increase ethanol substitution for gasoline by favouring the production and use of neat ethanol cars (running on pure hydrous ethanol). There was an effort to accelerate ethanol production, stimulating the installation of autonomous distilleries that produced only hydrous ethanol, direct from sugar cane juice. This was accomplished mainly through low-interest loans.
3 Production of ethanol is also preferred on rainy days during the crushing season because of the higher associated impurities (e.g. mud), and when mechanical harvesting is involved. Ethanol production is less stringent with respect to the quality of the feedstock compared to sugar production.
4 Prior to 1975, ethanol production in Brazil used final molasses as feedstock (so-called residual ethanol), and ethanol was used mostly for industrial applications or blended with gasoline.
5 There are 297 distilleries currently in operation in the country; 120 are annexed distilleries and the remaining 177 are autonomous distilleries. Ethanol is produced from molasses only, out of which around 50 per cent is used for alcohol-based chemical industries and the balance is used for alcoholic beverages.
6 Recently one mill added a new distillery with a capacity of 700 kl/day, which will be accomplished in a single distillation column set.
7 The cane preparation index gives an indication of the efficiency of the cane preparation workstation, which includes the knifing and shredding of cane.
8 There are other variations of these two processes. The first is known as COMBAT in which fermentation is performed continuously in a large first vat up to a certain point and then completed in several smaller vats in a batch process. The second is called BATCOM: the batch fermentation process takes place in a series of first vats and is then completed in several continuous process vats located downstream. Fermentation efficiency can reach 92 per cent but it is difficult to maintain this high value due to infections and process parameters fluctuations. Together with juice extraction, these are by far the most important sources of sucrose losses in ethanol production from sugar cane.
9 The TRS represents all sugars in the cane converted to reducing (or inverted) sugars. The TRS equivalent of the final products is calculated using the stoichiometric relation of the conversion reactions equation. The mass of dried yeast produced as a by-product must also be converted to TRS; due to its minor contribution to the total TRS equivalent – it is simply multiplied by a fixed factor, normally chosen at around 1.3, to convert

the yeast mass to ethanol equivalent. Mills that sell syrup, molasses and other sugar-containing materials must convert these products into TRS equivalent using the Brix, purity and RS content data; the TRS equivalent can then be converted to sugar or ethanol equivalent.

10 Generally, vinasse has a light brown colour and a low total solid content (from 2–4 per cent) when it is derived from sugar cane juice, and a black reddish colour with 5–10 per cent solid content when it is produced from cane molasses. About 10–16 litres of vinasse are produced per litre of alcohol produced.

11 Potassium is the main mineral nutrient, followed by calcium, magnesium and nitrogen; the high sulphate content causes odour problems with the fermentation of vinasse after application in the field. The pH of around 4 can cause corrosion problems in pipelines, pumps and other vinasse-handling equipment, but does not cause acidification of the soil due to secondary chemical reactions.

12 Vinasse is transported to the spraying areas by pipes (fibreglass-reinforced plastics), open channels and trucks, and the sprinkler systems use high-pressure pumps and self-propelled hose reels. The average maximum economic transportation distance is around 12 km and depends on several factors, such as topography, cane field productivity, soil moisture, etc. As a general rule, doses of vinasse below 300 m^3/ha are considered safe.

13 There is one Biostil plant in operation in Australia (CSR Sarina Distillery), one in Colombia and 25 in India. No published data have been found for those plants but apparently they are in normal operation.

14 Although all these alternatives are already known, only two are currently practised in Brazil for vinasse disposal on any significant scale: ferti-irrigation and biodigestion (although the latter has failed to take off in Brazil, due to the high investment costs and lack of legislation requiring such a treatment). However, in India about 150 of the 297 working distilleries have installed biodigesters to treat vinasse and produce biogas for use as boiler fuel. Also about 50 distilleries are composting vinasse mixed with filter mud and using it as a fertilizer.

15 Depending on the reactor temperature, biodigestion can be classified into three types: psycrophilic (with temperatures below 20 °C), mesophilic (20–45 °C) and thermophilic (above 45 °C). Thermophilic biodigestion has the advantage of presenting the highest conversion rates and a lower retention time, but the microorganisms are more sensitive to process parameter variation, especially temperature. The mesophilic type is a slower process but is more robust with respect to operating conditions, which makes it the preferred option. Psycrophilic biodigestion is not of interest with stillage.

16 Calculation: 70,000 Nm^3/day × 1/24 day/hr × 21 MJ/Nm^3 = 61,250 MJ/hr; and 61,250 MJ/hr/3,600 MJ/MWh = 17 MW (fuel basis), at 30 per cent efficiency that results in 5.1 MWe. Four hundred thousand litres of ethanol per day requires 4,700 tc/day or 196 tc/hr; therefore 5,100 kW/196 = 26 kWh/tc.

References

ANFAVEA (2010) Brazilian Automotive Industry Yearbook, Associação Nacional dos Fabricantes de Veículos Automotores, São Paulo.

Bollók, M., Réczey, K. and Zacchi, G. (2000) "Simultaneous saccharification and fermentation of steam-pretreated spruce to ethanol", *Applied Biochemistry and Biotechnology*, 84–86, pp. 69–80.

Bosch (2011) Bosch Chainless Diffuser, Bosch Global Engineering Projects Ltd, www.boschprojects.com/site/files/6885/Diffuser.pdf.

CGEE (2009) Bioetanol Combustível: Uma Oportunidade para o Brasil, Centro de Gestão e Estudos Estratégicos, Brasília.

CETESB (2006) Companhia Ambiental do Estado de São Paulo, Standard P4.231, December, Vinhaça – Critérios e Procedimentos para Aplicação no SoloAgrícola, São Paulo.

Chematur (2011) Chematur Engineering AB, Biostil 2000 – Sugar, available at www.chematur.se/sok/download/Biostil_2000_rev_Sugar_0904.pdf, accessed 3 March 2011.

CONAB (2010) Companhia Nacional de Abastecimento, Acompanhamento da Safra Brasileira – Cana-de-Açúcar Safra 2009/2010, Terceiro Levantamento, Brasília.

Cornland, D.W., Johnson, F.X., Yamba, F., Chidumayo, E.N., Morales, M.M., Kalumiana, O. and Mtonga-Chidumayo, S.B. (2001) "Sugar Cane Resources for Sustainable Development: A Case Study in Luena, Zambia", Stockholm Environment Institute.

CTC (1999) "Fundamentos dos Processos de Fabricação de Açúcar e Alcool, Caderno Copersucar", Série Industrial n° 020, Revisão 08/99, Centro de Tecnologia Copersucar.

CTC (2003) Power Point presentation for an internal seminar, Centro de Tecnologia Copersucar.

Donzelli, J.L., Penati, C.P. and Souza, A.V. (2003) "Vinasse: A Liquid Fertilizer", ISSCT Co-Products Workshop, Piracicaba, July.

EPE (2010) Plano Decenal de Energia 2019, Empresa de Pesquisa Energética, Rio de Janeiro.

Fernandes, A.C. (2005) Personal communication.

Freire, W.J. and Cortez, L.A.B. (2000) *Vinasse from Sugar Cane*, Livraria e Editora Agropecuária, São Paulo.

IEA (2008) "Energy Technology Perspectives 2008: Scenarios & Strategies to 2050", International Energy Agency.

Johnson, F.X. and Matsika, E. (2006) "Bio-energy trade and regional development: the case of bio-ethanol in Southern Africa", *Energy for Sustainable Development*, vol. 10, no. 1 (March), pp. 42–54.

Mosier, N., Wyman, C., Dale, B., Elander, R., Lee, Y.Y., Holtzapple, M. and Ladisch, M. (2004) "Features of promising technologies for pre-treatment of lignocellulosic biomass", *Bioresource Technology*, vol. 96, no. 2005, pp. 686–673.

Moura, A.G. (2006) Introduction of the Systems with Thermal Integration in the Brazilian Distilleries: Concentration of Vinasse and Splitfeed, ISSCT Co-Products Workshop, Maceió, AL, Brazil, 12–14 November.

NIPE (2005) "Estudo sobre as Possibilidades e Impactos da Produção de Grandes Quantidades de Etanol Visando a Substituição Parcial de Gasolina no Mundo", Report to Centro de Gestão e Estudos Estratégicos (CGEE), Nucleo Interdisciplinar de Planejamento Energético da Universidade Estadual de Campinas (UNICAMP), Campinas, SP, Brazil.

Oliverio, J.L. (2007) "As Usinas de Açúcar e Álcool: O Estado da Arte da Tecnologia", International Congress of Agroenergy and Biofuels, Teresina, PI, Brazil, 13 June.

Oliverio, J.L. (2008) "Cogeração, Uma Nova Fonte de Renda para Usinas de Açúcar e Etanol", SIMTEC 2008, Piracicaba, SP, Brazil, 4 July.

Paturau, J.M. (1989) *By-Products of the Cane Sugar Industry*, 2nd edn, Elsevier Scientific Publishing Company, Amsterdam and New York.

Rao, P.J.M. (2001) *Energy Conservation and Alternative Sources of Energy in Sugar Factories and Distilleries*, ISPCK Publishers and Distributors, New Delhi.

Rein, P. (2007) *Cane Sugar Engineering*, Verlag Dr Albert Bartens KG, Berlin.

REN21 (2010) Renewable Energy Policy Network for the Twenty-First Century, Renewables 2010 Global Status Report.

Rossell, C.E. (1988) "Sugar cane processing to ethanol for fuel purposes", in M.A. Clark and M.A. Godshall (eds), *Chemistry and Processing of Sugarbeet and Sugar cane*, Elsevier, Amsterdam.

Rossell, C.E. (2006) Personal communication.

Seebaluck, V., Mohee, R., Sobhanbabu, P.R.K., Rosillo-Calle, F., Leal, M.R.L.V. and Johnson, F.X. (2008) "Bioenergy for Sustainable Development and Global Competitiveness: the case of Sugar Cane in Southern Africa", Thematic Report 2 – Industry. Cane Resources Network for Southern Africa (CARENSA)/Stockholm Environment Institute, www.carensa.net.

Seemann, F. (2003) Energy Reduction in Distillation, ISSCT Co-Products Workshop, Piracicaba, SP, Brazil, 14–18 July.

SMA (2004) "Balance of the Greenhouse Gas Emissions in the Production and Use of Ethanol in Brazil", Report to the Secretary of the Environment of the State of São Paulo (SMA), São Paulo.

Tewari, P.K., Batra, V.S. and Balakrishnan, M. (2007) "Water management initiatives in sugarcane molasses based distilleries in India", *Resources, Conservation and Recycling*, 52 (2007), pp. 351–367.

UNICA (2005) "Sugar Cane Energy, Twelve Studies: The Sugar Cane Agroindustry in Brazil and its Sustainability, Coordinated by Macedo, I.C.", São Paulo.

7

OTHER CO-PRODUCT OPTIONS FROM CANE RESOURCES

Kassiap Deepchand and P.J. Manohar Rao

Background

Other chapters of this volume have placed emphasis on large-scale commercial production of the most common traditional products recovered from the sugar cane stalk, including raw and refined sugar energy (steam and electricity) from bagasse and ethanol. In this chapter, we present a broader review of the many existing and potential sugar cane resource streams and co-products from which value-added products can be developed, including the use of other fractions of the sugar cane biomass (e.g. green tops and leaves, dry cane trash). The final products include speciality sugars, sucro-chemicals, alco-chemicals, pulp and paper, furfural and other manufactured goods, relying on commercially proven technologies. We also present the status of research and development on the use of cane tops and leaves and cane trash. Finally, we present a review on the sugar cane-based agro-industrial complexes in a selected number of countries. These complexes integrate sugar and a number of value-added products while optimizing use of energy (steam and electricity obtained bagasse combustion) as well as water; the overall objective is to maximize the revenue stream from sugar cane biomass, thereby enhancing the sustainability of the sugar cane industry.

The sugar cane biomass

Sugar cane biomass comprises of the cane stalk (itself composed of fibre and juice), the green tops and leaves (CTL), the dry trash, the underground rhizome and the roots. The dry matter content of these fractions of the cane biomass based on an average cane yield of 79 tonnes per hectare obtained in 2001 in Mauritius is given in Table 7.1.

TABLE 7.1 Dry matter content of sugar cane biomass in Mauritius (2001)

Biomass fraction	Dry matter (tonnes/hectare)	
	Elemental	Total
a Cane stalk	29.94	
b – Fibre	11.51	
c – Juice	13.43	29.94
d Cane tops and leaves (CTL)	5.53	
e Trash	9.26	
f Rhizome and roots	5.30	20.09
		45.03

Source: Based on assumptions in Deepchand, 1986a.

Notes
a Based on a cane yield of 79 tonnes/hectare obtained in 2001.
b Based on a fibre % cane of 14.87.
c Based on a Brix[1] % absolute juice[2] of 17.00.
d Based on a dry matter content of 28% and a CTL % harvested cane of 25%.
e Based on records at the MSIRI.

Sugar cane processing and production of co-products

Under normal sugar cane industry conditions, the mature cane stalks, freed of the green leaves and immature top (cut at its natural breaking point) as well as the dry leaves (or trash), are harvested and delivered to the cane processing plants. The processing of the stalks for the recovery of sugar concurrently leads to the generation of a series of co-products at different stages of the sugar-manufacturing process. These are: bagasse (obtained after juice extraction in the milling department); final molasses (obtained after massecuite[3] exhaustion in the boiling house); filter mud (obtained from the rotary vacuum filter of the juice clarification station); and bagasse furnace ash and fly ash (obtained from the boiler house). Figure 7.1 outlines the generation of these co-products in the sugar-manufacturing process. It will be recalled that bagasse is burnt to generate high-pressure steam, which, when allowed to expand in condensing extraction turbo-alternators, enables generation of electricity and process steam for use in sugar and co-product recovery processes. In a well-designed, modern sugar cum power plant, a significant excess amount of electricity is generated and exported to the grid (see Chapter 5).

Value-added sugars

The process of recovering sugar from the cane stalk has been extensively addressed in Chapter 5. A number of countries, especially those forming part of the African Caribbean Pacific (ACP) region, had until recently been relying on refineries in Europe to refine their sugar that was destined for consumption in the European market. The bulk of the sugar (raw) sold under the ACP/EU Sugar Protocol was

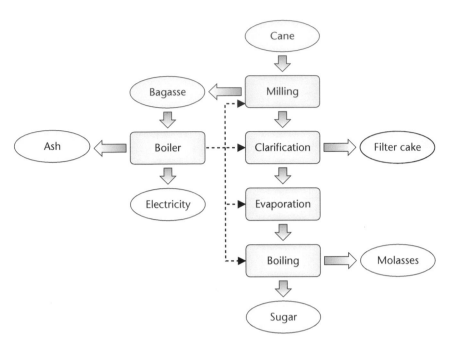

FIGURE 7.1   Sugar cane processing and generation of co-products

then converted into value-added refined white sugar in these refineries. The major refineries are now no longer in operation and those countries that have opted to continue cane and sugar production have invested in refineries locally, which enables them to tap an additional income from the sugar value chain.

An additional means of deriving income is the production of so-called speciality sugars, which are obtained through adoption of technologies within the traditional cane-to-sugar process (MSA, 2003). Many of the technologies are country-specific or even factory-specific within a country and have been developed in-house. The idea of producing speciality sugars was about creating a final product that would preserve the natural goodness of the cane juice with its richness of flavour as well as its molasses and mineral content. Details of production are kept as closely guarded secrets. Such speciality sugars meet quality and food-grade norms (HACCP and ISO 9002, see Chapter 12) and are sold in niche markets that have been developed over the past few decades, mainly in Europe and the US. In Mauritius, for example, there is a market for around 15 speciality sugars, each one "crafted" as an ingredient to meet specific demands for both household and industrial uses. Typical speciality sugars are demerara, golden granulated, dark soft brown, light soft brown, dark muscovado, light muscovado and molasses sugar. These sugars have varying appearance, texture and flavour. They have different physical characteristics in terms of colour, grain size and insoluble matter content as well as different chemical composition in relation to polarity[4] and moisture content. The production technique is generally tailor-made according to the type of speciality sugar to be produced.

Sucro-chemicals

Sucro-chemicals are organic products made from feedstocks derived from sucrose extracted from sugar cane or beet. The market for such products is currently limited as the price and conversion ratio are often not competitive with existing products that are obtained mainly from petrochemicals. Cost and availability as well as lack of appropriate technologies are barriers to their successful development. In addition the markets are generally dominated by large multinationals, and this presents another barrier to new and small-scale business investments.

A wide range of products can potentially be obtained from sucrose as feedstock and a non-exhaustive list includes oxalic acid, lactic acid, glycerol, fatty acids, esters, polydextrose and diethyl oxalate (Paturau, 1989; Rao, 1997; Ministry of Agro-Industry, 2006).

Oxalic acid

Oxalic acid is widely used in the pharmaceutical industry for the production of antibiotics and in the textile industry as a neutralizing and acidifying agent as a substitute to the most common mineral acids. Its tin and antimonyl salts are used in the textile, printing and dyeing industries. The acid is also used as a whitening agent for leather and a bleaching agent for plant fibres; straw for hats, for example.

Manufacture of oxalic acid is based on the reaction between glucose or any sugar-containing raw material and nitric acid, during which oxalic acid and nitrogen oxides are formed. The nitrogen oxides evolved during the reaction are reconverted to nitric acid and recirculated, thus rendering the process economical. The reaction takes place in the presence of vanadium pentoxide as a catalyst and the yield is around 80 per cent. The plant consumes a negligible amount of steam and does not require any specialized type of equipment or technology.

Ethanol/glycerol

Fermentation of sucrose to ethanol is a well-established process but the economics are enhanced if other high value-added co-products or speciality chemicals are obtained along with the ethanol. One such example is the production of glycerol as a co-product of the fermentation process. Glycerol is used in a number of ways, including in the pharmaceutical, cosmetics and tobacco industries. The demand for glycerol from this source has increased considerably in recent years in order to replace glycerol originating from animal fats.

Lactic acid

Most of the world's lactic acid is produced by fermentation and the process is well established. Purac in the Netherlands was the leading company producing lactic acid until the mid-1990s, when Du Pont in the US invested in a massive new lactic acid plant. Lactic acid is used in the food, cosmetics and pharmaceutical industries, but this product can also be used in high value-added polymers such as polylactic acid and polylactide, which are biodegradable and bioresorbable when used in the medical field as suture material, medical implants and controlled drug delivery devices. Some companies are aiming at the manufacture of polylactides for use as a commodity plastic.

Fatty acid esters

Fatty acid esters can be of a low degree of esterification (30 per cent) or a high degree of esterification (70 per cent). Fatty acids derived from sucrose are produced on an industrial scale by the transesterification of sucrose to fatty acid methyl ester. The monoesters are isolated after several purification stages. These compounds show favourable surface-active properties and are used in the detergent, cosmetics, pharmaceutical and food industries. The palmitoyl/stearoyl sucrose esters have received FDA approval and have been used in food application in Japan (Rao, 1997).

The esters can also be used in the building industry to remove metabolic ions such as calcium and magnesium from washing fluids, and as a bleaching booster in the detergent industry. The products currently used are derived from petroleum resources and carbohydrate-derived products can conveniently substitute them in an eco-friendly manner in that such products are non-toxic in nature. One example of an efficient carbohydrate-based bleaching booster is sucrose polyacetate (SUPA) developed by Eridamia Beghin-Say, a French company.

Molasses-derived products

Molasses is a major co-product obtained after massecuite exhaustion in the centrifugation process of the boiling house. It is available at a rate of about 3 per cent of the weight of cane processed. Alternative technologies for utilization of this resource are:

- Development of improved strains of yeast for obtaining a high yield of ethyl alcohol by fermentation of molasses and processes like the Melle–Boinot process (patented in France in the year 1949), adopted by almost all distilleries worldwide to obtain a high yield of alcohol.
- Development of continuous fermentation processes by firms like Alfa Laval (Sweden), Vogelbusch (Austria) and others to increase the yield of alcohol while reducing the time for fermentation, and reducing the quantity of effluent.
- Development of mechanical vapour recompressors and multi-pressure distillation systems to reduce steam (energy) consumption during distillation.
- Development of process and equipment for the production of biogas from distillery effluents (vinasse or stillage).
- Development of improved processes for the production from molasses of yeast, monosodium glutamate, lysine, citric acid, lactic acid, acetone-butanol, ephedrine, etcetera.

Most of the above technologies so far developed have been fully exploited in various countries by the setting up of commercial-scale plants to produce value-added products from co-products of sugar cane.

Alcohol products

Molasses is mainly used for alcohol production, which is classified in different grades: industrial alcohol (96.5 degrees Gay-Lussac), superfine alcohol (96.5 °GL) and absolute anhydrous alcohol (99.8 °GL). Depending on their properties and characteristics, they are used for the production of different value-added products, among which rum production is a key commercial product.

Rum is by definition drinking or potable alcohol. It is produced from sugar cane molasses or cane juice. Molasses is diluted to around 15 degrees Brix with pure water, some vinasse is added and the pH is maintained at around 5.6. The total fermentable sugar content is around 10–12 per cent and the liquor is fed with yeast. The fermenting temperature is maintained at around 26 °C. After fermentation, the rum is distilled in a batch still. The best rum is obtained when distillation is slow and uniform, the distillate having an alcohol content of 75–80 °GL. The product is distilled with water and marketed at a concentration varying between 32 °GL and 33 °GL. This product is manufactured in many cane-producing countries, fetching a local market as well as an export market in some cases; the export market is, however, more oriented towards higher-quality potable alcohol derived from cane juice such as agricultural rum.

Agricultural rum, also known as Rhum Agricole, is another value-added product obtained from the juice fraction of cane biomass. Countries in the Caribbean region (in particular Jamaica, which has been a pioneer in this activity) are known for having developed a number of alcoholic beverages meant for direct consumption. This "rhum" is generally obtained from cane juice, which can be either mixed juice[5] or first-expressed juice.[6] The alcohol derived from these feedstocks has natural organoleptic (sensory) properties that are much different than rum derived from cane molasses. This activity is gaining in importance in most sugar cane-producing countries in an attempt to increase the revenue from the sugar cane industry. Such projects can be undertaken both on a small or a large scale, depending on the market. In Mauritius, besides the production of molasses-based rum, a number of small-scale projects for the production of agricultural rum have recently been developed that generate higher income and have a growing export market.

Molasses-based ethyl alcohol projects

A major portion of molasses produced in different countries is used for the production of ethyl alcohol by fermentation. Brazil is the world's largest producer of ethyl alcohol from molasses. In 2008 the annual production was 18 billion litres, produced in huge distilleries of 1–3.6 million litres alcohol production per day. In

India, there are around 300 distilleries with a total annual production of about 2.0 billion litres (one-ninth of Brazil's production).

In recent years, some firms like Alfa Laval and Vogelbusch have introduced improved technologies for alcohol production, such as: (1) continuous fermentation; (2) multi-pressure distillation; (3) generation biogas from distillery effluents (vinasse); (4) production of compost from distillery effluent. Vogelbusch has a technical collaboration with Praj Industries Ltd (India), which is among the companies in the forefront of development of ethanol projects.

Other molasses-based products

Other potential co-products from molasses include: yeast, monosodium glutamate, lysine, citric acid, lactic acid, oxalic acid, ephedrine hydrochloride and glycerine. Though there are a few plants for producing such products, it is notable that already in 1983 Cuba had ten plants for the production of torula yeast (otherwise known as fodder yeast) to be used as cattle feed. The production of milk increased many-fold after feeding torula yeast to cattle (Rao, 1997).

Monosodium glutamate and lysine are produced in many countries, such as in Japan by firms such as Ajinomoto Co. and Kyowa Kogyo. Some countries produce citric acid, lactic acid and oxalic acid from molasses, using indigenous technologies. Ephedrine is a drug used for the treatment of bronchial asthma, hay fever and whooping cough. Molasses is used to develop a special type of yeast, which is used to convert benzaldehyde to phenylacetylcarbinol (PAC). This is extracted by benzene and subjected to hydrogenation to obtain ephedrine hydrochloride. There are two plants in India and the process is a highly guarded secret. Vogelbusch has developed a process for the manufacture of glycerine from molasses, and this process has been tried in a sugar mill in China (Rao, 1997).

De-sugarization of molasses

Many processes have been developed in countries like Finland, Germany and Belgium to recover sugar from beet molasses, and there are commercial-scale plants working in various countries. But a successful process to recover sugar from cane molasses has not yet been developed. In recent years, a French firm known as Applexion has developed a new process for de-sugarization of cane molasses, with trials conducted by the Audubon Sugar Institute, Baton Rouge (Louisiana, USA) (Rao, 2004). The sugar mills in Mauritius also adopted this process to conduct pilot plant-scale trials, but some technical difficulties were encountered and a large volume of non-sugar was obtained. A proper disposal method has still to be developed.

Fibrous co-products from sugar cane

Fibre resources from the sugar cane industry can be classified into field resources and factory resources, depending on the location where they are generated. Bagasse (a factory resource) is the fibrous residue left after juice extraction and is a main co-product of sugar cane processing. Bagasse application for power and process heat generation was discussed in Chapter 5, and this represents its key traditional commercial functions. The following sections present the scope, opportunities, challenges and some international experiences regarding other co-products from the fibre resources of sugar cane.

Co-products from factory-based fibre resources

Bagasse is the major factory-based fibre resource in sugar mills. A number of value-added co-products can be produced from it due to its attractive physico-chemical properties. Alternative technologies developed for the emerging utilization of bagasse are as follows:

- Moist and wet depithing technology for the complete removal of pith from bagasse, which is essential to produce high-quality pulp, paper and particleboards.
- Development of the "Ritter biological process" and other similar processes to arrest the deterioration of bagasse during long storage periods at paper mills.
- Modern pulping processes, such as mechanical and thermo-mechanical pulping processes, to produce high quality newsprints and printing papers from bagasse.
- Efficient chemical bleaching of pulp, such as the CEHD process (chlorine, alkali extraction, chlorine dioxide and hypochlorite), to obtain white pulp.
- Development of the high-speed "Fourdrinier" paper machine running at 5,000 feet per minute with improved "wet end" and "dry end" and calendaring rollers.
- Development of the "Clupak process", consisting of disturbing the fibre formation in a zig-zag manner to produce Kraft paper with a high tensile strength.
- Development of technologies for efficient depithing and uniform drying of depithed bagasse to a moisture content of 2–4 per cent to obtain high-quality particleboards.
- Development of high-pressure boilers up to 110 bars, perfect boiler water treatment technologies and special types of steam turbines like extraction cum condensing/back pressure, to generate both electric power and process heat at low cost.
- Development of an improved "dry process" for the manufacture of furfural from bagasse.

Some of the important applications and value-added products from bagasse are in: (1) pulp and paper; (2) market pulp; (3) dissolving pulp; (4) particleboards and fibre-boards; (5) corrugated boards and boxes; (6) furfural; (7) xylitol; (8) biogas and producer gas; and (9) charcoal and activated carbon.

Pulp and paper

Bagasse has been successfully used for the production of different grades of paper all over the world. Paper is usually produced from fibrous materials like bamboo, cotton stalks, rice straw, softwood and hardwood. Bagasse as a feedstock has an advantage in that it does not require debarking or chipping as wood does. However, depithing has to be carried out when using bagasse, and this process consumes additional water and energy. In India, about 8 per cent of total bagasse production is being used for paper production (Seebaluck *et al.*, 2008).

Tamil Nadu Newsprint and Papers Ltd (TNPL) in India is regarded as the world's largest producer of bagasse-based paper, with a daily production capacity of 600 tonnes of paper. It is also a leader in technology for the manufacture of newsprint from bagasse. TNPL has the most advanced paper mill in India with a unique bagasse procurement, storing, preserving, handling, processing and pulping system. It produces 180,000 tonnes of paper and newsprint per annum, consuming 800,000 tonnes of bagasse in the process. TNPL adopted the Beloit–SPB process for the manufacture of newsprint using mechanical pulp. The use of bagasse as a raw material helps prevent the denudation of 30,000 acres of forestland a year in India (Seebaluck *et al.*, 2008). There are five other bagasse-based paper mills also operating in India. Bagasse-based paper plants have also been successfully implemented in Argentina, Peru, Brazil, Thailand, South Africa and Iran.

Market pulp

Market pulp is defined as pulp sold on the open market and excludes any pulp used for captive consumption in a paper mill or its associated uses for the manufacture of paper. Market pulp is predominantly used as a major component in the manufacture of paper and paper products (newsprint, writing and printing papers, catalogue papers, tissues, towelling, wrapping and bag paper, etc.). Until recently, market pulp was predominantly from Canada, the USA, Sweden, Finland and Norway, with these countries producing 85 per cent of market pulp from wood, but commercial-scale plants based on bagasse have been established and are operating successfully in Mexico, Latin America, Taiwan and Australia.

Particleboards

Particleboard is a panel manufactured from lignocellulosic materials (normally wood), primarily in the form of particles mixed with a synthetic resin or other suitable binder and bonded together under heat and pressure in a hot press during which the entire inter-particle bond is created by the binder.

Use of bagasse in the manufacture of particleboards has a number of advantages. Since bagasse is in the form of particles, it does not need any equipment or machinery for particle preparation. The only operation required is the depithing of bagasse to separate the pith from the fibre that is to be used as the raw material in the manufacture of particleboards. The pith can be used as a fuel in boilers, while the fibre is sent for further processing. Most bagasse-based particleboard plants are located in Latin American countries, such as Cuba, Argentina, Costa Rica and Puerto Rico. There are several operational bagasse particleboard plants in India, China and Pakistan. China is a country with as many as 24 plants producing particleboards using bagasse. Of these, 14 have an annual capacity of 5,000 m^3 and ten have an annual capacity of 10,000 m^3 (Rao, 2004).

Fibreboards

Fibreboard is a panel manufactured from any lignocellulosic material, mainly by the interfacing of fibres, consolidated under heat and pressure in a hot press to varying densities. Bagasse is a potential raw material and there are several commercial plants successfully using bagasse as feedstock.

Corrugated boards and boxes

In the early days, hardwood and straw pulps were used for making corrugated boards. However, in recent years, several commercial-scale plants based on bagasse pulp have been set up in many countries. In some plants, 100 per cent bagasse is employed successfully. The corrugated board planks produced are converted into boxes of different sizes through simple mechanical operations.

Moulded bagasse products

Bagasse can be admixed with thermosetting synthetic resins (phenol formaldehyde or urea formaldehyde) and can be hot pressed into any desired shape with the use of an appropriate mould. Products such as boxes for packing fragile instruments, kitchen furniture, tiles, trays, window and door frames are manufactured in this way. The process consists of depithing bagasse, drying the depithed particles and reducing the size of the particles using hammer mills. The particles are then separated into different sizes to suit the product to be manufactured. The Taiwan Sugar Research Institute has developed many types of moulded products from bagasse. Biodegradable food trays have been developed using about 76 per cent bagasse pulp, 12 per cent waterproof agent and heat-resistant binder, 7 per cent dispersing agent and 5 per cent flocculants. In India, Sitapur Ply Wood Products Ltd (Sitapur, UP) manufactures different types of plywood products like door and window frames based on bagasse (Rao, 1997).

Furfural

Furfural is an aldehyde derivative of pentosans found along with cellulose in most agricultural commodities. It is also known as furfuraldehyde. A whole range of derivatives can be produced from furfural, such as furfural alcohol, tetrahydrofurfuryl alcohol, furfural resins, furans, tetrahydrofuran and pyrols. The process of manufacturing furfural is based on the hydrolysis of pentosans in agricultural residues in the presence of an acidic catalyst, such as sulphuric acid. A plant at Belle Glade in Florida, USA makes use of bagasse from an adjoining sugar mill to produce furfural, whereas most other plants use oat hulls, corncobs, cottonseed hull, and rice hulls as raw material. The world's largest furfural plant based on bagasse is found at the Central La Romana sugar mill in the Dominican Republic, with an annual capacity of 50,000 tonnes of furfural. In India, a few plants have been established to produce furfural from bagasse (for example, Southern Agrifurane Industries Ltd and Oswal Agrofurane Ltd); there is also a large plant in South Africa.

Bagasse co-products still under research

The technologies for the production of above-mentioned products using bagasse are commercially proven and successfully operating in various countries. However, bagasse application in the manufacture of many other products is still being explored, and some of these are reviewed in the sections below.

Dissolving pulp (rayon grade)

Dissolving pulp, which is also known as rayon pulp, is highly purified alpha-cellulose. It is mainly used for making viscose yarn (rayon) for fabrics. It is also used for medicinal tablets, pharmaceuticals and other cellulose ethers and esters, plastics, explosives, cellophane and photographic films. There is no commercial-scale plant at present to produce such pulp based on bagasse. Currently, commercial plants are based on wood. South India Viscose Industries Ltd in India has installed a $0.7\,m^3$ circulation-type pilot digester with liquid pre-heater. Dissolving pulp from bagasse can substitute softwoods that are currently being used as raw material in the plant.

Carboxymethyl cellulose

Carboxymethyl cellulose (CMC) is an ether of cellulose and monochloroacetic acid. Depending on its degree of purity, it is used for a variety of applications, including in oil drilling, the ceramic industry (for a shiny finish), the paint industry (as a thickener or stabilizer) and the wood industry (as a thickener). CMC is normally produced from waste cotton, rags, cotton cuttings, high-grade wood pulp, etcetera. The Cuban Research Institute for Sugar Cane Derivatives (ICIDCA) has conducted research on the production of CMC from bagasse, with results that suggest potential applications that may become economically attractive (Seebaluck et al., 2008).

Xylitol

Xylitol is a five-carbon sugar alcohol and is used in diabetic foods as a substitute for sugar. It is also used as a non-sugar sweetener in children's chewing vitamins, gums, tablets, jams, puddings, ice creams and other products. The main raw material used for xylitol manufacture is birch wood, but other raw materials such as rice and cottonseed hulls, bagasse, corn stalks and coconut shells are also used. Pilot projects for xylitol production from bagasse are being implemented in many countries, but no commercial production has been reported. The Taiwan Sugar Research Institute has carried out work in this area with a view to improving the techno-economic feasibility of the process.

Biogas and producer gas

The production of biogas from cellulosic materials like bagasse – which also contain lignin – is complex compared to its production from distillery wastes. There is no reported commercial-scale plant producing biogas from bagasse. In India, a pilot plant was developed and operated by the National Sugar Institute in Kanpur (in collaboration with Hungary), for the production of biogas and biofertilizer from bagasse (Seebaluck et al., 2008).

Activated carbon

Bagasse is pyrolysed at a temperature of 450 °C for a period of 20 hours in the absence of atmospheric air. The activated carbon produced is purified by washing it with dilute and hot hydrochloric acid with a retention time of one hour. The Mexican Institute of Steel Research has successfully produced charcoal from bagasse using this process (Seebaluck et al., 2008).

Plastics and other petrochemicals-based products

Research is currently being undertaken on further developing and promoting biodegradable plastics made from sugar and bagasse that break down into water and carbon dioxide within six months. It takes 17 kg of bagasse or 3 kg of sugar to make 1 kg of biodegradable plastic (Seebaluck et al., 2008). Bagasse-based plastics are still uncommon. However, their chemical resistance, quality and biodegradability are promising despite the high costs associated with their development. In the near future, environmental requirements may accelerate a wider dissemination of sugar cane-based plastics. Bagasse could also compete, to some extent, with other petrochemical products such as adhesives, synthetic fibres, herbicides and insecticides, as well as substances such as ethyl ether, acetic acid, ethyl acetate and diethylamines. Again, the main advantage of bagasse in these applications is its biodegradability.

Co-products of field-based fibre resources

The other fractions of sugar cane biomass available in significant amounts comprise the green tops and leaves (CTL) and dry trash. CTL is made up of the top immature stem tip and the green leaves of the cane plant. As new leaves emerge, the older ones dry off and are referred to as trash. This material is comprised of the leaf sheath and the leaf blade, which is normally detached from the stalk as it grows and is found at each node of the stalk. Under a normal manual cane-harvesting system in unburnt cane fields, the dry trash is detached from the cane stalk prior to harvest and the CTL is cut from the stalk at its natural breaking point. Hence in such manual systems the two distinct fibrous fractions of the cane biomass can be left in the field separate from each other, in case a separate collection and processing for each one is envisaged.

When unburnt cane fields are mechanically harvested, the CTL is topped off and the cane stalks with the adhering trash are chopped into billets. The trash is blown off the billeted cane and spread as a blanket on the fields together with the CTL. These cane biomass fractions are either left *tel quel* or windrowed in the cane interlines, depending on soil and agro-climatic conditions in a given location.

In India, Thailand and some other regions where sugar cane is harvested by semi-mechanical green cane harvesting methods, large quantities of green tops and leafy trash are left over in the fields at the time of harvesting. Such resources are not available in countries where cane fields are burnt before harvesting. In most sub-Saharan African countries cane is burnt before harvesting to get rid of snakes in the fields and make it safer for manual labourers to work. However, the industry has recently started to promote green cane harvesting in a global effort to reduce greenhouse gas emissions.

It will be noted from Table 7.1 that tops and green leaves comprise 5.5 per cent, dried leaves (trash) 9.3 per cent and clean cane stalks 30 per cent of the dry sugar cane biomass. The various applications of these resources are as follows.

Green tops as cattle feed

Green tops of sugar cane are high in fibre but low in protein. Using the green tops as cattle feed is the most common practice adopted in countries (mainly India) where cane is manually harvested. In Cuba, the Ministry of Sugar has developed a mechanical process for the production of dehydrated sugar cane tops (DSCT) as there was a good potential for export (Seebaluck *et al.*, 2008). Green tops and leaves can also be mixed with other ingredients such as molasses, filter cake or molasses-urea, corn, cornmeal, yellow meal and bagasse pith to improve the nutritive value.

As fodder, CTL is of low nutritive value compared to other forage crops. However, it is still considered significant by the peasant cow keepers, given that this fodder is available during the winter months when cane is normally harvested. It thus alleviates fodder shortage and ensures continuous cow keeping. Cow keepers adopting feedlot systems also ensile this fodder after chopping and mixing with molasses for off-season use. In some cases, other feed ingredients are added to improve the nutritive value of the ensiled fodder.

Green tops for field applications

The cane tops and leaves are generally left in the field after harvesting to prevent weed growth and soil erosion. It also helps in fertilization and irrigation. It adds organic matter to the soil and retains moisture in areas with low rainfall or irrigation.

Cane trash for field applications

Trash is used mainly in the sugar cane fields as mulch or is incorporated into the soil. This can help increase soil organic matter and available nitrogen (N) and phosphorus (P). It also enhances soil diversity in terms of microbial and earthworm populations. When used as mulch, it prevents evaporation of moisture from the soil, controls the growth of weeds and significantly reduces the risk of soil erosion. But in fields with high cane yields and, in particular, those fields with dual row planting, windrowing the cane trash in the cane inter-rows with an excessive amount of trash poses a problem. However, in spite of the above uses there is consensus that there remains considerable excess for diversion to other uses, in particular towards power generation.

Attempts have been made in past studies at the University of Mauritius to determine the availability of CTL, its characteristics and potential uses. Around 24 tonnes of CTL at a moisture content of 72 per cent are available per hectare of land under cane. On a dry matter basis CTL contains 6.5 per cent crude protein, 30.2 per cent crude fibre, 3.3 per cent crude fat and 5.4 per cent ash. On average, a dry matter content of 28 per cent, or around 6–7 tonnes of dry matter, is thus produced on an annually renewable basis per hectare of land under cane (Deepchand, 1986b).

Based on its characteristics, CTL and/or trash can be used as a raw material in the manufacture of a wide range of products to provide food, fuel, feed, chemicals, energy, pharmaceuticals, pulp and paper. A summary of such applications is presented in Figure 7.2.

Cane trash as boiler fuel

Cane trash contains valuable cellulose, which can be used either as fuel in boilers or as fibrous raw material in the board or paper-manufacturing industry. Trash has a higher calorific value and lower moisture content than bagasse. In India, Thailand, Jamaica and Hawaii (USA), trash balers are employed to compact the trash for use as fuel in the sugar mill boilers. Sugar mills in India have started realizing the benefits of cane trash utilization as boiler fuel, and as a result trash usage as fuel is gradually increasing. However, the low bulk density and high content of potassium reduces the attractiveness of cane trash as a fuel, as it increases transportation costs and causes damage to boiler tubes respectively. However, mixing trash in appropriate quantities (up to 30%) with bagasse and using better construction materials for boiler tubes can surmount these problems and ensure optimum utilization of the green fuel in sugar mill boilers.

In Brazil the Sugarcane Renewable Electricity (SUCRE) project will build on an earlier project, BRA/96/G31 – Biomass Power Generation: Sugar Cane Bagasse and Trash, supported by UNDP and the EU (CTC, 2005), to catalyse the transformation of the sugar cane industry in Brazil into one for which the supply of renewable electricity from sugar cane biomass to the grid becomes a significant and core aspect of their business, alongside sugar and ethanol production. To maximize the

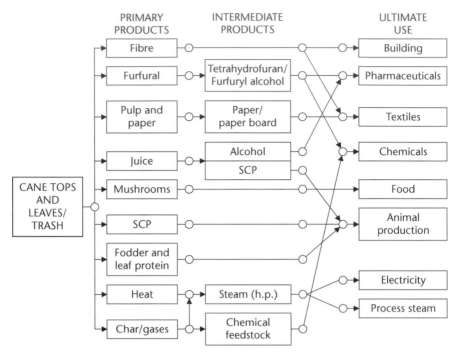

FIGURE 7.2   Use of CTL and trash as raw materials for the manufacture of co-products

potential for electricity generation from sugar cane, the project will facilitate the expanded use of bagasse and launch the widespread use of sugar cane "trash", the tops and leaves of the sugar cane plant that have historically been burned on the cane field as a waste product (UNDP, 2009). The typical quantity of trash available in cane fields is roughly equal to the amount of bagasse produced. Thus, considering trash and bagasse, the biomass resource from sugar cane is effectively doubled with the additional trash harvested solely used for surplus electricity generation.

Other potential products from CTL and trash

Laboratory-scale trials were undertaken to produce a selected number of items, and these included energy in the form electricity, leaf protein, single-cell protein, ethanol, pulp and paper and char. Fresh CTL prepared in conventional shredders and subjected to pressure using three-roller mills enabled separation of a juice fraction from a fibrous fraction. The fibrous fraction, which had a particle size distribution and moisture content similar to bagasse, was burnt in existing boilers to generate steam and electricity. This fibrous fraction had a lower calorific value, 90 per cent of that of bagasse on a dry matter basis. Coupling electricity generation from both bagasse and CTL can almost double the potential of exportable electricity per hectare of land under cane, yielding 100 tonnes of cane and 25 tonnes of CTL together with trash (Deepchand, 2001).

The juice fraction obtained from the fresh CTL when heated to 80 °C and sub-jected to centrifugation yielded a coagulate and a supernatant (whey) that had around 4.5 per cent fermentable sugars. The coagulate when dried contained around 31 per cent crude protein compared to that from Lucerne (*Medicago sativa*) – 56.6 per cent. Examination of the amino acid pattern of this leaf protein from CTL revealed that it was rich in all essential amino acids except methionine (Deep-chand, 1985a). The lysine content was double that of the FAO reference protein. The whey containing the fermentable sugars has been shown to sustain yeast activi-ties through aerobic fermentation to the same extent as cane molasses at similar concentration. Alternatively, the whey was anaerobically fermented to produce ethanol and around 21.8 g/litre of whey or an equivalent amount of 7.7 kg of alcohol at 60 °GL/tonne CTL were obtained. Such whey can conveniently be mixed with molasses instead of dilution water to enhance ethanol production from molasses (Deepchand, 1985b).

Laboratory-scale trials were conducted for the use of CTL for pulp and paper manufacture using standard laboratory procedure with different concentrations of alkali and different mixtures with paper and cotton rags. The paper obtained from CTL "tel quel" and mixed with cotton rags or paper had a smoother feel and was of better colour than that obtained from bagasse. The thickness, burst strength, tensile strength and weight of paper from CTL were determined and compared with standards. These parameters were below standard in general but could be improved with further investigation and fine-tuning (Deepchand, 1987a).

Thermochemical conversion of the fibrous fraction of sugar cane biomass to char and/or gases is another energy option that is being considered in the context of value addition in the sugar cane industry. The char and gases can be substituted for imported fossil fuels to generate electricity or alternatively in the production of methanol and ammonia. A first study was undertaken to characterize the sugar cane fibrous products for carbonaceous char production by means of thermogravimetry (Deepchand, 1987b). It was found that, compared to cellulose and wood, the fibrous products had lower volatiles and higher ash and char contents. This gave an indication that the sugar cane fibrous product had a higher lignin content than wood and cellulose. Examination of the char also showed that the skeletal structure of the pyrolysed samples of cane-derived fibrous product were well preserved and this could be beneficial to the overall rate of gasification, especially at temperatures above 1,000 °C when pore diffusion is likely to be rate-controlling.

Handling of cane field residues

In European countries, hay balers are common for baling grass, wheat straw and maize stalks; these hay balers can also be used for baling trash in sugar cane fields. In 1984 one such hay baler was imported into India from Sperry New Holland, Belgium, and trials of baling trash were conducted in sugar cane fields. The results were encouraging. Two machinery companies in India have started manufacturing trash balers. Twelve Indian sugar mills are baling trash and bringing it to the mills to be used as boiler fuel, particularly off-season. The experience gained so far shows that, on average, the trash available on a dry matter basis in the fields is about 8–10 tonnes per hectare. This can be baled by one baler in three or four hours. The calorific value of trash is comparable to that of mill-wet bagasse. Hence, every tonne of trash baled in the fields and brought to mills can replace roughly one tonne of mill-wet bagasse when used as fuel. Sugar mills in Thailand, Jamaica, Hawaii, Philippines and Mauritius conducted similar trials from 1987 onwards and the results were encouraging.

This is a low-cost project, as a trash baler costs about US$10,000 while the fuel value of baled trash is very high. Sugar cane-producing countries can thus potentially adopt this practice of baling trash in the fields for use as fuel. In Brazil, a large-scale demonstration of the physical and economic viability of sugar cane trash collection on a commercial scale for the generation of electricity is being undertaken (UNDP, 2009).

Filter cake

Filter cake or filter mud is one of the co-products that contains impurities precipitated from cane juice after the addition of lime and a flocculating agent in the juice-clarification stage of the sugar recovery process. The mud also contains colloidal organic anions that are precipitated during clarification. Bagacillo (fine particles of bagasse) is added to the clarifier muds, which allow efficient filtration of the impurities on rotary vacuum filters. Filter cake content varies between 3 per cent and 5 per cent of cane by weight on a wet basis; its composition is given in Table 7.2.

Products that can potentially be derived from filter cake include manure and compost, biogas and cane wax. Filter cake can also be used as cattle feed. However, the most popular commercial use is its application to cane fields as an organic fertilizer. Alternatively, it can be mixed with vinasse to produce high-value compost.

TABLE 7.2 Composition of filter cake

Constituents	% wet basis
Sucrose	2.1
Moisture	78.2
Wax	2.0
Fats	1.0
Nitrogen	0.4
P_2O_5	0.4
K_2O	0.4
CaO	0.4
Fibre	4.3

Source: Paturau, 1989.

Agro-industrial complexes based on sugar cane-derived products

By definition, a sugar cane complex or sugar agro-industrial complex is a cluster of industries based on sugar cane and its many co-products, such as those discussed previously. Such a complex utilizes sugar cane and all its co-products in an integrated manner with the sugar factory being the mother unit or nucleus to produce sugar and other production units associated with various co-products. All these plants serve as feedstock for the production of as many value-added products as possible, while also modulating input requirements (especially energy and water) in various forms (live steam, exhaust steam, vapour at various pressures and temperatures, and electricity) from the combustion of the fibrous fraction of cane (bagasse) in an integrated manner. In some cases, coal is used to complement the bagasse in dual-fired boilers. Even water is available at various temperatures and degrees of purity and is used in the process in an integrated manner.

A review of existing cane sugar industries worldwide has revealed that there are many sugar factories with attached co-product plants based on bagasse and molasses as feedstock; however, there are very few factories with integrated sugar cane complexes as defined above. According to Rao (1997), the Jiangmen Sugar and Chemical Complex in China is the most advanced sugar agro-industrial complex. Other advanced complexes are: (1) Socieda Paramonga Ltd (Peru) and (2) Sociétés des Sucreries et de Distilleries d'Egypte. Other countries where sugar cane agro-industrial complexes have been developed include Cuba, India, Taiwan, South Africa and Mexico. The flow diagram of the Jiangmen complex in China is given in Figure 7.3.

In the Jiangmen complex in China, there are 20 different industries producing over 28 different products from sugar cane. In addition to the various sugars (raw, refined, various grades), the complex includes production units for various pulp and paper products, yeasts and alcohols, biochemicals, condensate water (for pharmaceutical products), building materials (e.g. bricks, cement) and CO_2-related products such as dry ice. This complex has proven that nothing is waste in any industry and even waste can be converted into a useful product by adopting appropriate technologies.

The Jiangmen complex covers an area of 710 hectares and has over 3,900 workers and staff members, including 1,000 women and 180 engineers and technologists. It is a huge complex by world standards.

The Indian sugar industry has more modest units that could potentially be considered for implementation in the African region once a market is identified for a given product. Three examples of such complexes in India are described below.

Chemical complex at Sangli SSK

This complex is located in Kohhapur district, Maharashtra. It operates an ethanol distillery with a 30,000 litre/day capacity based on molasses obtained from an adjoining sugar factory with a daily cane-crushing capacity of 5,000 tonnes of cane. The ethanol is used as feedstock for the production of potable liquor, acetaldehyde (ten tonnes/day) and acetic acid (14 tonnes/day). These alcohol-derived chemicals use 5.6 million litres of ethanol out of the 9 million litres produced annually.

The chemical complex at Sanjivani

Located in Ahmednagar, this is an alcohol-derived chemical complex and is typical of what can be found in a number of sugar factories in India. The complex has a

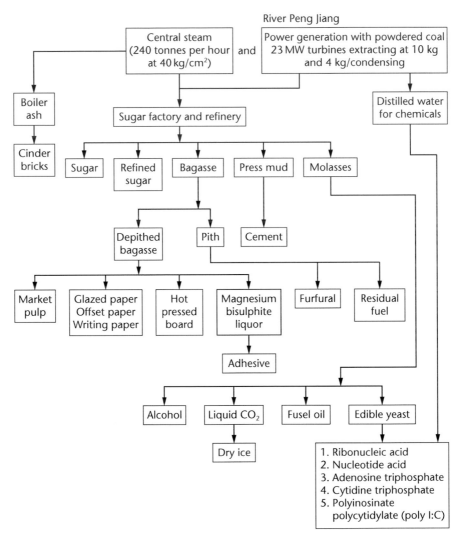

FIGURE 7.3 The Jiangmen sugar-alcohol-paper-cement-chemical complex in China (source: Rao, 2004)

sugar factory with a licensed capacity of 5,000 tonnes of cane per day, processing one million tonnes of cane annually into sugar and various sucro-chemicals and alco-chemicals, as given in Table 7.3.

The technologies for manufacture have been indigenously developed and the plants and machineries were deployed using locally available labour and materials.

Chemical complex at Tanaku

This chemical complex is located next to a cane sugar factory with a daily cane processing capacity of 5,500 tonnes. It has three integrated co-product plants: an ethanol distillery to produce industrial alcohol; an acetic acid/acetic anhydride/ ethyl acetate plant; and an aspirin plant. The distillery produces all the ethanol required for the production of acetic acid, acetic anhydride and ethyl acetate, as well as the carbon dioxide required for the production of aspirin and salicyclic acid. Biogas is also produced from the vinasse fed to a digester 20 m high and 26 m in diameter. Around $0.54\,m^3$ of biogas is produced from 1 kg of BOD.

Status of co-products in Africa and potential

African sugar production amounts to around nine million tonnes annually. A significant proportion of the sugar produced is consumed in the internal market. Some companies – mainly multinationals – are involved in the production of value-added refined sugars and a variety of speciality sugars. Most of the sugar-producing countries convert their molasses into ethanol, mainly for use in potable (spirits) and industrial applications, with only a few thus far engaging in ethanol as a fuel (see Chapter 8 for a detailed discussion of sugar and alcohol markets in Africa).

Downstream products are currently produced mainly in facilities operated by the Illovo group (Illovo, 2009) include:

• furfural (used mainly in lube oil refineries for the purification of oils);

TABLE 7.3 Production at the chemical complex at Sanjivani

Product	Production (tonnes/year)
Oxalic acid	720
Diethyl oxalate	600
Acetic acid	4,500
Ethyl acetate	1,800
Alcohol	13,500
Country liquor	As per Government quota
Butyl acetate	1,000
Sulfamethoxazole	200
Acetic anhydride	2,000

- furfural alcohol (used mainly to produce a resin in the foundry industry as a binder for foundry sands);
- diacetyl and 2,3-pentanedione (both used as high-quality natural flavourants);
- AgriGuard (an agricultural nematocide);
- BioMass sugar (a sugar cane-based fertilizer);
- ethyl alcohol and lactulose (a natural laxative).

The majority of the above products are marketed as high-value sugar cane-derived chemicals in niche markets. Furfural and derivatives are produced at the Sezela mills in KwaZulu Natal. High-quality ethyl alcohol, from which various grades of alcohol are made, is produced at the Merebank plant in Durban and the Glendale distillery on the north-east coast. Lactulose is also manufactured at Merebank (Illovo, 2009).

There is scope for more extensive production of value-added co-products from sugar cane on the African continent considering the successful production of sugar cane-derived fibrous products as well as chemicals in India and China. However, this requires appropriate policy measures, entrepreneurial skills and indigenously built capacities.

Concluding remarks

Exploiting the various fractions of sugar cane and sugar cane-derived products (including cane stalk, cane trash, cane tops and leaves, bagasse, molasses and filter cake) in an integrated manner for recovery of as many value-added products as possible can contribute significantly to the creation of a sustainable sugar cane agro-industry. Although the technologies are available to produce a huge number of products from sugar cane and its co-products, key factors such as market access, competitive pricing and specific local conditions will dictate the pace of investments and the development of new products.

It is clear that the full potential of the sugar cane plant has not yet been fully tapped and more focused research is needed to look into the possibilities of substituting renewable feedstocks based on sugar cane, given the context of dwindling fossil fuel reserves.

Cost-effective utilization of cane fibre resources for the production of value-added products is yet to be commercialized on a large scale. The large-scale production of fibre-based products is hampered by poor returns on investments, the availability of cheaper alternative resources, high technology costs, low volumes of production, lack of process development for commercial production, and lack of research and product development work.

The African sugar industry has so far concentrated on sugar production, with only a rather limited development of value-added products, many of which have already been commercially exploited in other cane-producing countries. There is wide scope for replication of such activities in Africa, both on a small and a large industrial scale, given the availability of cane resources, the expansion of the industry, the market potential with increasing development in the continent, and more importantly the need to improve the sustainability of the industry.

Notes

1 Brix is the apparent percentage of soluble solid matter (sucrose and soluble non-sucrose) determined densitometrically.
2 Absolute juice is the hypothetical juice, comprising all the dissolved solids plus the total water in the cane.
3 Massecuite is the mixture of crystals and sugar liquor discharged from a vacuum pan in which sugar crystals are formed.
4 Polarity is the apparent sucrose content of a sugar product (determined by the polarimetric method).
5 Mixed juice is the combined juice obtained from all the mills of a milling tandem that is sent to the boiling house.
6 First-expressed juice is obtained from the first two rollers of a milling tandem (it is a juice of high purity).

References

CTC (2005) "Biomass Power Generation: Sugar Cane Bagasse and Trash", Copersucar Technology Center, Project BRA/96/G31, Report to UNDP/MCT/GEF.
Deepchand, K. (1985a) "System for the production of electricity, leaf protein and single cell protein from sugar cane tops and leaves", *Solar Energy – Journal of the International Solar Energy Society*, vol. 35, no. 6, pp. 477–482.
Deepchand, K. (1985b) "Utilisation of cane tops and leaves for energy and food – An integrated system", *Biomass – International Journal*, 7, pp. 247–255.
Deepchand, K. (1986a) "Economics of electricity production from sugar cane tops and leaves – a preliminary study", *International Sugar Journal*, vol. 88, no. 1055, pp. 210–216.
Deepchand, K. (1986b) "Characteristics, present use and potential of sugar cane tops and leaves", *Agricultural Wastes – An International Journal*, vol. 15, no. 2, pp. 139–148.
Deepchand, K. (1987a) "Sugar cane tops and leaves as a raw material for pulp and paper", *Biological Wastes – An International Journal*, vol. 19, no. 1, pp. 69–73.
Deepchand, K. (1987b) "Study on the pyrolysis behaviour of sugar cane fibrous products", *Biological Wastes – An International Journal*, 20, pp. 203–208.
Deepchand, K. (2001) "Commercial scale cogeneration of bagasse energy in Mauritius", *Journal of Energy for Sustainable Development*, vol. 5, no. 1, pp. 8–15.
Illovo (2009) "Group Profile", www.illovo.co.za/About_Us/Group_Information/Group_ Profile.
Mauritius Sugar Authority (2003) "Sugar Cane Co-Products Development in Mauritius", report of MSA Committee on Sugar Cane Co-Products Development in Mauritius (under chairmanship of Dr K. Deepchand), Port Louis, Mauritius, p. 69.
Ministry of Agro-Industry (2006) Multi-Annual Adaptation Strategy Plan (2006–2015), Port Louis, Mauritius.
Paturau, J.M. (1989) *By-products of the Cane Sugar Industry*, Elsevier Scientific Publishing Company, Amsterdam.
Rao, P.J.M. (1997) *Industrial Utilisation of Sugar Cane and its Co-Products*, ISPCK Publishers and Distributors, New Delhi.

Rao, P.J.M. (2004) Technological Development and Opportunities for Sugarcane Co-Products, Workshop on Sugarcane as a Renewable Energy Resource, University of Mauritius, 14–20 November.

Seebaluck, V., Mohee, R., Sobhanbabu, P.R.K., Rosillo-Calle, F., Leal, M.R.L.V. and Johnson, F.X. (2008) "Bioenergy for Sustainable Development and Global Competitiveness: The Case of Sugarcane in Southern Africa – Thematic Report 2: Industry", Cane Resources Network for Southern Africa/Stockholm Environment Institute special report series.

UNDP (2009) "Sugarcane Renewable Electricity ('SUCRE')", UNDP project document.

PART III
Markets

8

SUGAR REFORMS, ETHANOL DEMAND AND MARKET RESTRUCTURING

Lindsay Jolly

Section 1: world market overview

Production

World sugar production has grown by 2.8 per cent annually since 2000, reaching 162.5 million tonnes in 2008, with sugar cane accounting for 80 per cent of the total and Brazil continuing to dominate the world market (Tables 8.1 and 8.2). Brazil accounts for 50 per cent of world sugar trade; thus, the share of sugar cane going to produce ethanol or sugar in Brazil is a leading determinant of global sugar production.

India is the second-largest cane sugar producer after Brazil and is the world's largest consumer of centrifuged and semi-processed sugar (gur and khandsari or jaggery). Due to a lack of alignment between sugar prices and government-controlled cane-pricing policies, the country's sugar balance moves regularly between surplus and deficit, with strong repercussions on the world sugar balance.

Sugar production responds strongly to: policy-induced adjustments in the EU and the countries of the North American Free Trade Agreement (NAFTA); the continuing sugar cycle in India; and the rate of growth in Brazil's fuel ethanol/sugar industry, which will continue to shape sugar-production growth worldwide over the longer term. Whilst still subject to weather shocks, world sugar production will increase further to meet steadily rising sugar demand. Whilst Brazil is anticipated to be the centre of growth, other countries/regions such as India and sub-Saharan Africa will also likely show strong production growth.

Evolution of consumption

Growth in sugar consumption has averaged 2.4 per cent annually over the past ten years, but fell in 2009 with the global recession and higher sugar prices. The stable

TABLE 8.1 Evolution of production by five largest producers

		2000	2001	2002	2003	2004	2005	2006	2007	2008
1	Brazil	16.46	20.34	23.57	25.73	27.29	26.13	31.62	33.20	32.29
2	India	20.25	19.91	19.52	21.70	14.43	15.21	22.64	29.09	25.94
3	EU	17.85	15.50	18.27	16.58	21.84	21.69	18.02	18.44	16.38
4	China	7.62	7.16	9.80	11.43	10.92	9.79	10.68	13.89	15.40
5	Thailand	6.16	5.37	6.44	7.74	7.46	4.59	5.65	7.15	7.77

TABLE 8.2 World beet and cane sugar production

	2000	2001	2002	2003	2004	2005	2006	2007	2008
Cane	94.68	87.75	106.54	113.71	109.91	103.74	115.54	130.27	129.64
Beet	35.35	32.90	35.55	34.42	37.36	37.62	36.61	36.07	32.86
Beet %	27.2	25.2	25.0	23.2	25.4	26.6	24.1	21.7	20.2

TABLE 8.3 The five largest cane and beet sugar producers

Largest cane sugar producers		Largest beet sugar producers	
(million metric tonnes, raw value)			
1 Brazil	32.29	1 EU-27	16.12
2 India	25.94	2 USA	3.88
3 China	14.43	3 Russian Federation	3.79
4 Thailand	7.77	4 Turkey	2.15
5 Mexico	5.94	5 Ukraine	1.70

long-term growth occurred at a time of relatively low world sugar prices, as well as continuing strong income growth and population growth particularly in the developing world, and is also related to the increasing prices of sugar substitutes such as corn sweeteners and high-fructose corn syrup. The responsiveness of consumption to income growth has been greater than that of population growth for developing countries. Furthermore, the trend of stagnating sugar demand in developed countries but growing consumption in developing countries has continued over the past decade. Sugar consumption is expected to rise steadily over coming years as income growth remains a key driver, as, to a lesser extent, do sugar prices. Brazil and India are among the top five sugar consumers as well as producers (Table 8.4).

International trade of sugar

The world's physical trade of sugar is increasing at a faster rate than the growth in global output. Between 1989 and 1998, global sugar exports accounted on average for 27 per cent of world output, compared to 31 per cent in the period 1999–2008.

TABLE 8.4 Evolution of consumption by five largest consumers in 2008

		2000	2001	2002	2003	2004	2005	2006	2007	2008
1	India	16.55	17.27	17.86	18.62	19.86	20.11	20.11	20.89	22.55
2	EU	14.11	13.59	14.37	14.14	17.69	16.76	17.53	19.31	20.47
3	China	8.50	8.90	9.98	11.07	11.61	11.79	11.98	13.83	14.73
4	Brazil	9.73	9.80	10.52	10.22	10.86	10.95	12.51	12.47	11.86
5	USA	9.05	9.14	9.08	8.84	9.00	9.25	9.23	9.11	9.81

According to ISO data, world exports of raw sugar reached a record of 28.4 million tonnes in 2008, or 59 per cent of world sugar trade, up from 20.4 million tonnes in 2000, or 56 per cent of the global trade. Global exports of white sugar amounted to 19.9 million tonnes in 2008, or 41 per cent of world sugar trade. Indeed, world white sugar trade has been growing more slowly than raw sugar trade – at an average compound growth rate of 2.7 per cent vs 4.2 per cent, respectively.

The past decade has seen a tremendous transformation in the composition of the world sugar trade. In the raw sugar market, Brazil's emergence as the key exporter has led to expanding exports to the Middle East and Northern Africa as well as Eastern Europe, which have now become the world's largest sugar trade routes (Tables 8.5, 8.6, 8.7).

In the white sugar market, the abrupt decrease in sugar exports from the EU due to sugar reforms has given way to a dynamic new set of trade flows involving Brazil, the Middle East, Africa and the Far East, partly as a result of several new large destination refineries in the Middle East and North Africa as well as in Asia, thus further boosting import demand for high quality raw sugar. Growth in the white

TABLE 8.5 The five largest net exporters, 2008 (million metric tonnes, raw value)

Total			Raw sugar			White sugar		
1	Brazil	20.14	1	Brazil	14.10	1	Brazil	6.04
2	Thailand	5.11	2	Australia	3.11	2	Thailand	2.34
3	India	4.23	3	Thailand	2.77	3	India	1.94
4	Australia	3.29	4	India	2.29	4	UAE	1.25
5	Guatemala	1.33	5	Cuba	0.85	5	EU-27	0.87

TABLE 8.6 The five largest net importers, 2008 (million metric tonnes, raw value)

Total			Raw sugar			White sugar		
1	Russian Fed.	2.52	1	EU-27	2.88	1	Iran	0.78
2	USA	2.37	2	Russian Fed.	2.42	2	Iraq	0.72
3	EU-27	2.01	3	USA	1.68	3	USA	0.70
4	Nigeria	1.57	4	Korea, Rep. of	1.65	4	Egypt	0.69
5	Iran	1.45	5	Malaysia	1.44	5	Indonesia	0.69

TABLE 8.7 Africa in the world sugar economy, 2008 (million tonnes, raw value)

	Africa	World	Share (2008)	Share (2000)
Production*	9.4	162.5	5.7%	7%
Consumption	15.6	163.1	9.6%	9%
Imports	9.1	48.3	18.7%	14%
Exports	2.9	48.3	6.0%	8%

Note
* Includes beet – 914,125 tonnes, and cane – 8,723,277 tonnes.

sugar trade is expected to recover over the coming years, however, as countries with new raw sugar-refining capacity start to build a surplus to export increasing quantities of white sugar to neighbouring markets.

Alternative sweeteners

Caloric sweeteners

From a period of sustained and strong market expansion during the 1990s, market growth has proven unreliable for high fructose syrup (HFS) producers during recent years. Producers have faced unstable feedstock costs and surging energy prices, resulting in significant swings in industry profitability. The share of HFS in the global sweeteners market fell from 10 per cent in 2000 to 9 per cent in 2009, chiefly reflecting a stagnant US market. Global HFS production in 2009 reached 12.6 million tonnes dry basis. US production has contracted but still accounts for nearly two-thirds of the total, whereas Europe's output increased due to higher production quotas provided under EU sugar reform. China has increased its output of corn starch sweeteners, including HFS.

Non-caloric sweeteners

Non-caloric sweeteners such as saccharine and sucralose, also known as high intensity sweeteners (HIS), represented around 10 per cent of the global sweeteners market in 2009. Annual growth in HIS was 4 per cent during 2004–2006, but a substantial decline in saccharine consumption in 2007 and 2008 contributed to a 2 per cent reduction, followed by a 4–5 per cent increase in 2009. Consumption growth has been strongest for sucralose and there are high hopes for strong growth of stevia-derived sweeteners in the USA and the EU, although sugar will continue to dominate both globally and regionally.

Principal sugar policy reforms

EU sugar reform in 2006/2007 has led to lower production and exports, while African, Caribbean and Pacific (ACP) countries have to negotiate new agreements with the EU to replace the previous ACP "Sugar Protocol". The EU has been transformed from a large exporter of mainly white sugar to the world's largest net importer of mainly raw sugar, reaching an all-time high of 3.3 million tonnes in 2008/2009. The USA, India and the UAE are also major importers.

The USA and Mexico became an integrated sweeteners market from 1 January 2008 under the NAFTA, after years of bilateral tensions featuring anti-dumping duties, discriminatory taxes and a WTO panel ruling. Whilst policy reform and liberalization has been a feature of the world sugar economy for some years now, future reforms depend on the outcome of the WTO negotiations on agriculture in the Doha Round of multilateral trade negotiations, whose completion date is unknown.

Global bioethanol market

Global fuel ethanol production and consumption is dominated by the USA (from corn), Brazil (from sugar cane) and, increasingly, the EU (produced from cereals and sugar beet). World fuel ethanol production has more than quadrupled in the past ten years, reaching an estimated 83 billion litres in 2010 (Figure 8.1). The world market is driven especially by new and expanding consumption mandates in the USA, the EU and several other countries in Asia (e.g. Thailand) and South America, along with anticipated continuing growth in Brazil's domestic ethanol consumption. Brazil's tight supplies have limited export availability and boosted ethanol prices recently, but growth in production and trade is expected in 2011–2012.

In the longer term, legislation such as the US Energy Independence and Security Act and the EU Renewable Energy Directive insure robust demand growth and further investment in production capacity. However, producers have to meet increasingly strict GHG reduction thresholds and sustainability criteria, thus pushing the market towards second-generation cellulosic ethanol within the next five to ten years.

Section 2: Africa overview

Sugar

Sugar industries in Africa embrace a wide range of production systems over a large spectrum of agro-climatic conditions, socio-economic conditions and ownership structures. These range from irrigated beet sugar production in North Africa (such as in Morocco and Egypt) to rainfed and irrigated sugar cane in the sub-Saharan region. There are both corporate-owned cane plantations and extensive small-grower schemes. There are large, modern, efficient factories but also small older plants. Refineries vary from "white end" extensions to raw cane sugar factories to the large stand-alone refinery units in Egypt, Algeria and Nigeria that refine imported raw sugar (Innes, 2010).

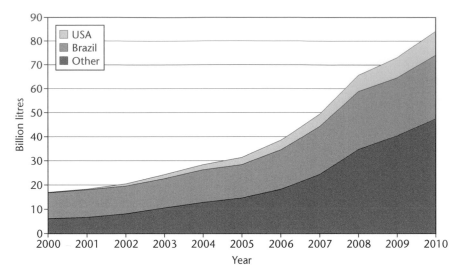

FIGURE 8.1 World fuel ethanol production

Africa in 2008 accounted for a lower share of global production and a slightly higher share of global consumption than the figures for 2000, with an accompanying increase and decrease in import and export shares, respectively (Tables 8.7, 8.8). Even with the continent's share of global consumption remaining relatively static, its import share has increased significantly because consumption growth (3.8 per cent annually) has consistently exceeded production growth (2.2 per cent). Key producers are South Africa, Egypt, Swaziland, Sudan and Kenya. Major consumers include Egypt, South Africa, Nigeria, Algeria and Morocco. The top net importers in 2008 were Nigeria, Algeria, Egypt, Morocco and Tunisia, whilst the top net exporters were Swaziland, Mauritius, South Africa, Mozambique and Zimbabwe.

Although the continent as a whole has not shown sustained production growth, there has been significant growth in Mozambique, Tanzania, Ethiopia, Zambia, Uganda, Cameroon and Sudan. On the other hand, consumption and imports have been increasing steadily, with total growth of 36 per cent for consumption and an astonishing 73 per cent for imports.

TABLE 8.8 Africa's sugar balance (thousand tonnes, raw value)

	2000	2001	2002	2003	2004	2005	2006	2007	2008
Production	9,254	9,065	9,871	9,253	9,418	9,848	9,701	9,831	9,637
Consumption	11,469	11,703	12,662	12,811	*13,537*	14,192	14,702	14,955	15,629
Imports	5,185	6,058	7,567	7,668	8,292	8,628	7,812	8,595	9,057
Exports	3,069	3,564	3,419	3,645	3,623	3,638	3,247	3,286	2,896

Alternative sweeteners

Africa is an insignificant producer of HFS on the world stage, with production of around 150,000 tonnes in Egypt, all from imported corn. One of the key conditions for the development of a commercially viable HFS industry is an ample supply of good-quality starch, which is most often not the case in the developing countries of Africa where cereals play a key role in basic food security.

Africa has no production of chemical sweeteners and only limited consumption of brand-name tabletop HIS, met from imports. However, the establishment of stevia plantations in East Africa will facilitate an important role in supplying the agricultural feedstock to meet the likely fast-growing future demand for this "natural" sweetener in developed countries.

Fuel ethanol

Governments within Africa are at varying stages of considering and implementing legislation for renewable energy, including biofuels. Biofuels like ethanol are seen as one potential instrument to effect poverty alleviation in Africa. Whilst having considerable potential for the production of fuel ethanol, both for home consumption and for export to the EU under preferential access, notably the EBA initiative, Africa is not yet far advanced in terms of government support and investor confidence. Even so, a growing number of countries are beginning to implement ethanol mandates, including Ethiopia, Kenya and Nigeria. The African biofuels industry seems likely to expand in the coming years. African biofuel producers, African governments and investors will likely partner in the creation of sustainable energy systems for local development as well as for export to the EU, China and other markets. Fuel ethanol prospects in Africa are discussed in more detail later in this chapter.

Section 3: Africa

Key drivers impacting production and export potential

In addition to selling their sugar on growing domestic markets, some African countries enjoy sales into high-priced regional markets that offer "natural" price protection due to their remoteness and considerable distances from key ports. But at the same time, other industries in Africa have to confront constrained access to export markets due to their landlocked status and distance from ports, which act to reduce the ex-mill prices considerably. This reduces in turn the attractiveness of export markets, including sales to the EU, which under two key preferential trading arrangements typically offer higher prices than the world market (discussed later in this chapter).

BOX 8.1: SALIENT FEATURES OF THE SUGAR INDUSTRY IN SOUTHERN AFRICA

Southern Africa is a surplus region for sugar, with all key producers having sufficient amounts for export (see Figure 8.2). There are about 30 sugar factories and two refineries in the region, producing 4.5 million tonnes of sugar. South Africa is by far the largest single producer, accounting for half of regional production in Southern Africa and 20 per cent of Africa's total. Other key industries are Swaziland (three mills); Zambia (one large and two smaller mills); Mauritius (six mills); Malawi (two mills); and Mozambique (four mills in operation). South Africa and Zimbabwe are the only countries in the region with stand-alone refineries, processing locally produced raw sugar into white sugar. Two annexed refineries have recently been constructed in Mauritius.

South Africa has 14 factories, of which two are owned by Tongaat Hulett, four by Illovo Sugar, three by Tsb Sugar and each of the other three by a different group. These companies purchase the majority of their cane from independent growers.

Agro-climatic characteristics are conducive to high sugar yields in Southern Africa (good soils, a defined wet season and seasonal temperature variation). The region boasts some of the world's highest cane yields and field and factory performance: yields of over 100 tonnes of cane per hectare over a crop cycle of up to ten ratoons are found in some industries. Malawi, Zimbabwe and Swaziland are among the lowest-cost sugar producers in the world.

There is considerable diversity in production systems. In South Africa there are both irrigated and rainfed systems, depending on climatic conditions. Water is often a limiting factor, and large-scale water infrastructure is used in some areas.

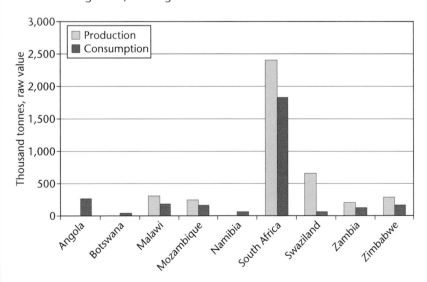

FIGURE 8.2 Southern Africa – sugar production and consumption

> Private ownership in Southern Africa is typical, with South African companies holding significant stakes in many mills in other countries. Outside South Africa, the structure of ownership is commonly "miller cum planter", whereby the mill grows a significant amount of its cane requirement on its own land and purchases the remainder from both large and small-scale "outgrowers".
>
> Many Southern African producers have relatively large domestic markets but also have significant access on a preferential basis to markets such as the EU and the USA, and to regional markets.
>
> *Source*: Innes, 2010.

Africa's sugar economy is diverse so it is not practical to cover the entire continent in this chapter. The focus here is on the sugar industries of Southern Africa[1] (see Boxes 8.1 and 8.2), which, although diverse, nevertheless share some common issues shaping their future development:

- broader integration within African markets through regional trade agreements;
- EU sugar policy reform and consequent new preferential trading arrangements; and
- possible commitments to the WTO.

These challenges are magnified in light of the wide variety of national sugar policies in the region. The outward-looking trade stance of some governments means that the success of their industries depends on their ability to respond to the above-mentioned challenges while maintaining or even improving their regional competitiveness. Each of these key drivers are now analysed in turn.

Broader integration of African markets via regional trade agreements

There are four key regional trade agreements (RTAs) in Southern Africa and East Africa: the South African Customs Union (SACU), the Southern African Development Community (SADC), the Common Market for East and Southern Africa (COMESA) and the East African Community (EAC). Several countries are members of more than one RTA (Table 8.9). Some changes are under way, such as COMESA becoming a free trade area and SADC moving to a free trade area and ultimately to a customs union.

Membership of RTAs shapes industry strategies and can result in both winners and losers, causing problems for some industries while providing viable sales outlets for others. The diversity of membership and multiple memberships proves a stumbling block by creating confusion in member status, appropriate tariffs and other trade regulations required to achieve a stable sugar market.

In short, RTAs are dynamic and constantly evolving. Sugar industries can capitalize on trade and investment opportunities by constantly monitoring the shifting

---

BOX 8.2: ECONOMIC INTEGRATION WITHIN SADC

In 2007 the SADC sugar industries formulated a Regional Sugar Strategy and accompanying Action Plan to take advantage of their cost competitiveness vis-à-vis the rest of the world. Objectives include: efficiency improvements; enhanced external market access; rationalization of the SADC market; promotion of the consumption of regional sugar; and promotion of alternative uses for sugar cane and sugar. The provisional target date for free trade in sugar was set at 2012, subject to review mid-decade.

South Africa is by far the largest surplus producer within SADC, while the non-SACU surplus countries are Malawi, Mozambique, Zambia and Zimbabwe. However, due to the transition to duty-free and quota-free access into the EU for the former ACP-EU Sugar Protocol beneficiaries by 2015, the definition of "surplus" for non-SACU SADC sugar producers accessing the SACU market is currently being reviewed.

During the continuing transition phase towards free trade, all SADC members will clearly aim for sugar to be traded in accordance with the terms of the Sugar Cooperation Agreement, and in a manner that enables individual countries to retain control of their domestic sugar policy and prices. Crucially though, however successful the sugar industries are in managing the transition to unrestricted free trade within the region, there will likely be strong and growing pressure to harmonize sugar pricing policies.

---

nature of RTAs in the region and their relationship with the EU and other major trading partners. The formation of EPAs (economic partnership agreements) with the EU cut across RTA membership lines, adding another dimension to complexity (Table 8.9).

Closer integration between SADC and COMESA is also likely, as the two organizations have overlapping memberships.[2] Nine of the COMESA member states formed a free trade area in 2000, with Rwanda and Burundi joining in 2004 and the Comoros and Libya in 2006. The main focus within COMESA is to strengthen and consolidate the FTA, deepen integration through the Customs Union and make progress on economic partnership agreement negotiations with the EU.

A recent proposal is the possible formation of a "super-block" comprising COMESA, EAC and SADC and SACU, first to create a "grand free trade area" and then ultimately a combined Customs Union. It would encompass 26 nations. A key motivation for such a sugar RTA is the fact that WTO rules prohibit dual Customs Union membership. When the COMESA Customs Union is launched and the SADC Union also, countries with overlapping membership would effectively have to decide which customs union to be part of.

TABLE 8.9 Southern and Eastern Africa EPA configurations and regional trade agreements

Country	SACU	SADC	COMESA	EAC	ESA EPA	Southern Africa/ SADC EPA	EAC EPA
Angola		X				X★	
Burundi			X	X			X★
Botswana	X	X				X	
Comoros			X		X★		
Djibouti			X		X★		
DRC		X	X				
Egypt			X				
Eritrea			X		X★		
Ethiopia			X		X★		
Kenya			X	X			X
Lesotho	X	X				X★	
Libya			X				
Madagascar		X	X		X★		
Malawi		X	X		X★		
Mauritius		X	X		X		
Mozambique					X★		
Namibia	X	X				X★	X
Rwanda			X	X	X★		
Seychelles		X	X				X
South Africa	X	X				X	
Sudan			X		X★		
Swaziland	X	X	X			X	
Tanzania		X		X			X
Uganda			X	X			X
Zambia		X	X		X★		
Zimbabwe		X	X		X★		

Note

★ Least developed country (LDC)

EU sugar policy reform

In November 2005, after many months of intensive negotiations, EU Agriculture Ministers reached agreement on reform of the EU sugar regime. A major feature of the reform was the replacement of intervention prices with reference prices, which would be reduced by 36 per cent over four years starting from 2006/2007 (Table 8.10).

In a nutshell, the key objective of the reform was to reduce sugar output by about 6 million tonnes, which was done via a voluntary restructuring scheme with associated compensatory payments. Sugar is now produced in 19 countries as opposed to 23 before the reform, and nearly 70 per cent of production is in the seven countries with the highest sugar yields. According to the Commission, domestic prices are showing a downward trend consistent with the objective of the reform.

TABLE 8.10 Evolution of EU reference prices for sugar

	2006/2007	2007/2008	2008/2009	2009/2010
White sugar, EUR/tonne	631.9	631.9	541.5	404.4
Cumulative reduction as %		0	14	36
Raw sugar, EUR/tonne	523.7	496.8	448.8	335.2
Cumulative reduction as %		5	14	36

Domestic output is now expected to remain considerably smaller than internal demand; export availability is declining dramatically while import demand is growing (see Table 8.11). Indeed, the EU became the world's leading sugar net importer in 2008/2009.

The market deficit is structural; with domestic production limited by quotas to around 14 million tonnes raw value. Annual imports of 4–4.5 million tonnes are implied, which is substantially more than the previous five-year average of 3.1 million tonnes. Therefore, one may conclude that, in terms of the EU sugar balance sheet, the main results brought about by the reform of the EU sugar regime are the end of large-scale white sugar exports to the world market and a hefty growth in EU import demand.

Changes in preferential sugar trade arrangements

The EU clearly continues to be an important market through countries' membership in economic partnership agreements (EPAs), or for least developed countries, through the Everything but Arms (EBA) Initiative, which confers unlimited, duty-free and quota-free access. Several countries also have access to the US market through its tariff-rate quota. The world market is commonly the residual market, which offers returns to the lowest-cost producers. South Africa is the African country with the greatest exposure to the world sugar market.

TABLE 8.11 EU-27 sugar balance (thousand million tonnes, raw value)

	2009/ 2010	2008/ 2009	2007/ 2008	2006/ 2007	2005/ 2006	2004/ 2005	2004/ 2003	2002/ 2003
Production*	17,355	15,445	17,833	18,739	20,598	21,926	20,210	20,886
Consumption	19,937	19,742	19,570	19,612	18,495	17,734	18,517	18,218
Exports**	2,025	1,278	1,666	1,592	8,077	6,042	4,418	6,228
Imports**	3,720	3,546	3,076	3,340	3,196	3,343	2,414	2,712

Source: ISO, undated.

Notes
* Including DOM raw cane sugar production.
** Including exports/imports of individual new country-members before the latest EU enlargements in 2004 and 2007.

BOX 8.3: PREFERENTIAL MARKETS: THE CASE OF SADC

The SADC region produces 5 million tonnes of sugar each year – around 3.5 per cent of world sugar output. With consumption at 3.4 million tonnes, this makes the region a net exporter of 1.6 million tonnes per annum. Around 63 per cent of SADC's sugar exports were sold to the EU and the USA under preferential access arrangements during 2006/2008, compared to only 40 per cent during 2003/2005.

SADC producers export sugar to the EU and the USA under three preferential access arrangements:

- economic partnership agreements, which replace the previous Sugar Protocol;
- The EU's Everything but Arms initiative, which granted LDC sugar producers phased duty-free access to the EU sugar market (qualifying SADC countries are: Angola, Democratic Republic of Congo, Malawi, Mozambique, Tanzania and Zambia).
- The USA, where market access is granted through a tariff-rate quota. Of the WTO bound minimum quota of 1.2 million tonnes, SADC producers are allocated only 0.1 million tonnes.

Table 8.12 presents an overview of the sugar supply/demand balance in SADC and SACU countries for the three-year period of 2006–2008. It shows that 90 per cent of the 1 million tonnes that SADC sells under preferential access arrangements is sold to the EU. Some countries rely heavily on the preferential EU and US markets (e.g. Mauritius) while others have low reliance, notably South Africa, which has no preferential access to the EU market.

The reform of the EU sugar regime is also transforming existing sugar trade arrangements with developing countries. After a transitional period ending in 2015, both LDC and non-LDC EPA countries exporting sugar to the EU will receive duty-free, quota-free access to the European market, but prices paid will be reduced at the same rates as prices paid for domestically produced sugar.

From ACP Sugar Protocol to EPA

In 2002, the Community and the ACP states officially opened negotiations on EPAs, which are to be with six ACP regions (the Caribbean (CARIFORUM); West Africa; East South Africa, Central Africa, Southern Africa and the Pacific). The EPAs are very detailed agreements, specifying treatment of thousands of separate products, services and investments. Although the deadline for EPAs expired on 21 December 2007, only one full regional agreement has been negotiated, with the

TABLE 8.12 SADC Supply/demand balances: average 2006/2008 (thousand tonnes, raw value)

	Production	Consumption	Preferential trade			Surplus/deficit
			EU EPA/EBA	US quota	Total	
Botswana	0.0	51.0	0.0	0.0	0.0	(51.0)
Namibia	0.0	57.0	0.0	0.0	0.0	(57.0)
South Africa	2 395.2	1 700.4	0.0	21.6	21.6	673.2
Swaziland	639.4	112.0ϕ	173.5	11.9	185.4	342.0
SACU	**3,034.6**	**1,920.4**	**173.5**	**33.5**	**207.0**	**907.2**
Angola	0.0	256.0	0.0	0.0	0.0	(256.0)
DR Congo	62.8	85.0	0.0	0.0	0.0	(22.2)
Madagascar	18.6	140.0	0.0**	0.0*	0.0	(121.4)
Malawi	273.3	170.0	63.6	6.7	70.3	33.0
Mauritius	482.3	41.2	478.8	2.8	481.6	(40.5)
Mozambique	245.5	161.6	77.0	8.5	85.5	(1.6)
Tanzania	270.0	312.5	15.4	0.0	15.4	(57.9)
Zambia	231.1	116.2	43.1	0.0	43.1	71.8
Zimbabwe	361.9	223.0	62.4	11.2	73.6	65.3
Non-SACU SADC	**1,945.5**	**1,505.5**	**740.3**	**29.2**	**769.5**	**(329.5)**
SADC	**4,980.1**	**3,425.9**	**913.8**	**62.7**	**976.5**	**577.7**

Notes

★ Madagascar has not filled its quota allocation of 7,258 tonnes.

★★ Madagascar did not ship any sugar to the EU.

ϕ ISO estimate. Swaziland sales to SACU amounted to an average of 323,000 tonnes.

There is no separate data available for Lesotho.

CARIFORUM states. In the case of the other five regions interim agreements have been signed, which establish market access but leave other aspects (services, investments and trade-related matters) to further negotiations. Sugar provisions are included in the EPA regulations, as fixed by the EU Council Regulations No. 1528/2007 and agreed to by all countries. Consequently, the sugar provisions will apply de facto to all EPA countries, even if the full regional EPAs have not been finalized yet.

BOX 8.4: THE ACP SUGAR PROTOCOL

The ACP Sugar Protocol via the Cotonou Agreement of 2000 allowed for annual duty-free imports into the EU of 1.3 million tonnes of sugar at the EU reference price. The Cotonou Agreement specified that a new trade regime had to be agreed by the end of 2007, coinciding with the expiry of a WTO waiver for the system on non-reciprocal trade preferences.

TABLE 8.13 ACP sugar quotas for 2008/2009 (tonnes, white sugar equivalent)

Barbados	32,097	Jamaica	122,234	Swaziland	117,845
Belize	46,680	Kenya	5,000	Tanzania	10,186
Congo	10,186	Madagascar	10,760	Trinidad & Tobago	43,751
Côte d'Ivoire	10,186	Malawi	20,824	Zambia	7,215
Fiji	165,348	Mauritius	491,031	Zimbabwe	30,225
Guyana	165,131	Mozambique	6,000	TOTAL	1,294,700

In order to address the impacts of EU sugar policy reform on the ACP sugar exporters market, the EU adopted and funded an eight-year assistance scheme of accompanying measures with €1.284 billion. There are 18 beneficiary countries within the ACPs, which were selected based on the importance of sugar in their economy and their dependence on the EU market. There are nine SADC countries among the beneficiaries: *Democratic Republic of Congo, Madagascar, Malawi, Mauritius, Mozambique, Swaziland, Tanzania, Zambia and Zimbabwe.*

In the case of Southern Africa (the SADC EPA), eight SADC countries decided to negotiate an EPA with the EU. In February 2010, following a high-level SADC/SACU meeting it was agreed that Botswana, Lesotho, Namibia, South Africa and Swaziland would move forward as one entity in the SADC EPA negotiations. Five of the SADC members decided to join the East South Africa (ESA) configuration, namely Madagascar, Malawi, Mauritius, Zambia and Zimbabwe (see Table 8.14). One member (Democratic Republic of Congo) has since joined the CEMAC (Communauté Economique des Etats de l'Afrique Centrale).[3] This complicates matters in the SADC as regional overlaps are seen as a stumbling block in deepening regional integration and negotiating trade arrangements. This may even work to the disadvantage of progress, with the negotiations of the EPAs including notification to the WTO.

South Africa has a Trade, Development and Cooperation Agreement (TDCA) with the EU that includes reciprocal market access arrangements (but which excludes sugar). It is still not clear how the TDCA will be aligned with the SADC-EU EPA. Even so, South Africa has requested inclusion with the SADC EPA negotiating group.

During the transitional period (2009–2015), shipments arriving in the EU from ACP countries will require licences and will be subject to a safeguard clause limiting the total allowable as duty free. From 1 October 2015, the regular EPA safeguard will apply, in particular where the value on the Community's market declines for two consecutive months to below 80 per cent of the EU market value that prevailed during the previous marketing year.

Sugar imports under Everything But Arms (EBA) Initiative

The Everything But Arms (EBA) Initiative was adopted in 2001, allowing 50 LDCs to export everything except arms (weapons) duty free to the EU. This initiative was later incorporated into the General System of Preferences Council Regulation. This foresees the special arrangements for LDCs being maintained for an unlimited period and not being subject to the periodic renewal of the Community's scheme of generalized preferences. As of 2009, sugar imports will be free of both tariffs and quotas. According to the framework agreement on EBA sugar, 25 countries were identified as the LDC sugar-supplying states, with 21 of these in Africa, although not all have yet exported through the EBA (see Table 8.15).

As of 1 October 2009, there is a safeguard clause for arrivals in the EU from a third country under the EBA arrangements: if imports increase by more than 25 per cent compared to the previous marketing year, the Commission may open

TABLE 8.14 EPA regions in Africa

West Africa (CEDEAO + Mauritania)	**Central Africa** (CEMAC + STP)
Benin*, Burkina Faso*, Cape Verde* Gambia*, Ghana, Guinea*, Guinea Bissau*, Côte d'Ivoire, Liberia*, Mali*, Mauritania*, Niger*, Nigeria, Senegal*, Sierra Leone*, Togo*	Cameroon, Central African Republic*, Chad*, *Congo*, DR Congo*, Equatorial Guinea*, Gabon, S. Tome*
	East Aftrican Community (EAC)
	Burundi*, *Kenya*, Rwanda*, *Tanzania*, *Uganda**
Southern Africa (SADC)	**Eastern and Southern Africa** (ESA)[4]
Angola*, Botswana, Lesotho*, *Mozambique**, Namibia, *Swaziland*, South Africa	Burundi*, Comoros, Djibouti*, Eritrea*, Ethiopia*, *Kenya*, Malawi* Mauritius, *Madagascar**, Rwanda*, Seychelles, Sudan*, Uganda*, Zambia*, Zimbabwe

Notes
In *italic* – holders of ACP sugar quotas.
* Least developed country (LDC).

TABLE 8.15 EBA sugar shipments (thousand tonnes)

	2001/2002	2002/2003	2003/2004	2004/2005	2005/2006	2006/2007	2007/2008	2008/2009
Bangladesh			8.99	0.33			11.65	11.17
Benin				6.06	8.30	7.79	8.59	7.35
Burkina Faso	7.07	7.24	7.67	8.02	5.24			
Cambodia								7.70
Chad							7.76	9.17
DR Congo				11.70	7.04	5.19	6.00	14.81
Ethiopia	14.30	14.69	15.25	13.80	13.67	24.04	21.15	21.22
Laos								7.08
Malawi	10.40	10.66	10.96	2.58	21.00	27.01	18.76	15.95
Mozambique	8.33	8.38	10.12	27.60	27.53	37.24	11.26	15.98
Nepal		8.97	8.67	9.00		12.95	10.36	12.27
Senegal						8.04	3.88	8.60
Sierra Leone					5.75	6.59	7.76	11.04
Sudan	16.26	17.04	16.98	16.36	14.25	28.11	27.28	26.51
Tanzania	9.07	9.32	9.94	1.67	7.57	8.81	22.52	14.33
Togo				6.55	3.67	3.45	5.82	7.33
Zambia	8.76	9.02	9.54	9.18	15.67	22.92	15.25	14.21
TOTAL	74.19	85.31	98.11	112.83	129.75	192.11	178.03	204.74

Source: ED&F Man, 2009.

proceedings to decide whether measures such as suspension or temporary with-drawal of trade concessions need to be applied. It is important to note that suspension or temporary withdrawal will affect only fraudulent transactions while genuine deliveries originating in LDCs cannot be subject to quantitative restrictions.

According to the Regulation, for the period from 1 October 2009 to 30 September 2012, the importer shall undertake to make purchases at a minimum value of not less than 90 per cent of the reference price (on a c.i.f. basis[5]). Thus, LDCs will be paid the same prices as EPA countries.

WTO Doha Round outcome

Should there be a final Doha Round agreement, then legally binding commitments to reduce tariffs and lower domestic support will be another factor ensuring closer alignment between domestic and world market sugar prices with the SADC, even though more than nine years of talks have yet to result in any agreement and many delegations have expressed frustration at the slow progress of the Doha Round.

A Doha Round outcome would require reduction commitments to be made on a product-specific basis. In a nutshell:

* A ceiling will be negotiated on <u>domestic support</u> for sugar (and every other commodity) in each country based on a historical base period.
* A date will be set for parallel elimination of all forms of <u>export support</u>, including sugar.
* Improvements in <u>market access</u> will be negotiated in each commodity sector, including sugar.

The impact on sugar in Africa and elsewhere will not only depend on the how-far-how-fast "hard numbers" for reducing tariffs, opening-up markets, lowering domestic support and phasing out export subsidies, but final agreement on base periods for reduction commitments and transition periods will be just as important. Moreover, there would be flexibility and lesser commitments for "sensitive products", which will most undoubtedly include sugar. For sensitive products, both developed and developing countries will be able to slate for shallower tariff cuts in return for expanded import quotas. Not only is the issue of sensitive products crucial to sugar, so too will be any final agreement on "special products" and the "special safeguard mechanism"; two types of flexibilities for developing countries alone. While the former would allow them to cut a limited number of tariffs more gently on the basis of food security, livelihood security and rural development concerns, the latter would provide them with a quick defence against import surges.

Fuel ethanol to impact sugar exports?

Governments within Sub-Sahara Africa are at varying stages of considering and implementing legislation for renewable energy, including biofuels. Moreover, the development community and governments generally believe that in order for economic growth to be sustainable, jobs must be generated and maintained within rural areas. Ethanol is seen as one potential instrument for poverty alleviation and rural development in sub-Saharan Africa. Indeed, there is a general presumption that ethanol can help displace imported petroleum and in some cases also substitute for traditional biomass.

The economics of ethanol production is heterogeneous and location-specific: feedstock availability and cost are the most important drivers. These factors in turn hinge on land availability and quality, agricultural productivity, labour costs and so on. Operating costs hinge upon plant location, size and technology. There may also need to be investment in storage infrastructure (Plate 16). The second key important determinant of viability is ethanol's competitiveness with rival fuels in the transport sector and, for Africa, in the household markets. This is a function of the cost, demand and availability of these other fuels, which fluctuate around global petroleum prices, foreign exchange rates and domestic refining capacity. An overriding consideration for fuel ethanol from sugar cane is opportunity cost – the value of the sugar foregone. For Southern African countries that are net exporters of sugar to regional and world markets, in some cases the opportunity cost of producing ethanol would be higher than for sugar production. As explained below this is generally the case for SADC sugar-producing countries under present and expected price levels for sugar and ethanol.

In Africa, markets for fuel ethanol are still very much at the drawing board stage. Malawi and Zimbabwe have small production capacities for fuel ethanol, but output growth is constrained by the lack of a legal mandate. Several LDC net sugar exporters, such as Angola, Ethiopia, Mozambique, Tanzania, Sudan and Zambia are reportedly seeing investment in new ethanol production from sugar cane molasses for exports.

Clearly, within Southern Africa there is considerable potential for investment in fuel ethanol production, not only to satisfy domestic markets but also to ship ethanol to the EU – a likely large importer over the medium to longer term. However, it is unlikely under present pricing levels within Sub-Saharan sugar industries for sucrose to be diverted (by using higher grades of molasses or cane juice) to ethanol production. Indeed, for the majority of the region the simple economics, with a world sugar price above 15 US cents and the opportunity of free market access for sugar into the EU from 2015, precludes the option of using cane juice for ethanol: the opportunity cost of doing so is likely to be prohibitive. Furthermore, domestic and regional prices are typically much higher than world prices. In a nutshell, absent significant policy-driven initiatives in ethanol production will be largely limited in the near term to production from molasses for local fuel consumption and some regional export markets.

Section 4: capacity expansion in Africa's preferential sugar exporters

The benefits to preferential exporters are undergoing structural changes with the reform of the EU sugar regime, negotiation of EPAs and with the transition arrangements under the EBA initiative. On the one hand, prices paid for sugar imported to the EU are set to decrease; indeed, a shortage of sugar occurred in the EU in the second half of 2010 as ACP/EBA suppliers chose to sell to the higher-priced world market. On the other hand, the level of access to the EU market will increase considerably. In short, those LDCs with efficient seaports and low production costs relative to the EU reference price (€404 in 2009/2010) will likely expand to supply the EU under the EBA and the EPAs.

Importantly, the two dominant multinational sugar producers – Illovo Sugar and the Tongaat Hulett Group – are pursuing sugar industry expansions in Malawi, Mozambique, Swaziland, Tanzania and Zambia. Illovo is Africa's biggest sugar producer and has extensive agricultural and manufacturing operations in six African countries. According to the Group's 2010 Annual Report, it manages agricultural estates in each of the countries in which it operates, which together produced 6.1 million tonnes of cane in the 2009/2010 season. In addition, independent growers supplied an aggregate of approximately 8 million tonnes of cane to Illovo's sugar factories, mainly to those in South Africa.

A recent independent survey of international sugar production costs for the period 2006/2007 to 2009/2010, which covered over 100 sugar-producing countries, indicated that, of the six countries in which Illovo operates, three are in the top ten lowest-cost cane sugar producers in the world and all six are within the top 30. Illovo Group is a major supplier of sugar to African consumer and industrial markets, particularly in its own countries of operation. In Malawi Illovo is the sole sugar producer, and in Zambia it manufactures 94 per cent of all local production. Illovo has significant and increasing access to preferential markets in the EU and the USA, whilst the operations outside South Africa also have access to the SACU market. The Group exports sugar into the world free market through the South African sugar industry.

According to its report, Illovo Group's cane production increased by more than one million tonnes to 6.1 million tonnes, and sugar output reached 1.685 million tonnes in 2009/2010 compared to 1.578 million tonnes produced in 2008/2009. These increases were driven primarily from the expansion of the Zambian operations where the existing cane supply was supplemented by nearly 3,000 hectares of newly developed estate cane fields. Growth in the Zambian cane supply and consolidation of factory operations will continue in the coming year, and a further significant increase in sugar production was achieved in 2010/2011 (see subsection on Zambia, below).

Cane production also increased in Malawi, Mozambique, Swaziland and Tanzania, following smaller expansions and irrigation upgrades in those countries. In addition, in Tanzania the introduction of new cane varieties has resulted in improved cane yields. The expansion of the Maragra factory in Mozambique has been completed, doubling capacity to 150,000 tonnes of sugar per annum.

A further significant group expansion is under way in Swaziland where the Ubombo factory expansion and cogeneration project is in progress, due for commissioning in 2011/2012. Annual sugar production at Ubombo is to increase by more than 40 per cent to over 300,000 tonnes per annum. Increased power generation using cane biomass to supplement boiler fuel will enable the Swaziland operation to become self-sufficient in electricity. In addition, following the conclusion of a power purchase agreement with the Swaziland Electricity Company, surplus power will in future be supplied into the national grid for 48 weeks of the year.

According to Tongaat Hulett's 2010 Annual Report, their cost-competitive portfolio of sugar operations and assets coupled with the leading "Hulett's" sugar brand, world-class technology and preferential market access, will enable them to capitalize on the changing sugar fundamentals in world markets. Production in 2009/2010 was 1.0 million tonnes of sugar, their milling capacity is 1.9 million tonnes, and expansion potential exists in the low-cost production areas of Zimbabwe and Mozambique. The company's sugar milling, refining and agricultural operations in KwaZulu-Natal and Zululand are also strategically positioned to improve its cost competitiveness.

Tongaat Hulett is also able to take advantage of the evolving renewable energy dynamics in its sugar operations by substituting the production of molasses and sugar for the international market with the production of bioethanol for local and regional markets. The opportunity exists through capital investments to increase efficiency in bagasse cogeneration (see Chapter 5), thereby producing a greater surplus of electricity for use by third parties and/or sale to the electricity grid.

In Mozambique, Tongaat Hulett's expanded sugar mills and estates surrounding Xinavane and Mafambisse are fully irrigated and are located in areas with ideal growing conditions, thus resulting in high cane and sucrose yields. These ideal agricultural conditions, combined with close proximity to ports and the technology availability from South Africa, makes it well positioned for future growth. Mozambique's LDC and ACP status allows its products to enter the EU on a duty-free basis, giving it an advantage over countries such as Brazil.

Sugar operations in Zimbabwe consist of Triangle and a 50.3 per cent stake in Hippo Valley Estates, representing a combined installed sugar-milling capacity of 600,000 tonnes. The Lowveld in Zimbabwe, with excellent topography, climate and established water storage and conveyance infrastructures for irrigation, is recognized as the lowest-cost sugar producer in Southern Africa, if not the entire world. Since 1980, the operations at Triangle have produced potable, industrial and/or fuel alcohol, with additional capital invested during 2007 for resumption in production of fuel-grade ethanol.[6]

Ethiopia (LDC and ACP)

Ethiopia has three state-owned sugar plantations, currently producing 280,000 tonnes of sugar a year, and the government has ambitious expansion plans that could allow greater duty-free exports into the EU. This would also allow for expanded ethanol production, which has attracted interest not only from the government for fuel blending but also for household cooking. The use of ethanol stoves has been championed by the award-winning Gaia Association, an Ethiopian NGO that is connected to the US-based Project Gaia. According to the Ministry of Trade and Industry, there are also plans for construction of a new sugar mill and distillery at Tendaho and expansion projects at three other sugar production sites. There is also a private initiative, the Hibir Sugar Share Company, which plans to construct a complex with annual capacity of 260,000 tonnes of sugar and 28 million litres of ethanol, as well as 14.4 megawatts of electric power capacity.[7]

Malawi (ACP and LDC)

Illovo Sugar Malawi Ltd, 76 per cent owned by South Africa-based Illovo Sugar, is the country's sole sugar producer. It operates two mills, one at Nchalo in the south and the other at Dwangwa in the centre, which produced 295,000 tonnes of sugar in 2009/2010. Both Nchalo and Dwangwa have been the focus of various agricultural and milling capacity expansion projects in recent years, leading to a current capacity of 330,000 tonnes of sugar per annum. Options for further expansion in Malawi are being evaluated, as considerable potential exists for expanding land under cane at the estate and also by outgrowers. Opportunities for power cogeneration are also being assessed. It is also important to note the ethanol production in Malawi, which includes potable, industrial and fuel grade, with blending of 10 per cent in petrol. Malawi has the longest continually running ethanol blending programme in Africa; it started operation in 1982, although the size of the blend has changed depending on amounts sold for potable, industrial and pharmaceutical markets (Plate 17).

Mali (LDC and ACP)

There is a large greenfield mill under development in Mali, where Illovo is partnering with the government to develop 14,500 hectares of sugar cane. Funding for the project is currently being evaluated and subjected to final due diligence from a consortium of concessional funders. The Malian operation, if completed, will ultimately produce 195,000 tonnes of sugar, 15,000 kilolitres of ethanol for fuel blending and generate sufficient electricity for the agricultural factory and operations, with additional capacity to export power into the national grid.

Mauritius (ACP)

Mauritius has historically been the largest ACP sugar exporter to the EU, traditionally selling more than 90 per cent of its output to the EU, and is thus significantly affected by EU sugar reform. Consequently, there are no expansion plans, and the industry continues to implement its ten-year Industry Action Plan, focusing on centralizing production at four sugar mills, down from ten sites in 2006. To make this possible, a voluntary retirement scheme was launched for workers older than

50. A related goal is to cluster together the country's 28,000 smallholder cane farmers to enhance economies of scale in cane and sugar production. Smallholders are utilizing 21,000 of the 72,000 ha of land under cane cultivation. The clustering project involves mechanization of all practices, including cane harvesting, irrigation and land preparation. The fields will be replanted with cane varieties with 20 per cent higher yields.

A cornerstone for the long-term survival of Mauritius's sugar industry is an agreement reached with Germany's Südzucker in May 2008. Südzucker AG signed a deal on delivery of around 400,000 tonnes of sugar annually from 2009 to 2015. The deal means Südzucker would take most of Mauritius's annual sugar output. Most of the raw sugar is refined at two new annexed refineries. A large part of the export is sold as speciality sugars, which fetch a higher price.

Mozambique (LDC and ACP)

Tongaat Hulett owns two mills in Mozambique – Mafambisse and Xinavane, which are targeted over the next two seasons to expand to the newly installed milling capacity of 300,000 tonnes per annum. Illovo Sugar plans to double output at its Maragra mill from 75,000 tonnes to 150,000 tonnes of sugar per year by 2012. The company has entered into a joint venture with the local community at Maragra to develop 4,000 ha of land to cane over the next two years, which is expected to produce an additional 400,000 tonnes of cane per annum. The Marromeu sugar mill in central Mozambique is also planning to expand: its majority Brazilian owner Açúcar Guarani announced in April 2009 that it aims to double the size of its sugar cane plantation in Mozambique to 30,000 ha in the course of the year and nearly triple its sugar production within three years. The additional sugar production in Mozambique will be sold in the domestic market with the balance being exported into the EU under the EBA initiative. The expansions will also result in a significant boost to molasses supplies, providing ample feedstock for a possible increase in ethanol production.

South Africa (non-LDC and non-ACP)

South Africa, whilst by far SADC's dominant sugar producer, has the least potential for expansion given its lack of LDC status, which prevents EBA access to the EU, and no EU access under the Sugar Protocol. Production costs are generally higher than in other SADC countries. As already noted its two major sugar companies, Illovo Sugar and Tongaatt Hulett have invested in other production centres within the SADC, which have lower production costs and access to the higher-value preferential markets in the EU and the USA. Production is variable because the bulk of the cane land is rainfed. Production over the past five years has ranged between 2.3 million and 2.7 million tonnes. The fortunes of South Africa's sugar industry are predominantly linked to the global sugar market and domestic market protections, but on a seasonal or annual basis are subject to the vagaries of weather.

Sudan (LDC and non-ACP)

Sudan is producing 900,000 tonnes of sugar, complemented by refining of imported raw sugar. Kenana – the country's largest producer – in September 2009 announced plans to more than triple sugar output within three years after prices for the sweetener rose to a record high. Plants under construction will take output to more than three million tonnes. Part of the additional production will be for the rapidly growing domestic market, some will be exported to match growing import demand and market opening in the EU, and some could also be exported to Middle Eastern markets. Kenana has also invested in ethanol production capacity, amounting now to 65 million litres annually, some of which is exported to the EU. Kenana's ethanol production is slated to reach 400 million litres of ethanol after the White Nile and Blue Nile Sugar Factories became operational. Sudan also has four public sugar mills – New Halfa, Genain, Assalaya and Sennar – whose output was pegged at 356,000 tonnes in 2008/2009. Plans for significant expansion of the sugar industry include the improvement of irrigation infrastructure as well as an increase of the area under cultivation.

Swaziland (ACP)

Swaziland's annual sugar quota to the EU under the ACP protocol in 2008/2009 was 170,000 tonnes, making the sugar sector a mainstay of the economy. The country is ranked among the top ten world producers and presently produces 650,000 tonnes of sugar annually on less than 50,000 ha of irrigated land. The country's largest sugar producer is the Royal Swaziland Sugar Corporation (RSSC), which produced 423,000 tonnes of the 2009/2010 total at its two mills – Mhlume and Simunye. These crush around 3.2 million tonnes of cane per annum. A 32 million litre capacity ethanol plant is situated adjacent to the Simunye mill producing industrial grade ethanol and potable spirit from all of the molasses at Mhlume and Simunye. RSSC manages some 15,000 ha of irrigated sugar cane on two estates leased from the Swazi nation and manages a further 5,000 ha on behalf of third parties, delivering some two million tonnes of cane per season. An additional 1.2 million tonnes of cane is sourced from independent (outgrower) farmers, who grow cane on around 11,300 ha.

Swaziland's production is expected to increase significantly, reflecting its relatively low production costs; perhaps by as much as 110,000 tonnes within the next few years. The aim of the National Adaptation Strategy is to minimize the adverse effects of the EU sugar regime reform. Short-term priorities include the establishment of a Restructuring and Diversification Management Unit, the restructuring of smallholder farmer financing and the use of grant finance to restructure existing smallholder debt, the maintenance of health and education facilities for the affected communities and the development of suitable models for the future provisions of social services presently financed by the industry. In the medium term, priorities are focused on programmes aimed at reducing the cost of smallholder production and reducing marketing costs, together with measures for the diversification into alternative crops and the promotion of small and medium enterprises.

In line with the expansion and restructuring, two new smallholder development schemes have been started. The Lower Usuthu Smallholder Irrigation Project was slated to produce an additional 100,000 tonnes of cane to be processed by Illovo's Ubombo sugar mill, boosting sugar production to around 300,000 tonnes per annum, together with an increase in power generation capacity utilizing bagasse. Under the current development plans, it is envisaged that an additional 1,900 ha of cane will be available for milling at Ubombo by the 2011/2012 season, growing to 5,000 ha thereafter and ultimately up to 9,000 ha of cane in the longer term.

There is also the Komati Downstream Development Project (KDDP). This project is assisting smallholder farmers' associations to cultivate sugar cane and other crops on 7,400 ha of irrigation fields (using Swaziland's share of water from the Maguga dam). The Mhlume sugar mill is expanding to accommodate the additional production of sugar cane.

Tanzania (ACP and LDC)

Tanzania's four main sugar mills have a production capacity of 400,000 tonnes of sugar per year, but did not manage to produce more than 263,000 tonnes in 2009/2010. Tanzania has four major sugar companies. Kilombero, a unit of South Africa's Illovo Sugar; two locally owned companies – Mtibwa and Kagera; and TPC, which is majority-owned by Mauritius's Sukari Investment Ltd. All four sugar mills have plans for investment and expansion, and Kagera Sugar Company and TPC have already completed a substantial part of their rehabilitation and capacity-expansion plans at the mills.

According to the adopted National Sugar Strategy, priorities have been given to improving efficiencies in smallholder farm management, developing "public good" types of infrastructure at the smallholder level and supporting environmental management. The strategy is meant to address inefficiencies in the smallholder production process and support the establishment of block farming at the smallholder level. Under the programme, farm tracks and sugar cane access roads for smallholders would be rehabilitated, while a farm trust would be set up to support infrastructure maintenance. The package also includes comprehensive support to achieve synchronization of farming among smallholders and improve agricultural practices and logistics management.

For the time being the government is giving priority to satisfying domestic demand, as it is currently a net importer of sugar. The annual expected sugar demand of 400,000 tonnes will far exceed the expected supply of 318,000 tonnes in 2011.

A large and ambitious sugar cane ethanol production scheme was pursued by the Swedish company SEKAB during 2007–2009 and was intended mainly for export to the EU. This ran into a variety of political-economic difficulties as well as the international financial crisis, although a restructured smaller-scale investment in Tanzania is continuing (see Chapter 17).

Uganda (ACP and LDC)

Continued growth of the country's leading sugar producers Kakira Sugar Works Ltd, Kinyara Sugar Works and Sugar Corp. of Uganda Ltd (SCOUL), as well as the entry of Mayuge Sugar Works Ltd and GM Sugar Works, is substantially boosting Uganda's sugar production. Output in 2010 reached 292,051 tonnes, a slight increase compared to the 287,400 tonnes produced in 2009. Ugandan sugar production has more than doubled in the past ten years, averaging an annual increase of 13 per cent. The three major companies have invested a combined US$100 million in the past three years and are all continuing their expansion programmes, which involve more sugar cane production – especially from outgrower farmers, higher milling capacity and installing cogeneration power capacity. Consumption still exceeds output and is expected to grow strongly over coming years, and thus the industry is expected to prioritize satisfying domestic demand rather than exporting to the EU.

Zambia (ACP and LDC)

Zambia Sugar Plc, 90 per cent owned by Illovo Sugar, accounts for about 90 per cent of the country's total output. The company's only mill – Nakambala – has undergone significant expansion over recent years. An agreement with the Zambian government in March 2007 led to expansion in existing sugar cane-growing area from 10,500 ha to 27,000 ha, and an increase in annual milling capacity from 200,000 tonnes to 440,000 tonnes of sugar a year. The expanded milling capacity was successfully commissioned in April 2009. The factory reached rated capacity during the 2009/2010 season and produced 315,000 tonnes of sugar, a 62 per cent increase compared to the 194,000 tonnes produced in 2008/2009.

Zambia Sugar also acquired Nanga Farms – a cane-growing company located adjacent to the Nakambala estate, producing 325,000 tonnes of cane per annum. This acquisition, together with expanding the existing area under estate cane, enabled Zambia Sugar to more than double estate cane supply to 1.7 million tonnes from 720,000 tonnes the year before. As a result, the estate supplied more than 65 per cent of the total cane supplied to the Nakambala mill in 2009/2010. Furthermore, outgrowers lifted their cane deliveries to just short of 910,000 tonnes, resulting in a total delivery to the expanded mill of 2.6 million tonnes, an improvement of 1.0 million tonnes year on year.

Zambia Sugar consolidated its expansion gains even further in 2010/2011, reaching a record 385,000 tonnes. Exports to the EU and regional markets increased to another record level of 233,000 tonnes, thus doubling the level of exports in the previous season.

Zimbabwe (non-LDC and ACP)

South Africa's Tongaat Hulett now owns both mills and estates – Triangle Sugar and Hippo Valley. The company is poised to expand output following last year's

US dollarization of the economy and the "return to more normal economic fundamentals relevant to the sugar business", including restoration of domestic sales prices to regional levels. A recovery programme is currently under way, focused on improving cane yields and the re-establishment of outgrower cane lands, so as to restore sugar production to the existing installed capacity of 600,000 tonnes per annum (based on the crush of 4.8 million tonnes of cane in a 38-week crushing season). Both sugar mills operate refineries with a combined capacity to produce 140,000 tonnes of white sugar per annum.

Meanwhile, the government of Zimbabwe initiated the Chisumbanje Sugar Milling/Ethanol Project, which involves the growing of sugar cane on 10,000 hectares under irrigation. A sugar mill with an annual capacity of 125,000 tonnes is planned for Chisumbanje. The government has set an implementation consortium, which is seeking project funding and foreign investment.

Despite a steady decline in output over recent years, exports to the European Union have increased. In 2008/2009, 146,000 tonnes were exported to the European Union (October/September). This compares with exports to the EU of 51,000 tonnes in 2007/2008 and 81,000 tonnes the year before.

Section 5: conclusions

In general, it can be concluded that in the case of preferential exporters, including those in Africa, the reform of the EU sugar regime works towards the same goal as within the European Community. On the one hand, the reform stimulates efficient producers to increase production, which can be directed to the still lucrative EU market with prices typically considerably higher than those of the world market (2010 presented an exception, with world sugar prices reaching 30-year highs), while market access is improving considerably. On the other hand, due to the severe cuts in reference prices, high-cost producers are expected to abandon sugar production or consolidate production to achieve efficiency goals.

The sugar industries of Africa will develop strategies according to the opportunities and challenges they face, but generally would be expected to seek to reduce production costs and improve their operations in an increasingly competitive market place. Several industries are already targeting specific market opportunities arising from trade agreements, including opportunities under preferential trade agreements with the EU under the EPAs and the EBA. There are significant expansions of sugar production capacity on the way in a number of African countries, with the most notable being those in Sudan (increase of 2.1 million tonnes), Ethiopia (increase of 1.3 million tonnes), Mozambique (increase of 0.4 million tonnes) and Zimbabwe (increase of 0.3 million tonnes). If a number of smaller projects are added, African production could rise by five million tonnes in the future. It has to be pointed out also that the 36 per cent reduction of the EU reference price for raw sugar since 1 October 2009 – ahead of the reform – could make it more attractive for African exporters to supply regional markets on the African continent that offer higher premiums.

Renewable energy has become a strategic element in the expansion options for African sugar producers. There are new opportunities for fuel ethanol production in response to a growing world ethanol market, including prospective national blending policies in Africa, but also in terms of export to the EU under the EBA initiative. Cooperation agreements have recently been announced between Brazil and Zambia, and Brazil and Mozambique. Investment in bagasse cogeneration has already been recognized in some countries (e.g. Mauritius and Swaziland) as a valuable addition to efficiency and low-cost production; where surplus electricity can be sold, the resulting revenue stream offers an additional bonus, since the efficiency gains also benefit overall factory performance and sugar output.

Crucially in terms of fuel ethanol production, the question of opportunity cost remains paramount. Africa's sugar producers are unlikely to forego sugar production to boost ethanol production unless they can achieve comparable returns between the two products. However, expanding supplies of molasses arising from likely increasing sugar production in several lower-cost industries does provide potential for higher ethanol production over the longer term, both for domestic and export markets.

Notes

1 Southern Africa includes: Malawi, Mozambique, Zambia, Zimbabwe, South Africa, Swaziland, Botswana, Lesotho and Namibia. SADC also includes: Angola, DRC, Madagascar, Mauritius, Seychelles and Tanzania.
2 Eight countries are members of both SADC and COMESA: Angola, Democratic Republic of Congo, Madagascar, Malawi, Mauritius, Swaziland, Zambia and Zimbabwe.
3 CEMAC consists of Cameroon, Central African Republic, Congo, Gabon, Equatorial Guinea and Chad (www.cemac.cf/).
4 The region includes countries of the East African Communities (EAC) – Burundi, Kenya, Rwanda, Tanzania, Uganda – a splinter group from the ESA.
5 The c.i.f. price (i.e. cost, insurance, freight) is the cost of a good delivered to the importer at the border of the importing country, thus including any insurance and freight charges up to that point.
6 Fuel-grade ethanol was produced in the 1980s and blended in petrol in Zimbabwe but was discontinued in the 1990s due to changes in markets and government policies, as well as the impacts of a severe drought.
7 Hibir Sugar Share Company (accessed January 2011): www.hibirsugarethiopia.com/.

References

ED & F Man (2009) "EBA Market Analysis", ED&F Man, London.
Hibir (2010) Hibir Sugar Share Company, www.hibirsugarethiopia.com/.
Innes, J. (2010) "Sugar in Africa: status and prospects", *International Sugar Journal*, vol. 112, no. 1337.
This chapter is based on statistics and forecasts of the International Sugar Organization (ISO) along with the following reports:
ISO (2010a) "Fuel Ethanol Prices and Drivers – a World Survey", MECAS (10)19.
ISO (2010b) "Industrial and Direct Consumption of Sugar – an International Survey", MECAS (10)18.

ISO (2010c) "World Sugar Demand: Outlook to 2020", MECAS (10)17.

ISO (2010d) "World Trade in Raw and White Sugar – Recent Trends and Prospects", MECAS (10)06.

ISO (2009a) "The International Physical Trade of Sugar – a Survey", MECAS (09)19.

ISO (2009b) "Domestic Sugar Prices", MECAS (09)18.

ISO (2009c) "Sugarcane Ethanol and Food Security", MECAS (09)07.

ISO (2009d) "Cogeneration – Opportunities in the World Sugar Industries", MECAS (09)05.

ISO (2008a) "EU Sugar Policy Reform – Ramifications for Preferential Exporters", MECAS (08)18.

ISO (2008b) "The International Trade of Fuel Ethanol: Outlook to 2015", MECAS (08)17.

ISO (2008c) "US Sugar under the 2008 Farm Bill: National and International Implications", MECAS (08)16.

ISO (2008d) "Sugarcane Smallholders in Sub-Saharan Africa: Status, Challenges and Strategies for Development", MECAS (08)05.

ISO (2007a) "Government Biofuels Policy and Sugar Crops: Outlook to 2015", MECAS (07)17.

ISO (2007b) "Southern Africa: Key Drivers Impacting Sugar Export Potential", MECAS (07)07.

ISO (2007c) "Cross-Border Investments in the World Sugar Industry", MECAS (07)06.

ISO (2006) "Analysis of Cane and Beet Payment Systems", MECAS (06)04.

9

IMPLEMENTATION, STRATEGIES AND POLICY OPTIONS FOR SUGAR CANE RESOURCES AND BIOENERGY MARKETS IN AFRICA

Francis D. Yamba, Francis X. Johnson, Gareth Brown and Jeremy Woods

Introduction

Modern use of biomass energy has increased in many parts of the world, with the key driving forces being soaring oil prices, commitments to address climate change and the need for rural development (Faaij and Domac, 2006). A reliable supply and stable demand of feedstocks are always required as a prerequisite to sustainable market development. The exploitable bioenergy potential of the sub-Saharan Africa region is significant and is estimated to be the highest of any world region (Smeets *et al.*, 2007).

Sugar cane has become an important source of bioenergy globally, and at the same time, the sugar industry has entered a period of transition. Subsidies within the European Union have been reduced, preferential market access to the African, Caribbean and Pacific (ACP) countries has been adjusted and prices and quotas for least developed countries (LDCs) are somewhat more constrained (Innes, 2010). There will remain, however, a considerable amount of preferential market access for sugar, so that the "free market" where world commodity prices apply will remain "thin" for some years to come (Alvarez and Polopolus, 2008; Chapter 8 in this Volume).

Two important co-products of the sugar cane industry are cogenerated electricity from bagasse and bioethanol. While the former may only have a domestic and regional market, bioethanol has an additional opportunity on international markets that can be exploited (Johnson and Matsika, 2006). The possibility for African LDCs to export ethanol to the EU has been recognized due to the fact that EU countries must meet mandates under the Renewable Energy Directive (EC, 2009), while at the same time many LDCs are exempt from agricultural import tariffs applied to products such as bioethanol (see Chapter 8).

The main driving forces for market diversification into renewable energy include uncertainties in oil and sugar prices, land and resource availability, power supply conditions, environmental policies and economic development policies. The challenges to diversification include lack of supporting national and international policies, difficulties with market access and technological constraints, as well as the more general barriers that stifle the development of renewable energy sources. This chapter will consider the implementation strategies available to sugar producers and the policy incentives that can stimulate greater production of renewable energy from sugar cane. The emphasis will be on bioethanol and cogeneration, although some of the same approaches can support other energy and non-energy products or markets that use biomass from sugar cane.

Market structure and co-products

The sugar cane plant has high photosynthetic efficiency and is currently the world's most commercially significant energy crop (El Bassam, 2010). It offers a cost-effective and versatile renewable resource, offering many alternatives for production of food, energy, fibre and feed. Sugar cane resources support a variety of uses and products based on different resource streams, which can be divided into sugars, molasses/juice and crop residues, as shown in Figure 9.1. Cogenerated electricity and ethanol are amongst the most important cane co-products in commercial terms, but the other by-products or co-products need to be considered as well.

Regulation is often required to facilitate the relationship between growers (farmers) and millers (factory operators) in the sugar industry. In most African countries, an estate provides the majority of sugar cane delivered to the factory, but small farmers generally provide a significant share. The way in which sugar industry revenues are shared can be crucial for market development. The division of revenues reflects the nature of the sugar industry, with growers and millers interdependent, and the welfare of both parties dependent on the performance of the other.

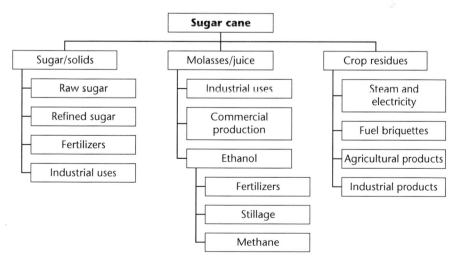

FIGURE 9.1   Sugar cane co-products

Sugar industry institutional arrangements – formal and informal – in each country are therefore important in understanding the way in which sugar industries might react to policy reform and deregulation outcomes, and to incentives for bioenergy production (Yamba *et al.*, 2008).

The economic reform in some African economies, accompanied by liberalization of the energy industries, has led to an energy supply shortfall in various countries, involving blackouts and impacts on reliability that can affect not only households but industrial customers, with considerable potential losses. The entry of sugar industries as independent power producers (IPPs) can contribute to stabilizing localized and national electricity shortfall. In the longer term, there is also concern that climate change may impact hydropower capacity in some areas; given the high reliance on hydropower in Africa, alternative energy sources that are available continuously will become more important.

The LDC sugar producers in Africa still benefit from domestic prices and preferential trade that are higher than world sugar market levels. Consequently, the opportunity cost of using sugar cane for ethanol production remains a key variable. The recent EU sugar reforms will favour the low-cost producers of Southern Africa when it comes to sugar exports (Innes, 2010). Whether these same producers will also choose to enter the ethanol market depends on other market factors as well as the policy incentives in the national and regional context. The nexus between sugar policy reform and incentives for bioenergy production will therefore have differential impacts across different producers, according to the strategies chosen by the producers and the market access and conditions.

Implementation strategies

There are a number of different implementation options that might be pursued by individual producers in expanding renewable energy from sugar cane in the region; these options can be viewed in terms of various "co-product strategies" that are based on differing levels of sucrose and fibre utilization (Yamba *et al.*, 2008). The more sucrose that is directed towards sugar, the less is available for ethanol production, while fibre utilization depends mainly on efficient factory configurations and bagasse and residue management strategies (as well as economic markets for cogenerated electricity). There are other energy-related co-product options, such as biogas production using the stillage (waste stream) from ethanol production, and there are also many non-energy co-products that might be included. These implementation strategies are discussed in this section.

Ethanol strategies

Since the sugar industry has traditionally focused on extracting as much crystallizable sugar (sucrose) from the cane as is economically possible, any sustainable strategy to co-produce ethanol and raw sugar is driven by market factors to apportion some of the sucrose resources to ethanol production. For economic and social sustainability, local demand for crystalline sugar must be met before any sugars can be diverted to ethanol production except for final molasses, the by-product from which the sucrose (around 8–10 per cent of the sucrose contained in cane) cannot be economically recovered.

The traditional strategy of only producing sugar is increasingly unfeasible as preferential export markets are removed (Innes, 2010). The alternative strategy of producing only ethanol in an autonomous distillery has been pursued in Brazil and elsewhere, and has the advantage of significant savings in capital investment costs since only juice preparation and extraction facilities are needed rather than a complete sugar factory. It is nevertheless difficult economically for producers to forego sugar production. A study carried out in Zambia (Cornland *et al.*, 2001) concluded that ethanol-only strategies are not viable unless sugar prices drop significantly. Furthermore, the feasibility of an autonomous distillery depends strongly upon both the size of the market and the size of the sugar cane estate. The plant size and feedstock production would need to be optimized accordingly. An autonomous distillery would need to be operated at a reasonable scale and would require a stable market at this higher level; such stability is difficult in the near term, given that no regional ethanol market currently exists. The introduction of E85 or flex-fuel vehicles coupled with policy measures to stimulate a regional market could improve the feasibility of ethanol-only options (via autonomous distilleries) in the future. The low rate of personal car ownership (and resulting low consumption of petrol) essentially means that the small scale of the national markets (except for South Africa) makes these markets less competitive in comparison to a regional market (Johnson and Matsika, 2006).

A third approach considers producing sugar and ethanol in fixed quantities by maximizing extraction of all the economically extractable sugars for sugar production following the standard 3-massecuite raw sugar production process, and leaving only C molasses (final molasses) available for ethanol production. In some cases, B molasses is used. This practice has been widespread and is based on annexed distilleries to make use of the surplus molasses stream while prioritizing sugar production. At Triangle Estates in Zimbabwe, C molasses was at some stage imported from the neighbouring Hippo Valley Estates and from as far as Zambia to supplement the B molasses from the sugar mill (Scurlock *et al.*, 1991). This approach based on fixed quantities remains viable if sugar prices are competitive, sugar markets are active and ethanol/oil prices are low. Where sugar markets become saturated and prices decline, and where ethanol markets emerge, then the industry would fail to capitalize on such opportunities by remaining in a fixed sugar/ethanol strategy.

The fourth strategy involves producing sugar and ethanol in flexible proportions. In this scenario, sugar is extracted up to the first or second stages, resulting in the production of A or B molasses, respectively. These molasses streams will have fermentable sugars that can still be economically extracted. The presence of additional fermentable sugars increases the efficiency of ethanol conversion. Consequently, if ethanol is expected to have a market value close to or greater than that of sugar, then it makes economic sense to prioritize ethanol production over some sugar production, by using A or B molasses as the ethanol feedstock. If market prices are fluctuating over time, a producer can benefit from having the flexibility to switch among these alternative balances of molasses use. The capital and operating costs of the additional processing stations for B and C molasses are not significant compared to the overall production costs (Seebaluck *et al.*, 2008). Consequently the decision as to whether to emphasize sugar or ethanol production can be made at the margin; such an approach has already been adopted by producers in the mature sugar/ethanol market of Brazil.

In the case of expected increases in ethanol market demand, the approach would involve a progression in the use of feedstocks for ethanol production from C molasses to B molasses to A molasses to cane juice, as required over time to meet increasing ethanol demand. Thus on an annual basis, the first step would be to compare ethanol yield from C molasses to target ethanol demand. If ethanol demand exceeds ethanol yields from C molasses alone, then B molasses is chosen as feedstock (for ethanol production in as many mills as required, starting with the largest mills, but only if national sugar demand is still met). If raw sugar demand is not met then additional feedstock needs to be developed. But as long as sugar demand is met, ethanol continues to be derived from higher-sugars feedstocks (up to A molasses and from cane juice). Where ethanol and sugar demand are not met, there is a need to assess additional sugar cane production required to meet demand and ascertain whether this would be feasible. This may include considering the need for additional factories. If expanding the cane-growing area is not feasible, alternative feedstocks (e.g. sweet sorghum, maize) or imports may be considered. The other crops could in most cases provide equivalent feedstock to the same ethanol distillery, although there are always some additional feedstock preparation costs when using multiple sources of raw material. Sweet sorghum is not suitable for crystalline sugar production but is suitable for ethanol production, and can therefore be viewed as a potential supplementary feedstock where ethanol production is valued more than sugar at the margin (Woods, 2001).

A scenario assessment for selected countries is shown in Table 9.1. The scenarios show the range of production options available. The simplest case is where all the final (or C) molasses is used for ethanol production, which might be the case when international sugar prices are extremely high and there are no countervailing policy incentives that can offset the high opportunity cost of foregoing sugar production. The zero sugar exports scenario is a purely theoretical exercise only, since no countries in the foreseeable future would be willing to forego sugar exports. The maximum ethanol production case is where cane juice and/or B and C molasses are used for ethanol production. A middle scenario can also be envisioned in which sugar exports move towards zero, thereby freeing up more sugars for ethanol production. Such a middle scenario seems more likely than the other two options, given the combination of higher sugar prices and the increasing demand for biofuels.

TABLE 9.1 Summary of scenario results for 2030 ethanol potentials

Country	From C molasses			With zero sugar exports			Maximal ethanol		
	Ethanol potential production (million litres)	Share of domestic gasoline consumption (energy basis)	Share of domestic gasoline consumption (volume basis)	Ethanol potential production (million litres)	Share of domestic gasoline consumption (energy basis)	Share of domestic gasoline consumption (volume basis)	Ethanol potential production (million litres)	Share of domestic gasoline consumption (energy basis)	Share of domestic gasoline consumption (volume basis)
Malawi	29.2	10.6	16.4	44.2	16.0	24.8	249.7	90.4	140.0
Mauritius**	79.9	19.6	30.4	726.1	178.5	276.5	683.3	168.0	260.2
Mozambique	28.2	14.1	21.9	163.1	81.8	126.6	241.2	120.9	187.2
South Africa	316.5	1.1	1.7	2133.9	7.4	11.5	2707.6	9.4	14.6
Swaziland	61.4	23.4	36.2	564.6	214.9	332.8	525.7	200.1	309.9
Zambia	27.7	5.2	8.1	160.0	30.2	46.8	236.6	44.7	69.2
Zimbabwe	63.0	5.1	7.9	390.5	31.8	49.3	538.9	43.9	68.0
Total	**605.8**	**1.9**	**3.0**	**4,182.4**	**13.2**	**20.5**	**5,183.0**	**16.4**	**25.4**

Source: Yamba et al., 2008.

Note

** Includes theoretical expansion options that are not achievable under current conditions.

Bagasse-based electricity generation strategies

Two main strategies for selling surplus electricity from a sugar factory are possible. The first option can be to sell to local off-grid customers, such as local industries or rural electricity cooperatives, thereby providing electrical services without the costs that accompany grid connections. For areas without a guaranteed market and with low local demand, investment in a cogeneration plant with the intention of selling electricity to the neighbouring communities may not be viable.

Mauritius is often held up as a model in promoting bagasse cogeneration, as it has successfully implemented cogeneration plants at many factories as part of the overall economic restructuring of its sugar industry, in which the larger and/or more competitive factories were consolidated, competitive feed-in tariffs were offered for electricity sold to the grid from sugar factories, and cooperation on management of the biomass resource base was established (Deepchand, 2001; see also Chapter 5 for more details).

The second option is to sell surplus electricity to established utilities companies or distributors, as an independent power producer (IPP). This requires appropriate national policies that allow IPPs to generate and supply the public network. However, in most African countries, the national utilities are not offering competitive feed-in tariffs to encourage this option. As an important input in the productive sector, many governments still maintain some form of control of electricity pricing to lower the cost of industrial production. This has generally stifled development of cogeneration in the sugar industry in the region despite sectoral reforms in the electricity industry. However, the rapidly changing electricity tariffs in the region will gradually make cogeneration economics more attractive in the short to medium term.

Cogeneration potentials were determined for selected countries based on sugar factory data, and by assuming constant growth scenarios, as given in Table 9.2. The simplest option requiring the lowest investment cost is to use improved back pressure turbines to generate a small surplus. A more state-of-the-art option would be to install condensing extraction steam turbines (CEST), while the technologically advanced option would be biomass gasifier combined cycle technology (BIG-CC). The CEST systems increase electricity output by a factor of seven, while BIG-CC systems result in a further quadrupling of output to 48 terawatt hours. BIG-CC systems are still not operated commercially, but within ten to 20 years the cost will come down and the value of the additional production will be higher, especially with increasing value for carbon credits. The proven success of CEST technology in Mauritius has provided a model baseline for evaluating cogeneration options in the sugar industry in Eastern and Southern Africa (Deepchand, 2001).

Other co-product markets

Besides production of electricity and ethanol, there are many other marketable by-products or co-products that can be produced in annexed plants to the sugar cane factory. The establishment of a centralized and integrated sugar cane processing

TABLE 9.2 Summary of 2030 cogeneration potentials in selected countries

Country	Electricity generation potential with improved back-pressure turbines (20 kWh/tonne cane; 20 bar)		Electricity generation potential with CEST technology (143 kWh/tonne cane; 82 bar)		Electricity generation potential with BIG-CC technology (650 kWh/tonne cane)	
	Cogeneration potential (GWh)	% of total electricity	Cogeneration potential (GWh)	% of total electricity	Cogeneration potential (GWh)	% of total electricity
Malawi	72	3.2	518	23.0	2,355	104.6
Mauritius	n/a	n/a	1,323	29.2	6,016	132.9
Mozambique	14	0.4	103	2.6	468	11.8
South Africa	865	0.3	6,185	2.1	28,111	9.6
Swaziland	141	8.6	1,006	61.7	4,574	280.3
Zambia	58	0.4	414	2.7	1,884	12.3
Zimbabwe	153	0.5	1,095	3.6	4,977	16.1
Total	1,489	0.4	10,645	3.0	48,385	13.7

Note

Mauritius has already installed several CEST systems, so the first scenario is not applicable.

complex creates further economic opportunities by increasing the viability for exploiting additional cane-based co-products. Whether this is done through extensions to existing factories or through entirely new facilities depends crucially on scale and market structure for the intended co-products. The initial investment costs will be lower via the expansion of existing factories, whereas a fully optimized production that is related to the specific product mix would require changes already at the design phase (Seebaluck *et al.*, 2008).

In a sugar agro-industrial complex, the sugar factory serves as the nucleus to produce sugar while other production units can process as many value-added products as the market allows, while also modulating the input requirements (especially energy and water) in various forms (live steam, exhaust steam, vapour at various pressures and temperatures and electricity) from the combustion of the fibrous fraction of cane (bagasse) in an integrated manner. Downstream products that are already produced in South Africa include furfural, various flavouring agents and lactulose (a natural laxative). These higher value-added products that are produced at lower volumes can serve as economic complements to high-volume energy co-products (see Chapter 7 for more details on co-products).

Barriers to market development

For many developing countries, where most sugar factories are found, the sugar industry has great potential to contribute to Africa's economic and energy development goals. At the same time, most African countries lack the transportation and communications infrastructure to expand into renewable energy markets, as well as the institutions needed to create reliable supply and demand that will attract investors and entrepreneurs. In this section, a review is provided of the key barriers to market formation, focusing on bioethanol and cogeneration.

Barriers to ethanol production

With the ever-increasing prices of petroleum-based fuels, the supply of which is finite in nature, bioethanol, which can be produced locally, presents the best sustainable development path for the oil industry. However, various barriers currently inhibit the penetration of ethanol as a commodity in Southern Africa. Presented in Table 9.3 are the key barriers.

Barriers to electricity production

This report has shown the high potential of cogeneration to contribute to energy security under the South African Power Pool (SAPP), and SADC as a whole. In a few years, the electricity demand for most countries will outstrip supply. Despite the aforesaid, production of electricity in sugar factories has not been fully exploited. Listed below are some barriers that have hindered the full potential of cogeneration in the sugar industry.

TABLE 9.3 Barriers to production and use of bioethanol

Category	Barriers
Policy	• Limited awareness of the benefits to accrue from investment in bioethanol-producing technologies. • Limited awareness of benefits of ethanol as a renewable energy source and its relationship to business by government, NGOs and private sector. • Tariff and non-tariff barriers to movement of energy products like ethanol. • Limited awareness of Clean Development Mechanism (CDM) objectives and its product/approval cycle in government, NGOs and private sector.
Technology	• Limited awareness of availability of information on ethanol-producing technologies. • Limited and sometimes non-existent knowledge of selection of appropriate ethanol-producing technologies as potential renewable energy sources. • Limited human resources in the development of bankable business proposals under sustainable energy path development. • Few support services for project idea note (PIN) and project design document (PDD) elaboration, for conducting feasibility studies, and for formulation of business plans related to renewable energy technologies (RETs).
Financial	• Lack of financial base from local investors to contribute to equity for project implementation. • Limited and sometimes non-existent awareness by local/regional institutions of the need to invest in ethanol projects. • Limited awareness of availability of international investment sources. • Under CDM, low market value of carbon credits and hence no financial attractiveness for business.
Legal	• Limited and sometimes non-existent awareness of legal issues regarding the development of RET projects at all levels (government, NGOs and private sector). • Limited and sometimes non-existent awareness of the Kyoto Protocol as an international law that promotes renewable energy sources under CDM. • Lack of capacity in most countries to negotiate for a Certified Emission Reduction Purchase Agreement (CERPA).

TABLE 9.4 Barriers to cogeneration in the sugar industry

Category	Barriers
Policy	• Inadequate political support for renewable energy.
	• Lack of data and information to support better policy making.
	• Most existing energy policies are inadequate, especially those on renewable sources.
	• Energy institutions are ineffective, mainly as a result of insufficient budget allocation to carry out the various activities.
	• There is no effective government institutional interaction in the energy sector.
	• Lack of private sector investment in electricity, mainly due to the commercially unviable low grid-based tariffs.
	• Limited awareness of the benefits to accrue from investment in advanced electricity-producing technologies.
	• Limited awareness of benefits of bagasse-generated electricity as a renewable energy source and its relationship to business under environmental programmes like CDM.
Technology and Human capacity	• Limited awareness of availability of information on renewable energy technologies.
	• Limited and sometimes non-existent knowledge of a selection of appropriate potential renewable energy technologies.
	• Limited human resources in the development of bankable business proposals under sustainable energy path development.
	• Limited human capacity in renewable energy technologies in key institutions like departments of energy and NGOs, who are key to the development and deployment of RETs.
	• Limited learning institutions providing in-depth system design in renewable energy technologies.
	• Research and development:
	• There is an inadequate number of projects to demonstrate the applicability of some of the available energy technologies.
	• Technological obstacles related to applied research, development and demonstration.

Financial	• Private sector participation in energy service development programmes has been lacking because rural energy projects tend to be less attractive and the investment incentives have not been operational.
	• Renewable energy technology costs have unrewarded environmental benefits; lack of appropriate and innovative financing mechanisms for long-term financial benefits.
	• Lack of fiscal incentives to enable renewable energy projects to become financially attractive due to distorted competition from big hydro schemes.
	• Relatively low grid-based tariffs make it difficult to commercialize renewable energy, making it unviable for sugar factories to produce export electricity for the public grid. Investment in advanced combustion technologies would be meaningful if revenue obtained from the electricity generation provided a positive return on investment. Otherwise, sugar factories will continue with inefficient bagasse combustion, while buying their additional electricity needs from the national grid.
	• In most SADC countries, high interest rates inhibit private investment in renewable energy technologies.
	• Lack of financial base from local investors to contribute to equity for project implementation.
	• Limited and sometimes non-existent awareness by local/regional institutions of the need to invest in renewable energy projects.
	• Limited awareness of availability of international investment sources.
	• Under CDM, low market value of carbon credits and hence no financial attractiveness for business.
Legal	• Similar to ethanol barriers.

International trade barriers

International trade in biofuels is currently limited by the fact that countries with large markets (the USA, Japan, the EU) maintain tariffs on these fuels, to protect their domestic industries and/or to assure that their substantial domestic subsidies are not used to support the industries of other nations. For instance, the USA applies an extra 54 US cents to each gallon (about 14 cents per litre) of imported ethanol on top of a 2.5 per cent tariff, bringing the cost of bioethanol in line with that produced domestically (IEA, 2004). Moreover, the tariff escalation systems that prevail in many industrialized countries encourage developing countries to export feedstock, such as unprocessed molasses and crude oils, while the final biofuel conversion – and associated value addition – takes place in the importing country. In addition, the lack of technical specifications for biomass and biomass import regulations can be a major barrier to trading. For example, in the EU denaturized ethanol above 80 per cent attracts an import levy of €102/m³, representing substantial additional costs (Junginger *et al.*, 2006). Denatured ethanol can be seen as a higher value-added product compared to undenatured ethanol since the former is designed to facilitate broader commodity trade.

The current lack of a clear classification of biofuels within the multilateral trading system constrains effective trade. At present there is no broad agreement on whether biofuels are industrial or agricultural goods. On the one hand, biofuels are traded as "other fuels", or as alcohol (in the case of ethanol) and are subject to general international trade rules under the World Trade Organization (WTO). Biofuels are potentially classifiable as environmental goods, as discussed in trade liberalization talks under the Doha Round, and such classification would facilitate their expanded use in environmental policy objectives as well as the current emphasis on energy security goals (Dufey, 2007).

Implementing a transition to sugar cane energy

The transition to greater utilization of energy from sugar cane requires a combination of investment in new technologies along with policy decisions about priorities for the development of the sector vis-à-vis sugar vs energy markets. In particular, the choice between sugar production and ethanol production will be influenced by key policies and regulations, while the value of additional electricity production depends on the structure and regulation of power production. The implementation is also significantly dependent on scale, as only the larger factories or mills will be able to pursue cost-effective strategies for ethanol and cogeneration. The optimal scale for ethanol production will generally be in the range of 1–2 million tonnes of sugar cane processed per year (see Chapter 6).

One way to encourage expansion is through transport fuel blending mandates, which exist already in countries such as Malawi and Kenya. On the power production side, the existence of a "renewable energy portfolio standard" can be used to encourage increased renewables in the power sector, such as is common in Europe.

Since bagasse cogeneration is generally quite cost effective in comparison to other renewable options it will fare well, especially since carbon finance fits fairly well into such projects (McNish et al., 2009). In this section, two algorithms are described that were developed to explore implementation strategies based on the possibility of coordinating national and regional policy mechanisms in promoting expansion options for ethanol and cogeneration (Yamba et al., 2008).

Algorithm for implementing ethanol expansion strategies

This process attempts to determine the most favourable transitions to ethanol production given the existing mill sets in the countries considered. It involves a progression in the use of feedstocks for ethanol production from C molasses to B molasses to A molasses to cane juice, as required over time to meet increasing ethanol demand (see Chapter 6 for a technical discussion on factory set-up and the three "strikes" in sugar production). The methodology assumes that the local sugar industry will select the most suitable mills for conversion to large-scale ethanol production using specific criteria based on local conditions.

For the region to meet projected sugar cane production and ethanol demand, possible feedstock and mill modification routes to achieve the transition were determined using a decision flow scheme, as described below. The scheme is based on the characteristics of the existing sugar mills and utilizes the three-strike model.

For each country, the procedure on an annual or seasonal basis for determining the potential for ethanol production is comprised of the following steps (see summary in Figure 9.2):

1. The ethanol yield from C molasses is determined from sugar cane data (using factory-specific data if available, or country averages if not).
2. Ethanol demand is compared with the calculated ethanol yield from C molasses. It should be noted that for South Africa some of the molasses is used for other purposes, mainly animal feed and potable alcohol.
3. If ethanol demand exceeds ethanol yields from C molasses alone, then B molasses are chosen as feedstock (for ethanol production in as many mills as required, starting with the largest mills, but only if national sugar demand is still met). If crystalline sugar demand is not met, then see (5) below.
4. As long as sugar demand is met, ethanol continues to be derived (and yields estimated) from higher-sugars feedstocks. The procedures for calculating ethanol potential from A molasses and from juice are similar to that for B molasses.
5. Where ethanol and sugar demand are not met it is necessary to calculate the additional sugar cane production required to meet demand and ascertain whether this would be feasible. This may include considering the need for additional factories.
6. If expanding the cane-growing area is not feasible, alternative feedstocks (e.g. sorghum, maize, etc.) may be considered. Imports may also be considered.

The same algorithm could be applied on a regional basis, with some modifications to consider greater intra-regional trade. Applying the algorithm on a regional basis would provide some better estimate of the additional gains from coordinating supply and demand.

Algorithm for implementing cogeneration strategies

Given electricity production targets, possible options for meeting these targets can be determined using a decision flow scheme. Following the same logic used for ethanol strategies, the largest mills are considered the most suitable for early modification to produce surplus electricity. The first step involves assessing electricity production from the sugar industry under existing conditions and comparing it with demand targets. If this not adequate, the second step is to assess electricity production assuming all factories optimized their existing boilers and back-pressure steam turbines, and improved efficiency in the use of steam. If demand is still not met, the next step is to assess potential electricity generation in a scenario where the largest mill is retrofitted with state-of-the-art condensing extraction steam turbines (CEST), and the highest boiler pressures deemed technically and economically viable for the mill. If demand is still not met, power production from the next-largest CEST-powered mill is estimated until demand is met or the mill set is exhausted. If after all viable mills are converted to CEST and demand is still not met, possible new cane production and factories are considered.

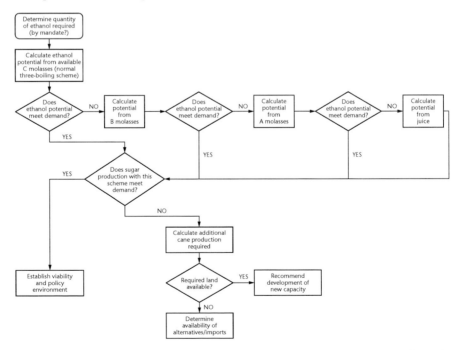

FIGURE 9.2   Algorithm for determining ethanol expansion options (*source*: Yamba *et al.*, 2008.)

For the projected sugar cane production and electricity demand, possible options for implementing the transitions to meet electricity production targets can thus be evaluated using a decision flow scheme as described below (Figure 9.3). Following the same logic used for ethanol scenarios, the largest mills are considered the most suitable for early modification to produce surplus electricity. The procedure involves the following steps on an annual allocation basis, for each country or region, depending on the application:

1. Determination of surplus electricity demand from the sugar industry.
2. Determination of surplus electricity that could be generated if all factories optimized their existing boilers and back pressure steam turbines, and improved efficiency in the use of steam.
3. If surplus electricity does not meet demand, potential electricity generation is estimated were the largest mill retrofitted to use condensing extraction steam

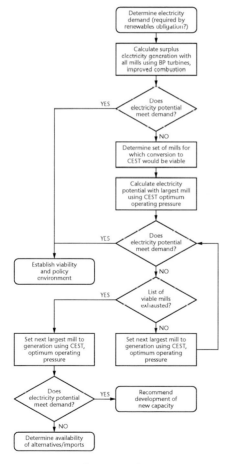

FIGURE 9.3 Algorithm for determining cogeneration expansion option (*source*: Yamba *et al.*, 2008.)

turbines (CEST), with the highest boiler pressures deemed technically and economically viable for that mill by local experts.

4.   If surplus electricity still does not meet demand, electricity production from the next largest CEST-powered mill is estimated until demand is met or mill set is exhausted.
5.   If after all viable mills are converted to using CEST and demand is still not met, possible new cane production and factories are considered.

Applications

The decision algorithms described above were applied to a dataset of factories from selected African countries in order to determine the range of potential for ethanol production and cogeneration (Yamba *et al.*, 2008). In order to apply such a decision algorithm, it is necessary to choose some objective; in this case the obvious cases are the use of a renewable fuel mandate for each of the two sectors (transport and power). A target of 25 per cent ethanol blending in gasoline and 10 per cent of electricity production offers one benchmark, which can be calculated from projected increases in gasoline and electricity demand to 2030; the algorithm can then be used to determine an implementation strategy based on the existing factories. The total ethanol required for the countries shown in Tables 9.3 and 9.4 would be 8 billion litres, while the required electricity production would be 37 TWh (Yamba *et al.*, 2008). The deficit for gasoline demand in South Africa would be met through excess ethanol production from the other countries, thus demonstrating the importance of a regional market in such policies. Although obviously power trade is more constrained than trade in transport fuels, there is nevertheless potential to address deficits in some regions through the SADC power pool (Yamba and Matsika, 2003).

It is important to note that the algorithms offer a fairly general approach to setting priorities in a manner that is economically rational (based especially on scale economics) but at the same time can respond to the energy-development-environment drivers that call for renewable energy targets. The general approach might be adapted to recognize infrastructure constraints and other limitations, and could also be adapted to consider the effects of trade liberalization on sugar and ethanol markets, to the extent that these will affect the scale economies and the location of production in the future.

Conclusions

This chapter has reviewed the implementation strategies and policy options for expanding ethanol and cogeneration from sugar cane, focusing especially on those factories that currently produce the majority of sugar in sub-Saharan Africa. The expansion of ethanol will depend on sugar prices and the willingness of producers to forego some sugar production; stronger policy incentives (e.g. blending mandates) as well as some financial incentives (e.g. carbon credits, loan guarantees) would be needed in order for national and regional markets to move towards a scale that is economically sustainable. Similarly on the power sector side, regulatory frameworks for independent power production in combination with incentives for renewable energy in the power sector will be needed in order to exploit the bagasse cogeneration potential, as has already been occurring in Mauritius. The inherent economics of the sugar industry are fairly competitive by global standards. Expanded production of ethanol and cogenerated electricity from sugar cane will require a greater willingness to apply policy instruments to shift the industry towards renewable energy production and thereby improve the competitiveness of both the sugar and energy sectors.

References

Alvarez, J. and Polopolus, L.C. (2008) "Domestic and International Competition in Sugar Markets", EDIS document SC 021, publication of Department of Food and Resource Economics, Florida Cooperative Extension Service, Institute of Food and Agricultural Sciences, University of Florida.

Cornland, D.W., Johnson, F.X., Yamba, F., Chidumayo, E.N., Morales, M.M., Kalumiana, O. and Mtonga-Chidumayo, S.B. (2001) "Sugar Cane Resources for Sustainable Development: A Case Study in Luena, Zambia", Stockholm Environment Institute (SEI).

Deepchand, K. (2001) "Bagasse Based Cogeneration in Mauritius – A Model for Eastern and Southern Africa", AFREPREN Occasional Paper No. 2, AFREPREN/FWD, Nairobi.

Dufey, A. (2007) "International Trade in Biofuels: Good for Development? And Good for Environment?", International Institute for Environment and Development (IIED) policy brief, January, http://pubs.iied.org/11068IIED.html.

Dufey, A., Baldock, D. and Farmer, M. (2004) "Impacts of Changes in Key EU Policies on Trade and Production Displacement of Sugar and Soy", Study commissioned by WWF to IIED, with the collaboration of IEEP.

EC (2009) Directive 2009/28/EC of the European Parliament and of the Council, of 23 April 2009, on the Promotion of the Use of Energy From Renewable Sources and Amending and Subsequently Repealing Directives 2001/77/EC and 2003/30/EC, European Commission, Brussels.

El Bassam, N. (2010) *Handbook of Bioenergy Crops: A Complete Reference to Species, Development and Applications*, Earthscan, London.

Faaij, A.P.C. and Domac, J. (2006) "Emerging international bio-energy markets and opportunities for socio-economic development", *Energy for Sustainable Development*, vol. 10, no. 1, pp. 7–19.

FAOSTAT (2011) "Sugarcane Crop Production Statistics", FAOSTAT, Statistical database, UN Food and Agriculture Organization, http://faostat.fao.org/site/567/default.aspx#ancor, accessed 21 February 2011.

IEA (2004) *Biofuels for Transport. An International Perspective*, OECD, Paris.

Innes, J. (2010) "Sugar in Africa: Status and Prospects", *International Sugar Journal*, vol. 112, no. 1337.

ISO (2010) "World Trade in Raw and White Sugar – Recent Trends and Prospects", MECAS (10)06.

Johnson, F.X. and Matsika, E. (2006) "Bio-energy trade and regional development: the case of bio-ethanol in Southern Africa", *Energy for Sustainable Development*, vol. 10, no. 1 (March), pp. 42–54.

Johnson, F.X. and Rosillo-Calle, F. (2007) "Biomass, Livelihoods and International Trade: Challenges and Opportunities for the EU and Southern Africa", SEI Climate and Energy Report 2007–01.

Johnson, F.X., Seebaluck, V., Watson, H. and Woods, J. (2008) "Renewable resources for industrial development and export diversification: the case of bioenergy from sugar cane in southern Africa", in *African Development Perspectives Yearbook*, University of Bremen Institute for World Economics and International Management, Lit-Verlag, Muenster.

Junginger, M., van Dam, J., Zarilli, S., Ali Mohamed, F., Marchal, D. and Faaij, A. (2010) "Opportunities and Barriers for International Bioenergy Trade", International Energy Agency Bioenergy, Task 40: Sustainable International Bioenergy Trade, www.bioenergytrade.org.

McNish, T., Jacobson, A., Kammen, D., Gopal, A. and Deshmukh, R. (2009) "Sweet carbon: An analysis of sugar industry carbon market opportunities under the clean development mechanism", *Energy Policy*, vol. 37, no. 12, pp. 5459–5468.

Scurlock, J., Rosenschein, A. and Hall, D.O. (1991) *Fuelling the Future. Power Alcohol in Zimbabwe*, Acts Press, Nairobi and Harare .

Seebaluck, V., Mohee, R., Sobhanbabu, P.R.K., Rosillo-Calle, F., Leal, M.R.L.V. and Johnson, F.X. (2008) "Bioenergy for Sustainable Development and Global Competitiveness: The Case of Sugar Cane in Southern Africa. Thematic Report 2 – Industry", Cane Resources Network for Southern Africa (CARENSA)/Stockholm Environment Institute, www.carensa.net.

Smeets, E.M.W., Faaij, A.P.C., Lewandowski, I.M. and Turkenburg, W.C. (2007) "A bottom-up assessment and review of global bio-energy potentials to 2050", *Progress in Energy and Combustion Science*, vol. 33, no. 1, pp. 56–106.

Woods, J. (2001) "The potential for energy production using sweet sorghum in southern Africa", *Energy for Sustainable Development*, vol. 5, no. 1 (March), pp. 31–38.

World Bank (2010) "Africa Development Indicators", http://data.worldbank.org/country.

Yamba, F.D. and Matsika, E. (2003) "Assessment of the Potential of State-of-the-art Biomass Technologies in Contributing to a Sustainable SADC Regional Mitigation Energy Scenario" Paper presented at the Risø Energy Conference held in Copenhagen, Denmark, May.

Yamba, F.D. and Matsika, E. (2004) "Factors and Barriers Influencing the Transfer and Diffusion of Biofuels Producing Based Technologies with Particular Reference to Southern Africa", Paper presented at the IPCC Expert Meeting on Industrial Technology Development, Transfer and Diffusion, Tokyo, 21–23 September.

Yamba, F.D., Brown, G., Johnson, F.X., Jolly, L. and Woods, J. (2008) "Markets, Technologies, Investment, Economics, Implementation Strategies, Barriers and Policy Issues. Thematic Report 3", Cane Resources Network for Southern Africa (CARENSA)/Stockholm Environment Institute, www.carensa.net.

10

INTERNATIONAL EXPERIENCES

Sugar cane bioenergy markets and trade

Frank Rosillo-Calle, Arnaldo Walter and
Antonio J. Gutiérrez-Trashorras

Introduction

Sugar cane has been a major world crop for centuries, primarily as a source of sugar and other derivatives, but nowadays its importance has increased significantly because of its potential contribution to energy supply. This chapter focuses exclusively on the potential of sugar cane for commercial energy production and the international ethanol trade, putting emphasis on the implications for expanding bioenergy from sugar cane in Africa. The chapter considers the global potential of ethanol and the contribution from cane resources, the environmental benefits from cane energy such as GHG mitigation and favourable energy balance, the current situation and future scenarios for ethanol production and use, the status and potential of sugar cane bagasse cogeneration for electricity export, and the role of the international bioethanol fuel trade in the consolidation of the biofuel market, particularly in respect to prospects in Africa.

Sugar cane as a source of energy

Sugar cane is an important crop globally and even more so in relation to the current global push for renewable energy. It is well known that with first-generation biofuels (G1) there is a potential direct competition with food crops (e.g. see Rosillo-Calle and Johnson, 2010), of which sugar cane offers the best alternative for various reasons. The factors concerning the suitability of cane as an energy crop are described below.

The overall potential of sugar cane

Despite its enormous worldwide importance (see Chapter 4) sugar cane occupies approximately 25 million hectares, a rather small land area compared with many other major crops; for example, wheat occupies some 230 million ha (in 2008), of which about five million ha are dedicated to ethanol production. For sugar cane, the potential to increase productivity and total production, without causing any serious impact on food crops and with relatively low investment, is quite large. For example, estimates for the future of sugar cane have indicated that the present efficiency of conversion of total primary energy from sugar cane can increase from about 30 per cent (the present value) to about 50 per cent, with a significant increase of ethanol production per tonne of sugar cane (up to 50 per cent more) and of the distillery gross annual revenues (Leal, 2009), although this may not be the case in all situations.

In Brazil, sugar cane occupies about 9 million ha, of which less than five million ha is used for ethanol production. In the 2009/2010 harvesting season between 40 per cent and 45 per cent of sugar cane was allocated to sugar production and between 55 per cent and 60 per cent to ethanol. Brazil has the world's most advanced ethanol-production industrial sector based on sugar cane. The main producing area is the State of São Paulo, which represents approximately 60 per cent of Brazilian sugar cane production.

Of the cane-producing countries, Brazil has been the most successful with sugar cane as an energy source, but there is also significant potential in many other countries – for example, Thailand, Pakistan, Mexico and Colombia, as well as African countries that have long practised sugar cane production, such as Mozambique, Kenya, South Africa, Sudan and Zimbabwe. In these countries ethanol could be initially produced from molasses and subsequently enlarged in dedicated units (see Chapter 6). As discussed further below, the potential of electricity production from sugar cane residues is also significant and can have positive economic impacts through reduced dependence on fossil fuels and the cost advantages of distributed generation.

GHG balance of sugar cane

Recent assessments have demonstrated the positive environmental benefits of sugar cane, as given in Table 10.1. For example, on a life-cycle basis it is estimated that sugar cane reduces GHG by 71 per cent compared to the gasoline life-cycle, in the case of ethanol production in Brazil and consumption in Europe (see Chapter 13). While for many other crops uncertainties remain as to the overall GHG benefit, this is not the case with sugar cane.

Energy balance of sugar cane

Unlike most feedstocks, sugar cane has a very positive energy balance with a ratio between 8 and 10 (Macedo *et al.*, 2004, 2008). That is, for each unit input of energy, around eight to ten times more energy is generated; this is the highest positive energy balance of any major energy crop today (see Chapter 13). Table 10.2 gives a comparative analysis of the energy balance for maize and sugar cane, the two main energy crops.

TABLE 10.1 Ethanol and biodiesel GHG emission reduction for selected paths[a]

Biofuel and path	GHG emission reduction (%)
Sugar beet ethanol	52
Wheat ethanol (process not specified)	16
Wheat ethanol (natural gas in CHP plant)	47
Wheat ethanol (straw as fuel in CHP plant)	69
Corn ethanol (natural gas in CHP plant)	49
Sugar cane ethanol	71[b]
Rape seed biodiesel	38
Sunflower biodiesel	51
Soybean biodiesel	31
Palm oil biodiesel (process not specified)	19
Palm oil biodiesel (with methane capture at oil mill)	56
Hydro-treated vegetable oil from rape seed	47
Lignocellulosic ethanol	70–85[c]
Fischer–Tropsch diesel	93–95[c]

Source: EC, 2009.

Notes

a   Renewable Energy Directive default values (land use change emissions not included).

b   Includes transport emissions from Brazil to the EU.

c   Range for different feedstocks.

Versatility and multi-product potential of sugar cane

Sugar cane mills/distilleries can generate all electricity and heat required in the process and also surplus electricity, which can be exported to the grid (see Chapter 5). In addition, many other products can be produced such as yeasts, carbon dioxide, animal feed and fertilizer. In fact, a sugar cane mill is quite similar to a "biorefinery". Although ethanol is mainly used as a fuel, there are many other important applications that should not be overlooked: it is a major feedstock in the chemical industry in India and it is also widely used around the world in the food (beverage), pharmaceutical and cosmetics industries (see Chapter 7). As fuel, it is used neat (i.e. by itself) or blended in different proportions as a substitute for gasoline in internal combustion engines and even mixed with diesel, be it in small percentages, or in the production of ETBE, a gasoline additive.

Current situation and future scenarios for ethanol production

Table 10.3 summarizes the main ethanol producers from 2006 to 2011. Global Renewable Fuels Alliance (GRFA, 2011) forecasts ethanol production in 2011 at approximately 88.7 billion litres, equivalent to one million barrels per day of crude oil, up 3 per cent from 2010, of which roughly 50 per cent comes from sugar cane.

TABLE 10.2 Energy balance of ethanol production from maize and sugar cane

Process	Maize[a] (GJ/ha/yr)	Cane[b] (GJ/ha/yr)
Energy consumption in agriculture	18.9	13.9
Biomass energy	149.5[c]	297.1[d]
Energy ratio in agriculture[e]	7.9	21.3
External energy consumption in distillery[f]	47.9	3.4
Ethanol energy content	67.1[g]	132.5[h]
Total energy ratio	1.21	8.32

Notes
a Source: Shapouri *et al.*, 2002.
b Source: Macedo *et al.*, 2004.
c Corn stover not included. See Shapouri *et al.* (2002) for further details.
d Tops and leaves not included.
e Energy ratio calculated as row 2 over row 1.
f External energy input coming from outside the system production boundaries.
g Does not include credit for co-products.
h Includes credit for 8 per cent bagasse surplus.

In Brazil, Colombia and India, the basic feedstock for ethanol production is sugar cane; China and Thailand use various feedstocks like maize and cassava, and the breakdown varies significantly from year to year. Feedstocks used in other countries/regions such as the EU also include cereals and sugar beet, though in smaller scale. Globally, about 60 per cent of ethanol production comes from sugar crops, 30 per cent from grains, 7 per cent consisting of synthetic ethanol, and 3 per cent from other raw materials. For decades, Brazil has been the world's largest producer and consumer of fuel ethanol, but it was surpassed by the USA in 2006. Together, these two countries produced 88 per cent of the world's fuel bioethanol. Although this chapter deals primarily with sugar cane, it is important to appreciate the overall world trends.

Worldwide, most governmental energy policies have focused on liquid biofuels due to the dangerous overdependence on oil in the transport sector. Although

TABLE 10.3 World ethanol fuel production, 2006 to 2011 (million litres)

Year	2006	2007	2008	2009	2010	2011[a]
Europe	1,627	1,882	2,814	3,683	4,615	5,467
Africa	0	49	72	108	165	170
N + S America	35,625	45,467	60,393	66,368	77,800	79,005
Asia/Pacific	1,940	2,142	2,743	2,888	3,183	4,077
World	*39,192*	*49,540*	*66,022*	*73,047*	*85,763*	*88,719*

Source: GRFA, 2011.

Note
a Estimated.

currently biofuels production and consumption is dominated by a handful of countries, this could change significantly by 2015 when biofuels could contribute to between 5 per cent and 10 per cent of transportation fuels on a global scale, if current policies are successfully implemented. However, large geographical differences will remain and so there is some degree of uncertainty (REN21, 2010).

Global estimates are not quantified in any great detail, and studies have focused on technology rather than economics or social considerations. Competition for other biomass uses like electricity and heat, which is a more effective use of biomass from an energy and economic standpoint, will also impact on the development of biofuels. There are many scenarios on the potential contribution of biofuels to transportation, but with considerable discrepancies. Sugar cane producers around the world, for different reasons, have failed to take full advantage of this crop as a source of ethanol and other energy applications.

Sugar cane and G2 biofuels

The continuity of sugar cane as the main source of ethanol will depend largely on the development of second-generation biofuels (G2). First-generation ethanol (G1) is made from sugars and starchy materials using the conventional technology and these feedstocks can compete directly with food production. G2 ethanol made from non-food crops such as cellulosic-based feedstocks does not compete directly with food crops, though purpose-grown energy crops rely on land, water and other inputs and in some cases may compete with food for land and water.

The IEA (2008) estimated energy demand from biomass in the transport sector in 2005 at 1.7 exajoules (EJ) per year (40 million tonnes of oil equivalent (toe)); global transportation energy demand in 2005 was about 2,140 million toe. It is projected to exceed 4,700 million toe in 2050 in the baseline scenario, an increase of 120 per cent; in the baseline, biofuels stay below 100 million toe in 2050, mostly G1 biofuels. In the study by IEA (2008), more optimistic scenarios have higher estimates for the share of biofuels. One scenario targets a reduction of 2050 emissions to 2005 level, and in this scenario biofuel use is estimated at 570 million toe in 2050 (17 per cent of total global transportation fuel). In this scenario, G2 biofuels dominate by 2050, with roughly equal shares of ethanol and biodiesel. Sugar cane is the only G1 crop that provides a significant production of ethanol after 2030. The second scenario relies more on the production of biofuels from gasified biomass (biomass to liquids – BtL) to replace petroleum diesel and estimates biofuels at 700 million toe, or 26 per cent of global transportation fuel in 2050 (2,656 million toe). G2 biofuels are estimated at 164 million toe (6.9 EJ) in 2030, and 612 million toe (25.7 EJ) in 2050 (IEA, 2008; Bradley et al., 2009). There are many other global predictions but with considerable discrepancies, partly because of the uncertainty surrounding the development of G2 and G3 technologies. Rapid changes could lead to large-scale use of ethanol fuel and also to the emergence of other alternatives (highly fuel-efficient vehicles, multi-fuel engines, hybrid vehicles, electric vehicles, hydrogen as fuel, etc.).

Current major producers and consumers of fuel ethanol from sugar cane

As mentioned, sugar cane is the most efficient and promising source of ethanol as far as G1 is concerned, dominated by Brazil which represents 90–95 per cent of the production; this is expected to be the case in the near-term future given its huge potential for further expansion. But there are also many other countries with a considerable potential, albeit on a much smaller scale. FAO projections indicate a rapid increase for sugar cane in Brazil, China and India, although there are other countries where production is also expected to pick up quickly such as Colombia, Thailand and some countries in Southern Africa and Central America. The following gives a brief description of the most significant countries likely to influence ethanol production and trade.

Brazil

In 2009/2010, 26.1 billion litres of ethanol was produced (–3.8 per cent compared to the previous year), which consisted of 7.0 billion litres (–26.8 per cent compared to 2008/2009) anhydrous and 19.1 billion litres (+8.7 per cent compared to 2008/2009) hydrous ethanol. The same year, 530 million tonnes of cane was harvested and sugar production accounted for 28.4 million tonnes (Agricultura, 2011). Around 90 per cent of all new light vehicles are FFVs and this has kept the domestic demand for ethanol very high. Due to the success of flex-fuel vehicles it is predicted that the domestic consumption of fuel ethanol could be about 40 billion litres in 2020 and close to 60 billion litres in 2030. On the other hand, the exports of ethanol that surpassed five billion litres in 2008 were deeply reduced and were below two billion litres in 2010, mostly because of overproduction in the USA, due to trade barriers and also because of unfair trade practices. Biodiesel production should surpass two billion litres in 2010 (an estimated fivefold increase over 2007); from January 2010, a 5 per cent blend is compulsory (UNICA, 2010; Datagro, 2011).

India

India is currently the world's second-largest producer of sugar cane, but unlike Brazil, sugar cane is mainly used for sugar production to satisfy its huge domestic market. However, India already has plans for expanded ethanol production. The government of India approved the National Policy on Biofuels in 2009, which proposes a target of 20 per cent blending of biodiesel and bioethanol by 2017. India's biofuel strategy continues to focus on the use of non-food resources; namely sugar molasses for production of ethanol and non-edible oils for the production of biodiesel. The government's current target of 5 per cent blending of ethanol in petrol has been successful in years of surplus sugar production, but unfilled when sugar production declines. Consequently, the government is currently unable to implement compulsory blending of 5 per cent ethanol in gasoline due to the short supply of sugar molasses in 2009/2010 and in 2008/2009 because of overall low sugar cane crop production in India (USDA, 2010). However, with a bumper sugar cane and sugar production outlook for 2010/2011, the government is likely to renew its focus and implement the mandatory 5 per cent ethanol blending in petrol. Industry sources report that the government is likely to take a decision on the purchase price of ethanol for the Ethanol Blending Program (EBP).

Currently, India produces conventional bioethanol from sugar molasses and production of advanced bioethanol is still at the research and development phase. India has 330 distilleries that produce four billion litres of rectified spirit (alcohol) per year. Of these, about 115 have the capacity to distil 1.8 billion litres of conventional ethanol per year; sufficient to meet the 5 per cent blending mandate (USDA, 2010). Table 10.4 summarizes ethanol production and uses from 2006 to 2009 and estimates for 2010 and 2011 (USDA, 2010).

China

China is the world's third-largest producer of sugar cane and ethanol. Table 10.5 shows fuel ethanol production from 2002 to 2011, which has steadily increased over this period.

The current use of biofuels in China is part of a strategy to decrease oil imports, foster agricultural and social development, and promote environmental sustainability.

Ethanol has been produced since 2001 from various feedstocks (maize, cane and wheat) and sometimes it is difficult to disaggregate these figures. Currently the situation is quite delicate since a combination of poor harvests and food price increases is causing concern and this will negatively affect ethanol expansion in the country.

TABLE 10.4 Conventional bioethanol production and consumption in India (thousands of litres), including estimates for 2010/2011

Year	2006	2007	2008	2009	2010	2011
Production	1,898	2,398	2,150	1,073	1,435	1,859
Imports	29	15	70	280	300	300
Total (*)	1,927	2,418	2,220	1,358	1,735	2,159
Consumption						
Industrial use	619	650	700	700	720	750
Potable liquor	745	800	850	880	950	1,010
Blended petrol	200	200	280	100	50	200
Other uses	75	100	110	100	110	110
Total consumption	1,639	1,750	1,940	1,780	1,830	1,970

Source: USDA, 2010 (summary of Table 1).

Note
(*) The total excludes opening stock, hence differences.

TABLE 10.5 Fuel ethanol production in China, 2002 to 2010

Year	Production (million litres)
2002	288
2003	799
2004	999
2005	1,200
2006	1,684
2007	1,480
2008	2,000
2009	2,050
2010	2,050

Source: Brown, 2011 (2002–2006) and RFA, 2011 (2007–2010).

Colombia

In March 2008 the Colombian government agreed on a policy framework for bio-fuels (biodiesel and ethanol). The first biodiesel blends commenced that year and covered the mandatory 5 per cent mix in 2009, and the minimum blend require-ment of biodiesel could reach 10 per cent by the end of 2010 (Rosillo-Calle *et al.*, 2009). In early 2011 ethanol installed capacity was about 1.3 million litres per day and one million L/day production, replacing about 8 per cent of gasoline. It is estimated that by early 2013 the installed capacity could reach 2.12 million litres and 1.9 million L/day production, sufficient to blend 15 per cent ethanol with gasoline (FNBC, 2011). Colombia is increasingly being recognized as a major pro-ducer of biofuels.

Thailand

In Thailand ethanol production is approximately 300 million litres annually, both from sugar cane and cassava. However, sugar cane is expanding and one of the reasons is the potential it offers for ethanol production.

The government has had a pro-biofuels policy since 2004, and energy policy makers plan aggressive promotional campaigns to double ethanol use by 2011 and increase gasohol consumption by drivers. The government had set a goal to increase ethanol content in motor fuels to two million litres per day in 2010 from an average of 1.5 million the previous year, and to raise this amount to three million in 2011 (Praiwan, 2010). In 2009, ethanol production was about 400 million litres, up 19 per cent from previous years due to new plants, primarily tapioca-based plants. However, ethanol production remained lower than the government's target of 3.0 million litres per day. In 2010 and 2011, ethanol production should continue to grow sharply by 20–22 per cent, in line with domestic consumption driven by the government's price-incentive policy (USDAFAS, 2010).

Mauritius

Historically sugar cane has played a key role in the economy of this island since the nineteenth century. Other activities, such as services and tourism, have been gradually replacing sugar cane as the backbone of the economy, but sugar cane still represents about 90 per cent of cultivated land. During the 1980s, sugar cane was responsible for 60 per cent of GDP, while today's sugar cane products represent 25 per cent of exports. Ethanol has not been produced in large quantities in the past but there have been rapid developments in recent years.

A policy of diversification introduced to avoid fluctuations and overdependence on sugar cane has impacted negatively on this sector. Nevertheless, sugar cane continues to play a major role in the economy in several ways: surplus electricity generation from bagasse provides about 16–19 per cent of the island's power requirements; there has been a shift from raw sugar production to higher refined-sugar quality; and ethanol production and use is projected to expand from 2012 onwards.

Southern Africa

Southern African sugar cane-producing countries (Congo, Malawi, Mozambique, Tanzania, Zambia and Zimbabwe) constitute a region that has been identified as having high potential for expansion of sugar cane and other energy crops (see Chapter 4). As a region characterized by severe poverty, the possibility to exploit a renewable energy resource offers valuable avenues for sustainable development and could support a more dynamic and competitive economy (Seebaluck et al., 2007). However, this region is far from taking full advantage of this potential.

Southern Africa can benefit from the experiences of Mauritius, India and Brazil with respect to electricity and ethanol production from sugar cane. Implementation strategies can be adopted using South-South technology transfer platforms. Bioethanol can replace significant shares of petrol, and export markets could support economic growth and global biofuel market development. The amount of land suitable and available is significant; production of cane based on recent yields would amount to over 60 times the current level (Seebaluck et al., 2007).

Others

Almost all sugar cane-producing countries are able to produce ethanol, although with considerable variation in potential due to land area, suitability of sugar cane, climatic conditions and stage of development of the sugar cane industry. The most important ones not previously discussed are Australia, Mexico, Pakistan and the USA, all of which already produce ethanol, but not necessarily for fuel purposes. All Central American and Caribbean countries produce sugar cane and have significant potential for further expansion, including in ethanol. However, except in the cases of Cuba, Guatemala and the Dominican Republic, total production and productivity are small and this limits expansion and large-scale use considerably unless external markets can be found. Nonetheless, sugar cane is expanding for three main reasons, namely:

- in most Caribbean countries, there are new opportunities to export ethanol to the USA, by taking advantage of trade agreements that allow exports without taxes;
- the need to reduce energy dependence (e.g. use of ethanol and bagasse cogenerated electricity);
- the need to modernize the sugar cane industry.

Cogeneration potential from sugar cane bagasse

The potential for cogeneration from sugar cane biomass has long been recognized and many studies have tried to throw some light into this vast and highly under-utilized potential (see Chapter 5). Table 10.6 summarizes an estimate of the world potential of sugar cane bagasse, which amounts to 424.2 million tonnes (equivalent to 90.14 million tonnes or 662.4 million barrels of crude oil) with a market value of US$34.4 billon (at US$52 per billion) (ISO, 2009). Based on this estimate, the continent with the largest

TABLE 10.6 Global potential of bagasse availability in 2007

Continent and main country	Mt (50% moisture)
Asia	198.2
• China	40.3
• India	94.8
• Pakistan	14.2
• Indonesia	9.2
• Thailand	23.3
Africa	28.0
• Egypt	3.8
• Kenya	1.7
• Mauritius	1.5
• South Africa	7.5
• Sudan	2.4
• Swaziland	2.0
America (South)	132.2
• Argentina	7.2
• Brazil	108.2
• Colombia	7.4
• Peru	3.0
America (Centre)	20.8
• Cuba	3.9
• Guatemala	7.7
America (North)	28.2
• Mexico	17.7
• USA	10.5
Oceania	16.0
• Australia	15.0
World total	424.2

Source: ISO, 2009.

potential is Asia, followed by South America. Africa, on the other hand, has a potential of just 28 million tonnes, but this estimate is based on current production, which is low in Africa, but will likely expand considerably in the future.

The ISO study (2009) has estimated the current bagasse-based cogeneration capacity in 13 of the most important producing countries at 6,483 megawatts, with potential to reach 26,751 MW by 2015 (Table 10.7).

Thus, it is surprising that given the considerable potential of cogeneration from sugar cane residues, capacity is still relatively small. There are various reasons, including lack of incentives, policies, regulatory framework, prices, barriers from conventional energy suppliers, lack of capital, technology, etcetera. This may explain why African countries, excluding Mauritius, are far behind in comparison to other countries. The following gives a brief description of the countries with the largest installed cogeneration capacity.

Brazil

It is estimated that sugar cane mills account for the generation of slightly more than 14,000 gigawatt hours, or 3.0 per cent of Brazil's total electricity production in 2009 (MME, 2010), and according to the Brazilian Electricity Agency (ANEEL) the installed capacity of electricity production from sugar cane bagasse by mid-2010 was 5.9 GW in 311 industrial mills, mostly in the State of São Paulo, the country's largest sugar cane producer. The installed capacity could reach 9–10 GW if all existing mills were to adopt the best practices in the sector, using only sugar cane bagasse and without major efforts for reducing steam demand in the production process. By recovering 40 per cent of the trash available in the field and reducing the steam demand, the potential would be higher. As sugar cane production could reach 730

TABLE 10.7 Bagasse-based cogeneration, current status and potentials in selected (six major) countries[1]

Country	No. of mills (2007/2008)	Mills connected to the grid	Cogeneration capacity (MW)	Plans for cogeneration expansion
Brazil	370	48	3,081	15 GW (by 2015)
Guatemala	14	9	497[a]	585 MW (by 2011)
India	492	107[b]	1,400	3 GW[a]; 10.5 GW (by 2015)
Mauritius	10	10	240[c]	n/a
Australia	27	n/a	392	n/a
South Africa	17	–	–	400 MW (by 2013)

Source: ISO, 2009.

Notes
a Including about 200 MW from bunker-based generation.
b There are 38 mills in the process of establishing cogeneration facilities.
c Including about 140 MW from coal-based generation.

million tonnes by 2012 and even surpass 1,000 million tonnes in 2017, the potential of electricity production from sugar cane bagasse could be two to three times higher in the mid-term (Walter, 2010). Despite considerable efforts by the Brazilian government (both federal and state) to support cogeneration of sugar cane residues (tops and leaves), there are still some bottlenecks that need to be addressed; for example, the high cost of interconnections and improvements on regulations.

Mauritius

During the 1990s, Mauritius, an island with no domestic fossil fuels and limited resources for hydroelectric power generation, took a proactive stand on bagasse-based cogeneration to address the future challenges in the sugar industry as well as in the energy sector. As a result exports of bagasse-based electricity to the national grid had nearly tripled between 1996 and 2005 and increased further in 2005 and 2007. In 2007 there were ten sugar mills generating electricity, with crushing capacities ranging from 100 to 350 tonnes of cane per hour. Three of the sugar mills' power plants operate throughout the year. Harvesting lasts for about six months, during which the cogeneration plants export bagasse-based electricity to the grid. However, to operate all year round, a complementary fuel (coal in this case) has to be used during the intra-harvest periods. In addition, cogeneration of the total sugar cane biomass (including tops and leaves) can further increase electricity generation potential by 350 GWh per year based on a 50 per cent collection of the cane trash from the fields. The government has already authorized further expansions by the sugar independent power producers (IPPs). According to the Multi-Annual Adaptation Strategy (2006), bagasse-based generation is targeted to grow by 300 GWh by 2015; a key feature of the strategy is the commissioning of five 42 MW/82 bar units (ISO, 2009).

India

Until the 1970s, sugar mills were using only low-pressure boilers. With the growing demand for electrical power and widening gap in the demand and supply of power, sugar mills in the mid-1990s incorporated high-pressure boilers and more efficient turbines to generate additional power for supply to the grid. Sugar factories started using 45 bar boilers, while the cogeneration units are currently using 67–105 bar boilers (ISO, 2009). Since the beginning of the 1990s, bagasse cogeneration has become a major activity of India's sugar industry. About one-third of operating sugar mills (145 out of a total of 492 in 2007) have installed, or are in the process of establishing, cogeneration facilities for surplus electricity production. The current capacity to cogenerate and export surplus power to the national grid in 107 mills is estimated by the industry at around 2,200 megawatts but will increase to 3,000 MW when new cogeneration units are assembled in about 40 mills. The full potential of the sector is projected at over 10,500 MW, including 3,500 MW for internal consumption and 7,000 MW for export to the national grid (ISO, 2009). However, Purohit and Michaelowa (2007) estimate the electricity cogeneration potential from bagasse in India at 5,575 MW (34 terawatt hours) in terms of plant capacity; only 20–30 per cent of all bagasse is used for these purposes and the remainder is wasted or incinerated for disposal. Therefore if all bagasse was to be used for energy and in highly energy-efficient boilers, this potential could reach about 11 GW.

Guatemala

In Guatemala, bagasse-based cogeneration has been a major activity of the sugar cane processing sector for nearly two decades. Unlike Brazil, the sector is supplying energy to the national grid all year around. The cogeneration programme started in Guatemala in 1994, when the biggest private electricity distributor (Empresa Eléctrica de Guatemala – EEGSA) and six sugar mills started surplus electricity production, with a total amount of 160.4 MW during the crushing season and 128.7 MW available during the intra-crop season.

To guarantee a firm supply of electricity operating the full year round, sugar mills have to operate the same steam boilers with two different types of fuel, sugar cane bagasse and bunker fuel. Most sugar mills use high-pressure flex-fuel steam boilers. In 2007 the sector had the capacity to generate 296.8 MW of electricity during the harvest season and 200.3 MW during the intra-crop period (ISO, 2009).

Until 1990 the utilities industry was strongly regulated and no sugar mills or IPPs from other sectors were allowed to export energy to the grid. According to the industry, apart from allowing sugar mills like IPPs to supply electricity to the national grid, no incentives have been provided by the government for bagasse-based cogeneration. The base price is established by the Administrador del Mercado Mayorista (AMM) and takes into account minimum offered prices from all the power-generating units and the required demand. Prices vary depending on hours when electricity is delivered and are higher during peak hours and lower during the rest of the day (ISO, 2009).

Other countries

Australia

In Queensland (the main Australian producer) in 2008, the sugar industry generated approximately 850 GWh, of which 370 GWh was exported. The total installed electricity-generating capacity of the industry was 392 MW (2009), with the largest facility in the industry located at CSR's Pioneer Mill in the Burdekin, rated at a capacity of 66 MW. In New South Wales there is an additional 80 MW of installed capacity (ISO, 2009).

Colombia

In 2008 the government passed Law 1215, which regulates the cogeneration of electricity where the CHP (combined heat and power generation) is an integral part of their production cycles. The law allows the IPPs (including sugar mills) to sell excess electricity to distributing companies. The proceeds from the sales to the national grid are subjected to a 20 per cent tax. Regulations are still being enacted that will determine technical requirements to be met by IPPs.

Currently, the sugar industry is preparing a number of cogeneration projects to be implemented in the near future with a total amount of investment at US$325 million. If current plans are implemented the total installed capacity will reach 266 MW, of which 155 MW will be in excess of the internal processing needs of the sugar mills (ISO, 2009).

Kenya

Although the country has still a small installed capacity, ethanol cogeneration is not new there. Ethanol has been produced for decades on a small scale. Kenya has a number of cogeneration projects at various stages ranging from feasibility assessments to commissioning. A cogeneration unit with the capacity to export 25 MW out of a total capacity of 32.5 MW was due for commissioning in Mumias mill in the first half of 2009. The project included the installation of a high-pressure (87 bar, 525 °C) steam boiler. Feasibility studies have been conducted at Nzoia, South Nyanza and Chemelil mills (ISO, 2009).

Thailand

At the beginning of the decade the Mitr Phol Sugar group started developing modern bagasse cogeneration in the plants with a 67 bar (510 °C) steam cycle. Two identical 41 MW extraction condensing steam turbo generators were installed in UFIC and Dan Chang mills. The power plants started their operation in 2004, with the capacity of each power plant increasing to 52.4 MW. Currently each plant is able to export on a firm basis throughout the year 37–39 MW to the Electricity Generating Authority of Thailand (EGAT) based on a long-term contract. Both plants are planning to use cane trash as supplementary fuel to compensate for any shortfalls in sugar cane output. The total investment cost of the two projects was €71 million excluding civil works, building foundations and financing costs; the expected payback period is about five years (ISO, 2009).

Uganda

Uganda provides a good example of successful cogeneration in the sugar industry of Southern Africa. Currently, in Kakira Sugar Works (KSW), the country's biggest sugar producer, out of the total installed capacity of 20 MW, 13 to 14 MW is exported to the national grid on a continuous basis except for 36 hours per month when the mill is not in operation due to regular maintenance. The plant was commissioned in December 2007. Kinyara Sugar Works, another sugar producer in Uganda, has implemented a smaller cogeneration project with an excess capacity of 5 MW to export electricity to the national grid (ISO, 2009).

South Africa

It is surprising that given the relative importance of sugar cane, no cogeneration facilities are currently producing surplus electricity in the country and this is despite its growing power requirements. Electricity-generating capacity in South Africa is about 40 GW; the estimated demand will be 42 GW by 2013. The White Paper on Renewable Energy sets a target of 10,000 GWh to be achieved by 2013, approximately 4 per cent of the projected electricity demand for that year. The sugar industry has reiterated its readiness to contribute to the bolstering of the country's faltering electricity-generation capacity. According to Tongaat Hulett, a leading sugar producing company in South Africa, the country's 14 sugar mills can contribute a combined total of 400 MW to the national electricity grid by 2013, with some mills already able to go on line in 2010 (Tongaat Hulett Sugar, 2011).

Summary

In summary, the potential for electricity generation from sugar cane bagasse (excluding tops and leaves, which are excluded) is far from being achieved, though existing development plans show that bagasse-based production of electricity for export to the national grid is fast becoming a major activity of sugar cane mills. Barriers to the realization of such potential include government policy with respect to renewable energy and how policy is translated into the regulatory setting for IPPs, lack of adequate fiscal incentives and lack of a clear long-term policy framework. Financial and tax incentives in line with those offered for other generators of renewable energy are of considerable importance, particularly in the initial stages when the necessary equipment has to be purchased and proper infrastructure has to be developed. Also, given the seasonal nature of bagasse and the need for uninterrupted supply throughout the year, it requires boilers capable of co-firing fossil fuels and other feedstocks like trash in the intra-crop periods.

Overall, from the sugar cane processing sector's perspective, there are three main benefits of bagasse-based cogeneration: (1) low or no fuel costs; (2) diversification of revenue streams; and (3) possibility of the application of the CDM of the Kyoto Protocol resulting in a monetary value for reduction in carbon dioxide emission. However, this is not necessarily the case for all countries; for example, in the case of Brazil this extra revenue almost makes little difference as the baseline of the electricity sector already has low greenhouse gas emissions (ISO, 2009).

Bagasse-based cogeneration also allows diversification of energy supply and provides additional income from export to the grid, particularly during the dry season, and also brings environmental benefits. In technical terms, the amount of energy that can be extracted from bagasse is largely dependent on the technology used for energy production. The use of high-efficiency boilers, generating extra-high pressures and temperatures (90–100 bar and 500–520 °C, respectively), allows production of more electricity in excess of a mill's captive consumption. The cost of boilers and their installation is relatively high, but, as shown by projects in Brazil and Thailand, capital investment costs may be covered by revenues from electricity exports to the national grid in a matter of three to five years (ISO, 2009). However, conditions vary from country to country and each specific circumstance needs to be taken into consideration.

The international market for ethanol fuel

The growing interest and international bioenergy trade prompted the International Energy Agency (IEA) Bioenergy Agreement to set up Task 40 in 2004 – "Sustainable International Bioenergy Trade: Securing supply and demand".[2] The areas of greatest interest and growth have been primarily in sustainable international bioenergy trade, development of bioenergy markets, optimization of supply chains, assurance and certification, formulation of policy and involvement of industry and traders. For example, woodchips, pellets, bioethanol, biodiesel and vegetable oils are all traded at national, regional and global levels. Table 10.8 summarizes the main international bioenergy trade activities in 2008.

Bioethanol, biodiesel and wood pellets are thus the three largest feedstocks being traded internationally. It is expected that international trade in this area will increase significantly in the next few years. Data related to fuel bioethanol trade is imprecise on account of the diverse end-uses of ethanol (i.e. fuel, chemical feedstock, pharmaceutical, beverages) and also because of the lack of proper codes for biofuels in the international trade statistics. In 2004, total trade of ethanol was estimated to be 2.4–3.2 million tonnes, with Brazil (1.9 million tonnes) as the main exporter, and the USA, Japan and the EU as the main importers. Estimates for ethanol trade (all

TABLE 10.8 Overview of global production and bioenergy trade of major commodities in 2008 (million tonnes; 1 tonne = 1,000 kg)

	Bioethanol[1]	*Biodiesel*[2]	*Wood pellets*[3]
Global production	52.90	10.60	11.50
Global net trade in 2008	3.72	2.90	4.0
Main exporters	Brazil (also China – small scale)	USA, Argentina, Indonesia and Malaysia	Canada, USA, Baltic countries, Finland and Russia
Main importers	USA, EU, Canada and Japan	EU, Japan	Belgium, Netherlands, Sweden and Italy

Source: Junginger *et al.*, 2010.

Notes
1 Bioethanol: 1 tonne = 1,266 litres; about 75% of bioethanol is traded as fuels, the rest for other industrial applications.
2 Biodiesel: 1 tonne is about 0.78 toe. Biodiesel traded is almost exclusively for use as fuel.
3 Traded wood pellets are exclusively for generating electricity and heat.

grades) provided by F.O. Licht indicate that trade grew steadily from about 2.4 million tonnes in 2000 to 4.7 million tonnes in 2005 (i.e. about 13 per cent of world production). Assuming that the increase in recent years has been mostly due to trade in fuel ethanol, it is reasonable to estimate that in 2005 trade covered about 10 per cent of fuel ethanol consumption (Rosillo-Calle and Walter, 2006). Such trends seem to have continued in recent years.

International trade in biofuels, however, fluctuates considerably, almost on a monthly basis. Up-to-date information on ethanol fuel trade can be obtained for Brazil and the USA (Agricultura, 2011; RFA, 2011). The fuel ethanol demand in the USA had risen from 7,892 million litres in 2002 to 36,476 million litres in 2008, with most of the production met locally but accompanied by a gradual increase in imports to satisfy the increasing demand (RFA, 2010). The main supplier was Brazil with 1.65 billion litres in 2006 and 0.7 billion litres in 2007 (directly exported to the USA).

The countries in the Caribbean Basin also import relatively high quantities of Brazilian ethanol but mostly for reprocessing (i.e. they convert Brazilian hydrated into anhydrous ethanol). They then re-export it to the United States, gaining value added and avoiding the 2.5 per cent duty and the 54 US cents per gallon tariff, thanks to the trade agreements and benefits granted by the Caribbean Basin Initiative (CBI). This process is limited by a quota, set at 7 per cent of USA ethanol consumption.

The USA also exports ethanol, mostly for non-fuel purposes, e.g. 157.8 million gallons (697.2 million litres) in 2008, and 150.2 million gallons (568.5 million litres) in 2007. Since January 2005, the monthly average for total ethanol exports has been 10.2 million gallons (38.6 million litres) (RFA, 2010).

Despite the relative success of the past few years, there remain serious barriers to overcome in expanding international trade on biofuels, including the following.

*   Trade barriers (especially unfair tariffs).
*   The lack of a clear classification for biofuels within the current harmonized system that restricts global trade.
*   Additional difficulties posed by certification and sustainability issues. For example, criteria related to environment and social issues could be too stringent or inappropriate for prevailing conditions in developing countries.
*   The costs of monitoring and certification are still unclear.
*   The lack of sufficient volume, which poses logistical problems.
*   Problems with statistical data (classification is not always clear).
*   The fact that biofuels have not reached maturity and are not yet widely traded as commodities.

Biofuels production in Africa

A controversial issue involves the question of whether a developing country with good potential for production of biofuels should prioritize the domestic market or should focus on exports as well. This section analyses the possible production of biofuels, and more specifically ethanol, from sugar cane, in some African countries. Three main points define the general context.

First, it is important to take into account that Africa has the potential to become an important producer and exporter of raw biomass and biofuels, and that the bulk of the production can occur on abandoned or underutilized and pastureland (Hoogwijk *et al.*, 2009). In some countries the conditions for bioenergy production are favourable due to the suitable climate and the availability of arable land and water resources. Second, most African countries depend heavily upon imports of crude oil and oil derivatives. Thus, the reduction of oil dependence is an important issue for many countries.

Third, most Africans live in rural areas, estimated as more than 60 per cent of the continent's land area, while a similar fraction of the population lives below the poverty line of US$1 per day (FAO, 2008). Mainly in Africa, people living in rural areas have constrained access to energy services and both agricultural productivity and life conditions are deeply impacted by constraints on water and energy supplies (i.e. electricity and motor fuels).

As a consequence of this set of factors, it seems that a logical priority would be the production of biofuels on a small scale, in order to match local demand and improve living conditions. More specifically, the priority should be small-scale projects and targets on local markets, aiming at rural electrification, water pumping and assuring the availability of transport fuels for agriculture. From this point of view, large-scale projects for biofuels exports should be a second-stage priority.

Nonetheless, in the case of ethanol production from sugar cane there are some particularities that should be highlighted. One very important issue is that ethanol from sugar cane is the best option among all first-generation biofuels, and the production in some African countries could be feasible and in accordance with international sustainability criteria. In addition, in countries with some tradition of sugar cane production (Mozambique, South Africa, Sudan, Tanzania and Zambia are among the major producer countries in Africa) the infrastructure and the sugar cane industry itself are more organized. Mozambique is an example of a country that has decided to use its capability and existing know-how for enlarging sugar cane production, aiming at ethanol production on a large scale, including exports. There is significant potential in Mozambique for bioenergy production and trade, due to the combination of excellent climatic conditions and a long coastline with several ports (Batidzirai *et al.*, 2006).

Regarding biofuels production, in most African countries it is more appropriate to focus first on their national markets rather than on exports, starting with blending at the regional level, if possible, as the risks would be reduced. As long as the conditions are improved, large-scale production for blending at national level and also for exporting could be considered.

Also regarding the sugar cane industry, and as previously described, there is significant potential for surplus electricity production from sugar cane residues, and this could positively impact on regions close to the mills and even the whole electricity system. Besides benefits due to the enlargement of electricity supply, these countries could reduce their dependence on imported fossil fuels and improve the efficiency of the electric system, through more distributed electricity generation.

Conclusions

This chapter has highlighted the increasing importance of sugar cane as source of energy, for example in the context of ethanol fuel and cogeneration. It is clear that such potential remains largely untouched. Among the G1 biofuels, sugar cane is recognized as the best option that is commercially mature: it has high productivity (yield per hectare), good potential for expansion, a very positive energy balance (ratio of 8 to 10), good potential for producing surplus electricity and good potential for avoided life-cycle GHG emissions compared to gasoline. Currently, a sugar cane mill is closest to the somewhat futuristic concept of the "biorefinery". The continuity of sugar cane as a major source of energy will depend largely on the development of G2 biofuels. Nevertheless, it is clear that even with G2, sugar cane will have an important role given its many advantages.

Many sugar cane producers around the world have failed to take full advantage of this crop as a source of energy. A clear example is bagasse cogeneration, as illustrated in this chapter. For example, the potential of electricity production using cogeneration systems is conservatively evaluated as being at least four times higher than the current installed capacity. Although this potential is not high enough to assure the expansion of the electricity sector in many countries, it can contribute significantly by facilitating the expansion of electricity production at reasonable costs while also reducing GHG emissions.

The international trade of bioenergy is still in its early stages, and existing trade barriers have constrained the expansion of this market, particularly in the case of ethanol fuel. In addition, bioenergy trade can be determined by different sustainability criteria, mostly imposed by developed countries. In light of this, a strategy based only on exports would be too risky for biofuels producers in developing countries. It is more appropriate to start the production of ethanol on a small to moderate scale, taking advantage of existing experience in blending ethanol with gasoline; the benefits of such a strategy are clear, especially in countries with high oil dependency.

Agricultural development and the expansion of a modern sugar cane industry are intertwined; a modern and dynamic agricultural sector is needed to increase sugar cane productivity and facilitate diversification, and to obtain multiple products, such as electricity and heat, animal feeds and ethanol. In many countries the expansion of the sugar cane agro-industry can serve as an excellent instrument for inducing regional economic development.

Notes

1 The other countries are Colombia, El Salvador, Kenya, Nicaragua, the Philippines, Thailand and Uganda.
2 Current member countries are Austria, Belgium, Brazil, Canada, Germany, Finland, Italy, Japan, the Netherlands, Norway, Sweden, the USA and the UK. In addition, the EU is also a full-paying member, as a non-country organization.

References

Agricultura (2011) Government of Brazil, Ministry of Agriculture, agricultural/statistical data, www.agricultura.gov.br./vegetal/estatisticas.
Batidzirai, B., Faaij, A. and Smeets, E. (2006) "Biomass and bioenergy supply from Mozambique", *Energy for Sustainable Development*, vol. 10, no. 1, pp. 54–81.
Bradley, D., Cuypers, D. and Pelkmans, L. (2009) "2nd Generation Biofuels and Trade – An Exploratory Study", www.bioenergytrade.org.
Brown, L.R. (2011) *The World on the Edge: How to Prevent Environmental and Economic Collapse*, Norton & Co., New York.
Datagro (2011) Datagro statistics and analysis, www.datagro.com.
EC (2009) Directive 2009/28/EC of the European Parliament and of the Council of 23 April 2009 on the Promotion of the Use of Energy from Renewable Sources, and Amending and Subsequently Repealing Directives 2001/77/EC and 2003/30/EC, www.eur-lex.europa.eu/LexUriServ/.
FAO (2008) "The State of Food and Agriculture", FAO, Rome.
FNBC (2011) "Federacion Nacional de Biocombustibles de Colombia", www.fedebio-combustibles.com.
GRFA (2011) "Global Renewable Fuels Alliance", www.globalrfa.org, accessed 11 February 2011.
Goldemberg, J. (2008) "The Brazilian biofuels industry", *Biotechnology for Biofuels*, vol. 1, no. 6, p. 4,096, www.intermonoxfam.org/es/page.asp?id=2143.
Hoogwijk, M., Faaij, A., Eikhout, B., de Vries, B. and Turkenburg, W. (2005) "Potential of biomass energy out to 2100, for four IPCC SRES land use scenarios", *Biomass and Bioenergy*, vol. 29, no. 4, pp. 225–257.
IEA (2006) "World Energy Outlook 2006", Paris.
IEA (2008) "Energy Technology Perspectives 2008 – Scenarios & Strategies to 2050", IEA-OECD, Paris.
IEA Bioenergy (2008) "From 1st- to 2nd-Generation Biofuel Technologies – An overview of Current Industry and RD&D Activities", IEA-OECD, Paris.
ISO (2009) "Cogeneration – Opportunities in the World Sugar Industries", MECAS (09)05 (April).
Junginger, M., van Dam, J., Zarrilli, S., Ali Mohamed, F., Marchal, D. and Faaij, A. (2010) "Opportunities and Barriers for International Bioenergy Trade", International Energy Agency Bioenergy, Task 40; Sustainable International Bioenergy Trade (www.bioenergytrade.org)
Leal, M.R.L.V. (2009) "Use of sugarcane bagasse and straw for ethanol and energy production", Internal Report, NIPE, University of Campinas, UNICAMP, Brazil.
Macedo, I.C., Leal, M.R.L.V. and Silva, J.E.A.R. (2004) "Assessment of Greenhouse Gas Emissions in the Production and Use of Fuel Ethanol in Brazil", Report prepared for the State of São Paulo Secretariat of the Environment, Piracicaba.
Macedo, I.C., Seabra, J.E.A. and Silva, J.E.A.R. (2008) "Green house gases emissions in the

production and use of ethanol from sugarcane in Brazil: The 2005/2006 averages and a prediction for 2020", *Biomass and Bioenergy*, vol. 32, no. 7, pp. 582–595.

MME (2010) Resenha Energética Brasileira, Ministério de Minas e Energia, www.mme. gov.br/mme/menu/todas_publicacoes.html.

NDRC (2008) "National Development and Reform Commission, People's Republic of China Document".

Praiwan, Y. (2010) "Ethanol use likely to double by 2011", *Bangkok Post Business*.

Purohit, P. and Michaelowa, A. (2007) "CDM potential of bagasse cogeneration in India", *Energy Policy*, vol. 35, pp. 4779–4798.

RFA (2011) Renewable Fuels Association statistics, www.ethanolrfa.org/statistics/.

REN21 (2010) "Renewables 2010 Global Status Report", www.ren21.net.

Rosillo-Calle, F., Pelkmans, L. and Walter, A. (2009) "A Global Overview of Vegetable Oil Markets and Trade, with Reference to Biodiesel", www.bioenergytrade.org/.

Rosillo-Calle, F. and Johnson, F.X. (2010) *Food versus Fuel: An Informed Introduction to Biofuels*, Zed Books, London.

Rosillo-Calle, F. and Walter, A. (2006) "Global market for bioethanol: historical trends and future prospects", *Energy for Sustainable Development*, special issue, vol. 10, no. 1 (March), pp. 20–32.

Shapouri, H., Duffield, J.A. and Wang, M. (2002) "The Energy Balance of Corn Ethanol: An Update", Agricultural Economic Report No. 814, US Department of Agriculture.

Seebaluck, V., Mohee, R., Sobhanbabu, P.R.K., Rosillo-Calle, F. and Leal, M.R.L.V. (2007) "Cane Resources Network for Southern Africa (CARENSA), Thematic Report 2 – Industry", European Commission DG-Research FP5 INCO-DEV.

Tongatt Hulett Sugar (2011) Annual Report, Tongatt Hulett Sugar, www.tongatt.co.za.

UNICA (2010) UNICA statistics and analysis, www.unica.com.br.

USDA (2010) "Gain Report", Global Agriculture Information Network, GAIN Report No. 1058, Delhi (US Department of Agriculture).

USDAFAS (2010) "Gain Report" USDA Foreign Agricultural Service, GAIN Report No. TH009, Biofuels Annual (US Department of Agriculture).

Walter, A. (2010) "Potential for electricity production from sugarcane residues", in L.B. Cortez (ed.), *Sugarcane Bioethanol*, Blucher, São Paulo, pp. 577–582.

PART IV

Impacts and sustainability

11

SUSTAINABILITY ASSESSMENT OF ENERGY PRODUCTION FROM SUGAR CANE RESOURCES

Maxwell Mapako, Francesca Farioli and Rocio A. Diaz-Chavez

Introduction

Firewood, dung and charcoal are common sources of household energy in many developing countries, while energy for transport is provided by imported petroleum products except for those few countries that are oil producers. The lack of access to modern energy services is acute in sub-Saharan Africa, where 89 per cent of the population still relies on traditional biomass energy (Gaye 2007; UNDP, 2010). The heavy reliance on traditional biomass results in health and environmental impacts, while in urban areas, imported fuels are often used; the low quality of energy services in rural areas and the reliance on imported fuels in urban areas both contribute to energy insecurity (see Chapter 14).

Modern bioenergy is a valuable domestic energy alternative in most rural areas of developing countries if adequate conditions (e.g. soil, water and land availability) are in place, or can be put in place without significant negative impacts. There are various feedstocks grown in Africa that could potentially be used for energy generation or as alternative fuel.

Sugar cane is an example of the potential for biomass modernization on a larger scale, with applications for energy use in the household, transport and heat and power sectors. The sugar cane plant is one of the most promising agricultural sources of biomass energy in the world. It is a highly efficient converter of solar energy, and has the highest energy-to-volume ratio among energy crops. It is found predominantly in developing countries, due to environmental requirements that restrict its growth to tropical and subtropical climates.

This chapter reviews the main socio-economic and environmental impacts of the use of sugar cane in Africa, focusing on its uses for bioenergy (rather than sugar and other non-energy products) and the implications of these positive and/or negative impacts in terms of their contribution to sustainability. The discussion focuses on Southern Africa, although there are some general aspects covered that are quite relevant elsewhere in sub-Saharan Africa and even more generally for all sugar cane-growing developing countries. Relevant assessments that were undertaken through the Cane Resources Network for Southern Africa (CARENSA) have been summarized where appropriate.

Sustainable development and bioenergy from sugar cane

Although sustainable access to energy was not treated as a priority in itself in the Millennium Development Goals (MDGs), most of the MDGs require energy for their realization. Sustainable human development requires a focus on improving the access of the poor to assets, goods and services, including food, water for drinking and irrigation, health care, sanitation, education and employment. Sub-Saharan African countries rank among the poorest in the world and a sustainable approach is particularly pertinent in this region.

There is no international consensus on the meaning of "sustainable development". The most common definition was developed by the World Commission on Environment and Development (WCED), which described sustainable development as: "*Development that meets the needs of the present without compromising the ability of future generations to meet their own needs*" (WCED, 1987). This general definition does not clarify the kinds of activities that would assist in attaining sustainable development goals. It contains two key concepts:

- the concept of "needs", in particular the essential needs of the world's poor, to whom priority should be given; and
- the idea of limitations imposed by the state of technology and social organization on the environment's ability to meet present and future needs.

While sustainable development is seen as desirable, there are differences in how it is understood, owing largely to differing interpretations of what is to be sustained. Some consider that sustainability applies to the resource base, while others focus on the well-being of people and their livelihoods deriving from the resource base. The fact that there are different types of capital stock that contribute to well-being has also led to a distinction between weak sustainability and strong sustainability (Neumayer, 2003).

Sustainability in the energy sector is mainly about energy sources or systems that have minimal environmental impacts (e.g. renewable energy and energy efficiency) and to what extent they encompass social dimensions of access to safe, affordable, modern energy services, using fewer resources. The energy consumption of developing countries will rise as they develop, but this must ideally become more efficient in terms of fuel type and energy utilization.

Conventional energy strategies that rely on supply-focused, fossil-intensive, large-scale approaches do not generally address the needs of the poor. Most African countries are heavily dependent on traditional uses of biomass to meet their energy requirements and have a small and inefficient modern energy sector.

Most of the impacts of sugar cane production are on the agricultural side, and would be incurred even when there is no surplus bioenergy production – though some bioenergy will always be used internally. Thus the energy production is essentially a bonus and from almost any perspective it will improve sustainability compared to production of sugar alone.

Around 100 countries in the world are sugar cane producers. The production and processing of sugar cane, as is the case with any other agricultural and industrial activity, has negative and positive impacts on the air, water, soil, flora, fauna, human population and global climate. These impacts vary from country to country since they are dependent on factors including local harvesting practices, environmental regulations, use of irrigation water, production model (small versus large producers) and level of mechanization (Woods *et al.*, 2008).

The Better Sugar Meeting Report (WWF, 2005) identified the key social and environmental impacts of sugar cane production and discussed how best to address these impacts through a collaborative approach. Nowadays most sugar cane producer countries have local standards for emissions and apply the ISO 14000 standards. Nevertheless, the World Wide Fund for Nature (WWF) and the International Finance Corporation (IFC) meeting on better sugar considered the development of better management practices, a forum to understand and reduce environmental and social impacts.

Socio-economic impacts from sugar cane production and use

Table 11.1 summarizes the main socio-economic issues relating to biofuels. The specific issues in any given location will depend on various factors, particularly the scale, production model and type of feedstock. It is clear that the agricultural side causes the most sweeping impacts in terms of number of persons affected, while the impacts on the industrial side can be significant in terms of their effect on infrastructure and quality of labour force.

Impact of ownership and control of land/resources

Poverty is endemic in sub-Saharan Africa, with human development indices for most of the countries stagnating or declining since 1990, making this region the poorest in the world. Since 1990, income poverty has fallen in all regions of the world except SSA (Handley *et al.*, 2009). It is therefore important to explore production models that provide local benefits to the communities. The rapid expansion in biofuels markets has led to greater foreign investor interest in Africa, while at the same time the different ownership models have been evolving, as discussed below.

Investor interest in biofuels

There has been considerable hype and investor interest in growing biofuels in Africa, especially from European companies interested in selling to the EU market,

TABLE 11.1 Overview of socio-economic impacts

Socio-economic impacts	Scale N, R, L	Biomass production (farm)	Biomass pre-treatment and conversion (factory)
Health and safety	L	E.g, pesticide application, use of harvesting machinery	Machinery risk, fire safety, contamination and hazardous substances
Freedom of association and collective bargaining	N, R, L	–	–
Working hours and remuneration/benefits	N, R, L	M, contract and law enforcement, CSR	M, contract and law enforcement, CSR
Migrant labour	R/L	M, contract and law enforcement, CSR	M, contract and law enforcement, CSR
Child/forced labour	L	M, contract and law enforcement, CSR	M, contract and law enforcement, CSR
Land ownership/access to land	L	M, law enforcement, CSR	M, law enforcement, CSR
Land use change	R, L	M, land use regulations, map zoning	M, land use regulations, map zoning
Food security – quantity and price★	N, R	M, land use regulations, map zoning	M, land use regulations, map zoning
Access to water resources	L	M, law enforcement, water governance	Law enforcement, water governance
Land/water contamination and associated health implications	N, L	N, law enforcement, international standards	N, law enforcement, international standards

Impact on landscape	L	–	M, land use regulations, map zoning	–	M, land use regulations, map zoning
Foreign control and imbalance of economic benefit	N, R, L	– –	M, land use regulations, CSR	– –	M, land use regulations, CSR
Gender (women's participation and support)		+ +	L and N, programmes to support community	+ +	L and N, programmes to support community
Community and cultural dilution	L	– –	L and N, programmes to support community	– –	L and N, programmes to support community
Rural employment and income generation	R, L	+ + +	L and N, programmes to support community	+ + +	L and N, programmes to support community
Infrastructure development	N, R, L	+ + +	N, R, L	+ + +	N, R, L

Notes

Code: + positive impact; – negative impact; M = mitigation measure; N = national; R = regional; L = local; CSR = corporate social responsibility.

although this has generally not translated into the expected large-scale activities. The current adverse economic climate may be partly to blame, as well as delays in formulating and implementing regulatory frameworks and limited infrastructure and discussions regarding the most suitable crops to use (GTZ, 2009). In some instances implementation has proceeded ahead of the development of the necessary regulatory frameworks, leading to calls for caution (Duku *et al.*, 2010).

Recent experience from Tanzania, one of the countries to see concerted biofuel investor interest in Africa, points to numerous risks and the need for caution (Sulle and Nelson, 2009; see also the brief case study in Chapter 17). Where the investor seeks to contract independent growers for biofuel crops, negative impacts on local land access can be minimized. The opposite case, where investors seek to acquire large tracts of land from communities as happened in Tanzania, has proved to be fraught with potential problems. Some of the major problems cited include the following.

- The targeted land is often already being exploited by local communities for forest-based economic activities.
- The value of the land may not have been accurately ascertained due to a combination of factors, including inability of the community and outsiders to determine such value, and investors taking advantage.
- Inadequate compensation procedures, which disadvantage villagers and lead to inequitable/questionable arrangements.
- Lack of capacity among communities to negotiate legal contracts with investors.
- Shortcomings in current biofuel guidelines.

The aforementioned problems are more likely to occur when investors are from outside the area of a given project, although it is also possible that local investors may contribute to such problems when participatory processes are not followed. The various possible arrangements for ownership are discussed further below.

Vertical integration/ownership

In this arrangement, the owner of the factory also owns (or has a long-term lease on) the land where sugar cane is grown. In such cases, the company (or the state, where applicable) can manage the land on the basis of maximizing output, regardless of the revenue streams for different products and inputs. Since sugar cane and ethanol production require a certain economy of scale, there will often be a tendency towards some degree of vertical integration so as to assure feedstock availability (see Chapter 6), although the actual ownership structures depend also on national and local regulations regarding land ownership. In practice, few African sugar cane operations are based on a pure vertically integrated model, although there are some state-owned factories that operate along similar lines. There is often a preference for some type of mixed model, as discussed further below.

PLATE 1 Hybridization of sugar cane parent varieties with excised flowering stalks under controlled conditions at the Mauritius Sugar Industry Research Institute (MSIRI)
PHOTO: Courtesy of the MSIRI (taken by Dr Kishore Ramdoyal)

PLATE 2: Test plots for different varieties of sugar cane at the South African Sugar Industry Research Institute (SASRI); note the high-fibre variety on the left
PHOTO: Francis X. Johnson (with permission of SASRI)

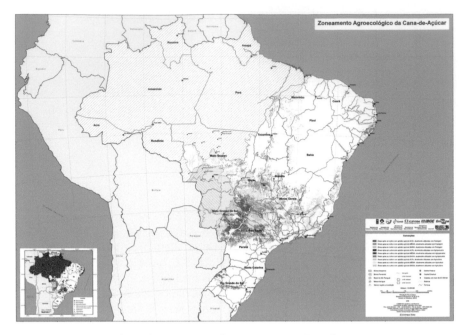

PLATE 3 Agro-ecological zoning map for sugar cane in Brazil

- Areas highly suitable for agricultural cultivation, currently pasturelands
- Areas moderately suitable for agricultural cultivation, currently pasturelands
- Areas of low suitability for agricultural cultivation, currently pasturelands
- Areas highly suitable for agricultural cultivation, currently mixed uses
- Areas moderately suitable for agricultural cultivation, currently mixed uses
- Areas of low suitability suitable for agricultural cultivation, currently mixed uses
- Areas highly suitable for agricultural cultivation, currently agricultural uses
- Areas moderately suitable for agricultural cultivation, currently agricultural uses
- Areas of low suitability for agricultural cultivation, currently agricultural uses

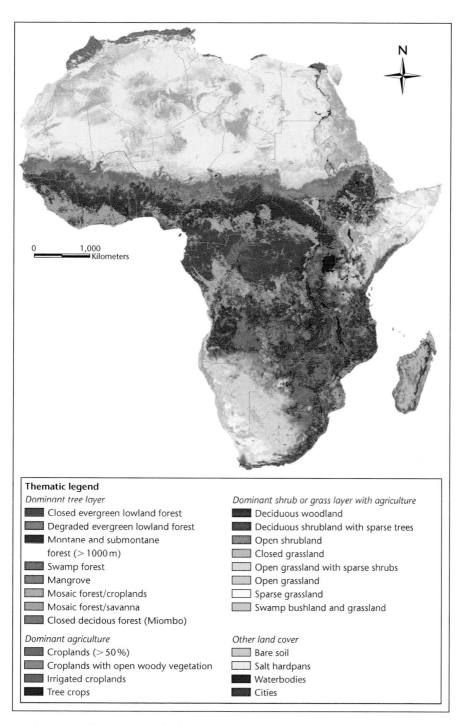

Thematic legend

Dominant tree layer
- Closed evergreen lowland forest
- Degraded evergreen lowland forest
- Montane and submontane forest (> 1000 m)
- Swamp forest
- Mangrove
- Mosaic forest/croplands
- Mosaic forest/savanna
- Closed decidous forest (Miombo)

Dominant agriculture
- Croplands (> 50%)
- Croplands with open woody vegetation
- Irrigated croplands
- Tree crops

Dominant shrub or grass layer with agriculture
- Deciduous woodland
- Deciduous shrubland with sparse trees
- Open shrubland
- Closed grassland
- Open grassland with sparse shrubs
- Open grassland
- Sparse grassland
- Swamp bushland and grassland

Other land cover
- Bare soil
- Salt hardpans
- Waterbodies
- Cities

PLATE 4 Land cover map of Africa; Year: 2000
Source: European Commission Joint Research Centre (JRC), 2003

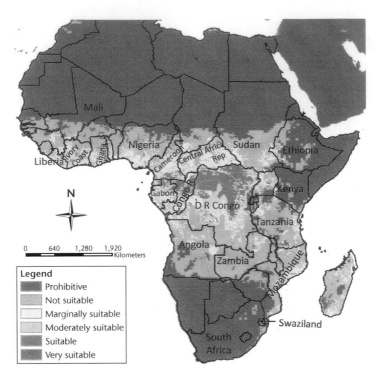

PLATE 5 Suitability of land for rain-fed sugar cane production in Africa
Source: Adapted from FAO (2004)

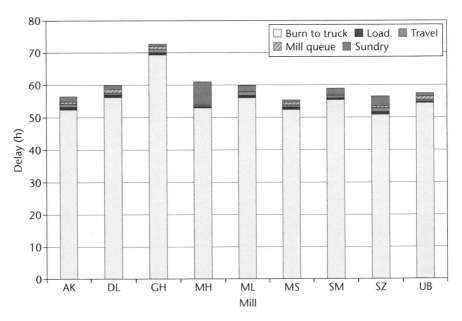

PLATE 6 Components of delay between burn and crush at different sugar mills

PLATE 7 Planting of cane *setts* (stem cuttings or stalk sections) for sugar cane fields in Mauritius
PHOTO: Vikram Seebaluck

PLATE 8 Burning of sugar cane in South Africa; the burning of green sugar cane prior to manual harvesting is practised throughout Africa, to facilitate harvesting and send a better cane quality with less trash to the factory, while it also clears out animals such as snakes in the fields. As mechanical harvesting is introduced, burning of sugar cane can gradually be phased out, such as is the case in Brazil
PHOTO: Maria M. Morales

PLATE 9 Manual cane cutting in Mauritius. The introduction of mechanized harvesting (see next photo) improves efficiency but reduces employment for the cane cutters
PHOTO: Vikram Seebaluck

PLATE 10 Full mechanical cane harvesting with chopper harvester and infield trailer in Mauritius; the chopper harvester is operated by large miller-planters, and is part of a modern system that even also utilizes GPS in order to ensure efficient and timely harvesting
PHOTO: Vikram Seebaluck

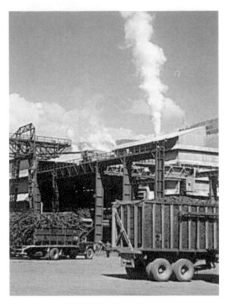

PLATE 11 Semi-mechanical loading of sugar cane at Dwangwa Estates, Malawi
PHOTO: Francis X. Johnson

PLATE 12 Sugar cane arriving by truck and trailer for processing at the sugar mill (factory), Nakambala, Zambia
PHOTO: Francis X. Johnson

PLATE 13 Transport of sugar cane to factory, Sri Ram Sugar Mills Pvt. Ltd., Rautahat, Nepal; animal power provides a significant share of transport to the sugar mill
PHOTO: Dilip Khatiwada

PLATE 14 Cogeneration plant, Kakira sugar mill, Uganda
PHOTO: Vikram Seebaluck

PLATE 15 Ethanol distillery, Sri Ram Sugar Mills Pvt. Ltd., Rautahat, Nepal
PHOTO: Dilip Khatiwada

PLATE 16 Ethanol storage tanks, Jalles Machado Sugar Cane Mill, São Paulo state, Brazil
PHOTO: Alexander Strapasson (Source. MAPA/SPAE/DCAA)

PLATE 17 Blending depot, Malawi: gasoline and ethanol are blended directly in the tank of the truck
PHOTO: Francis X. Johnson

PLATE 18 E10 blend at pump in filling station (pilot project for use of bioethanol in the land transportation sector), Mauritius
PHOTO: Vikram Seebaluck

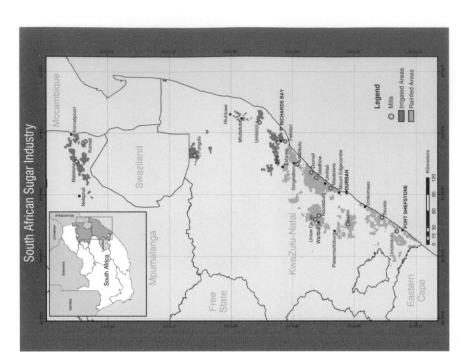

PLATE 20 Map of the South African Sugar Industry indicating location of sugar mills and rain-fed and irrigated sugar cane production areas

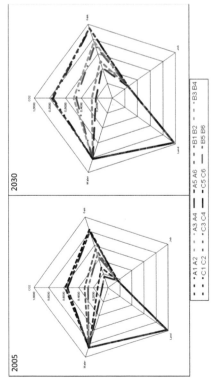

PLATE 19a Integration of sustainability assessment results for South Africa

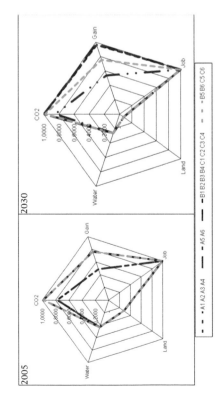

PLATE 19b Integration of sustainability assessment results for Mauritius

Source: Woods et al., 2008

Contract farming

Contract farming, also commonly known in Africa as outgrower schemes, is a business model based on an agreement between farmers and processing and/or marketing firms for the production and supply of agricultural products under forward agreements, frequently at predetermined prices (Eaton and Shepherd, 2001). It ranges from informal purchase agreements through to highly specified schemes, and varies considerably in the extent to which inputs, costs, risks and benefits are shared between growers/landholders and companies. Contract farming has been shown to work well with high-value-added crops, and products that require strict control over quality and quantity due to the high degree of processing, such as sugar cane. Several studies carried out in the 1990s in Africa and more recently in India indicate a positive effect of contract farming on smallholder income (e.g. Glover and Kusterer, 1990; Jaffe and Little, 1994; Porter and Phillips-Howard, 1997).

Several experiences can be reported in Africa showing the benefits deriving from use of contract farming or other types of inclusive business models where smallholders can maintain access to their land, participate on more equal terms in profits generation from biofuel production and gain access to more lucrative but remote markets for high-value crops, by reducing market risks for them and hence increase their income stability. Naturally there are advantages and disadvantages from both farmer and company perspectives, as summarized in Table 11.2.

Hybrid model

This model typically consists of a nucleus estate directly controlled by the company, which would have direct control over part of its supply and hold facilities for harvesting and processing. In addition there will be contract farming arrangements whereby the company provides inputs to family farmers and the latter sell their produce to the company, thereby augmenting supplies and feeding the processing facilities. An example of hybrid initiative which combines production from large plantations with significant contributions from small-scale farmers is the case of Tanzania's Kilombero Sugar Company. It owns an estate of 8,000 hectares, with contract farmers operating over 12,000 ha. The hybrid model of organization is quite common in Africa, especially in Eastern and Southern Africa, although the share of land held by small farmers tends to be a smaller fraction than in the case of Kilombero.

Revenue-sharing

In contrast to the "classic" contract farming model, whereby contract farmers merely sell their cane to the mill and the mill processes and sells the produce with no farmer participation, the model used in Tanzania's sugar industry rewards contract farmers not with a fixed price, but with a share of the revenues generated by the sale of processed sugar. Farmers can receive up to 55 per cent of the total proceeds, with the company getting the rest.

TABLE 11.2 Advantages and disadvantages of contract farming

	Advantages	Disadvantages
Farmers	• Inputs, production services often supplied by company. • Credit advances from company. • Contract farming often introduces new technology and also enables farmers to learn new skills. • Farmers' price risk is often reduced as many contracts specify prices in advance. • Contract farming can open up new markets that would otherwise be unavailable to small farmers.	• Particularly when growing new crops, farmers face risks of both market failure and production problems. • Inefficient management or marketing problems can mean that quotas are manipulated so that not all contracted production is purchased. • Companies may be unreliable or exploit monopoly position. • Company staff may be corrupt, particularly with quotas.
Company	• Contract farming with small farmers is more politically acceptable than, for example, production on estates. • Working with small farmers overcomes land constraints. • Production is more reliable than open-market purchases and company faces less risk by not being responsible for production. • More consistent quality can be obtained than if purchases were made on the open market.	• Contracted farmers may face land constraints due to lack of secure tenure, jeopardizing sustainable long-term operations. • Social and cultural constraints may affect farmers' ability to produce to company's specifications. • Poor management and lack of consultation with farmers may lead to farmer discontent. • Farmers may sell outside contract (extra-contractual marketing), thus reducing processing factory throughput and lowering efficiency. • Farmers may divert inputs supplied on credit to other purposes, thereby reducing yields.

The case shows that the higher the production by the large-scale estate, the higher the value gained by local communities through the contract farming scheme, because a share of the profits is retained locally and local communities benefit. This case shows that small-scale contract farming schemes can be a successful way of capturing foreign investment to increase significantly the income and productivity of rural communities in developing countries when conditions are favourable. Such a system is suitable in countries with multiple small-scale landowners.

In Tanzania, collective bargaining through the Tanzania Sugarcane Growers Association (TASGA) has helped outgrowers get a better deal through the Cane Supply Agreement (CSA) and an acceptable cane price formula, and through the negotiation of better credit access. As the hybrid model involves the establishment of a nucleus estate, it is subject to many of the problems characterizing pure, large-scale land acquisitions. Unless there is sufficient unused land available, establishing the nucleus estate is likely to be a challenging task, and may encroach upon existing land rights. Other limitations of the model include the following (Cotula and Leonard, 2010; Bekunda et al., 2009):

- Business ownership: the businesses are controlled by the companies – the growers have no equity stake in the sugar mills and have no say over their management.
- Limited capacity among association leaders at the grassroots level, which affects the inclusion of farmers as well as the association's ability to negotiate a better deal with sugar-producing companies.

The government of Tanzania encourages investors to specify the role of contract farmers in the production chain. It also encourages contract farmers to be more involved in value-adding activities related to biofuels and to form associations and cooperatives that may enter into contract agreements. Governments could facilitate this development by encouraging industry cooperation through starting up a national biofuel producers' association.

Scale of production

The consideration of the scale for bioenergy crops production (as well as for other commodities) may have implications at the local level, not just on environmental aspects but particularly on the socio-economic aspects. Figure 11.1 shows the different implications of scale for competitiveness and risk. For instance, at the large-scale level, greater emphasis on smallholder-led production will add value to local communities while retaining the export potential similar to centralized production in mill-owned estates. On the other hand, small-scale production, either through single or multi-product cropping, tends to be more complex and carry greater risks, with social issues coming into play to a greater extent. At the same time, such small-scale schemes can reduce market barriers through lower initial capital requirements and thereby bring greater local added value.

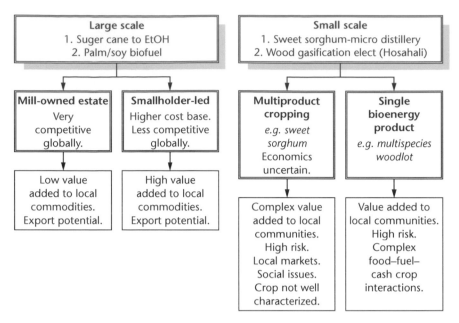

| Large scale |
| 1. Suger cane to EtOH |
| 2. Palm/soy biofuel |

| Small scale |
| 1. Sweet sorghum-micro distillery |
| 2. Wood gasification elect (Hosahali) |

| Mill-owned estate | Smallholder-led |
| Very competitive globally. | Higher cost base. Less competitive globally. |

| Multiproduct cropping | Single bioenergy product |
| *e.g. sweet sorghum* Economics uncertain. | *e.g. multispecies woodlot* |

| Low value added to local commodities. Export potential. | High value added to local commodities. Export potential. |

| Complex value added to local communities. High risk. Local markets. Social issues. Crop not well characterized. | Value added to local communities. High risk. Complex food–fuel–cash crop interactions. |

FIGURE 11.1   Bioenergy development options and impacts of scale of operation (source: Woods, 2006)

Co-products

Co-product markets will improve the economics for sugar cane, since there are various energy and non-energy products (see Chapters 5, 6 and 7) that can add value and diversify market options. Rural bioenergy projects can create new income opportunities for rural farmers. Producing biomass provides a new source of revenue and helps farmers to diversify. Depending on what other crops they are able to grow, such projects can reduce vulnerability related to crop failures or declining commodity crop prices.

Table 11.3 summarizes the drivers, co-products, strategies, impacts and indicators in the case of ethanol and electricity as co-products. The production of ethanol could be approached from a small-scale decentralized angle or could be large scale and centralized. It must be noted that small-scale production (i.e. less than 1 million litres per year) of ethanol will be more costly than large-scale production and there will consequently need to be cost savings in other parts of the supply chain such as distribution, or through a price premium such as may be available for non-fuel ethanol (see also Chapters 6, 9 and 10). The different production models have different impacts on local land access.

Co-benefits

Economic co-benefits of either a biomass power facility or an ethanol production

TABLE 11.3 Summary of drivers, co-products and impacts

Drivers	Co-product	Strategies	Impacts	Indicators
High oil prices, pressure on foreign currency reserves, limited energy access, need for greater energy security.	Ethanol: large scale. Ethanol: large/small scale. Ethanol: small scale.	Fuel blending; export; decentralized production, local appliances, kerosene substitute in liquid and gel form.	Foreign exchange savings, lead emissions, improved access to modern energy, cleaner indoor air, health risk – abuse, land use changes.	Qty petrol imported, % ethanol in blend, lead level in soil/air, lower particulates, % reliance on traditional fuels, incidence of upper respiratory complaints.
Power shortage in SADC region, need to diversify due to lower sugar prices, environmental concerns.	Electricity (bagasse), biogas[1] (anaerobic digestion).	Sell to grid, local mini-grid, direct use of gas, sell CO_2, generate electricity.	Facilitates productive activities (welding, power tools, cooling…), safe biofertilizer, more biomass available, production of N_2 source.	Range of income-generating activities, incomes, quantity of biomass harvested, quantity of biomass digested, quantity of gas, quantity of manure produced.

Source: Woods et al., 2008.

facility result from feedstock handling and processing activities, plant construction and operation, and product marketing. All contribute income to the local economy, primarily due to employment. Cogeneration of steam and electricity with innovative technology brings considerable efficiency and environmental benefits. Blending ethanol with fossil fuel or ethanol for domestic use and export brings considerable economic benefits by reducing the fuel import bill, saving precious foreign currency. Cogeneration projects reduce or delay the need for investment through the additional revenue derived from selling excess energy to the grid or other external consumer. Bagasse cogeneration also allows diversification in electricity generation, reducing possible disruptions and improving reliability (Woods *et al.*, 2008).

Small-scale uses and alternative markets

Ethanol can be turned into ethanol gel, which is now on sale as a cooking fuel. This viscous gel will not spill and is thus much safer than kerosene, which causes serious domestic fires (PASASA, 2011). It is also possible to use ethanol in its liquid form as cooking fuel in suitably designed appliances. There are some stoves specifically designed for liquid ethanol, and thermal efficiencies vary depending on fuel/stove combinations; tests conducted in South Africa showed that many can reach around 60 per cent (Lloyd and Visagie, 2007). There are also opportunities to supply the potable, industrial and pharmaceutical alcohol markets. Such opportunities can potentially stimulate local economies.

An example of local small-scale production

A small-scale case study around St Mary's and Epworth townships in Zimbabwe looked at the production of ethanol with indigenous fermentation and distillation knowledge using comparatively rudimentary equipment. The fact that sugar cane is sometimes used in the brewing of illicit spirits is of interest because this demonstrates indigenous knowledge in local production of ethanol from sugar cane as well as entrepreneurial skills. In addition to potable ethanol for spirits, another local market exists in the potential to substitute ethanol for kerosene, which is widely used as an illuminating and cooking fuel by low-income households (Woods *et al.*, 2008).

Nine brewers of illicit spirits were interviewed in two low-income suburbs (St Mary's and Epworth) of Harare, Zimbabwe. Samples of the distilled brew sold as spirits were measured at just over 54 per cent ethanol content (BUN, 2004). Dilution with tap water (25 per cent added) prior to sale resulted in approximately 43 per cent ethanol content in the final product. The profits were found to be generally very good, especially where the spirits were sold in small quantities at higher prices per litre than bulk sales. The monthly net income of the brewers was found to be comparable to a secondary school teacher's monthly salary in 2002 (BUN, 2004).

Environmental impacts

The main environmental impacts relate to water consumption, water pollution, soil impacts and air pollution from some harvesting practices. Compared to many other commodity crops, pesticide use is relatively low, and chemical application is mainly restricted to herbicides (Leal, 2006). Other impacts are associated with land use changes resulting from the expansion of sugar cane-growing area.

The impacts associated with bioenergy/sugar cane systems can be viewed in comparison to the likely alternative land use activities. That is, the relative impact of producing bioenergy feedstocks depends not only on how the biomass is produced, but also on how the land might have been used otherwise (including the opportunity cost of alternative uses). An underlying theme is that biomass must be produced in a manner that is sensitive to local ecological conditions. To the extent possible, crop types should be favoured that match native ecosystem types; for example, by selecting perennial grasses in prairie or savannah regions, and trees in woodland regions. Table 11.4 summarizes selected environmental impacts and selected characteristics.

Water impacts

Agriculture is by far the biggest user of fresh water, with a world average of 70 per cent of total human use, while industry and households consume 20 per cent and 10 per cent respectively (WWF, 2005). From the 1,500 million ha of world sugar cane crop area, around 275 million ha (18 per cent) is irrigated and account for 58 per cent of total agricultural production (Macedo, 2005); 202 million ha of irrigated cropland is located in developing countries, and this figure is expected to reach 242 million ha by 2030 (Leal, 2006). It has been estimated that by 2020 water demand will exceed availability in many parts of South Africa, with overconsumption and contamination as the major causes (Cheesman, 2005).

Water impacts associated with sugar cane crops can be divided into those linked to water consumption and water flow for the crops, and those related to water pollution, including water quality and effluent run-off problems, as well as those associated with the processing of the sugar cane either for sugar or for ethanol. Sugar cane is a deep-rooted crop that requires a large amount of water, and is extremely sensitive to soil water deficits; however, the efficiency of water use in terms of biomass production is better than that of many other crops (Woods et al., 2008).

Even in areas where sugar cane is not irrigated the crop can have a great impact on river flow, as it reduces run-off from the catchment into rivers and draws heavily on ground water resources. In South Africa, some 18 per cent of the 432,000 ha under sugar cane is irrigated, mostly in Mpumalanga Lowveld, but to a lesser extent also in northern Kwazulu-Natal (Pongola and Umfolozi Flats).

Watercourses can be polluted by agrochemicals and sediments; in some cases these impacts can extend to downstream ecosystems (IIED, 2004). The main pollutants are waterborne organic matter and solids, which can affect groundwater, rivers and

TABLE 11.4 Environmental impacts

Environmental impacts	Scale N, R, L	Biomass production (farm)	Biomass pre-treatment and conversion (factory)
GHG (overall benefit according to feedstock and local conditions)	Global	+ + +	+ + +
Water quality	R, L	– – M, local regulations	– – M, local regulations, better technology and management systems for waste water
Water quantity	R, L	– – M, local regulations	– – M, local regulations, better technology and management systems
Soil quality (pesticides, fertilizers)	L	– L, M, good agricultural practices	
Air quality	L	– L, M, good agricultural practices	– – M, local regulations, better technology and management systems
Land use change	N, R, L	– – M, land use regulations, map zoning	
Landscape	R, L	– M, land use regulations, map zoning	
Biodiversity	L	– M, land use regulations, map zoning; help to improve biodiversity and habitats; international conventions	
Protected areas	R, L	– – M, land use regulations, map zoning; protection of habitats and landscapes	
Monocultures	L	– L, M, good agricultural practices	
Invasive species	N, R, L	– L, M, good agricultural practices; international conventions	

Notes

Code: + positive impact; – negative impact; M = mitigation measure; N = national; R = regional; L = local; CSR = corporate social responsibility.

wetlands (IIED, 2004). Sugar mills generate about 1,000 litres of waste water per tonne of cane crushed. Sugar mill effluent has a high BOD (biological oxygen demand); effluents are also high in suspended solids and ammonium. Such was the case for three sugar factories next to the River Nyando in Kenya, which led to a decline in the quality of the source of drinking water to many families along the river's courses to Lake Victoria, and to nutrient over-enrichment of Lake Victoria (IIED, 2004).

Soil impacts

Residue burning and tillage, use of fertilizers and pesticides, the uptake and use of soil nutrients and changes in soil carbon content are the main impacts on soil. Specific impacts will vary with local context. Soil erosion has been associated with sugar cane growing, particularly where cane is cultivated on slopes. Because of monoculture and the use of various chemicals, processes such as leaching, eutrophication and turbidity contribute to off-farm impacts in the long run (WWF, 2005).

Positive impacts on soils include the generation of considerable organic matter that can enhance soil fertility. Sugar is not an annual crop and does not require annual ploughing, thus limiting soil erosion (WWF, 2005). Pre-harvest burning, or harvesting green, has its pros and cons. Continuous pre-harvest burning of sugar cane can contribute to soil quality deterioration (Cheesman, 2005) and reduction of soil organic carbon and available nitrogen (N) and phosphorus (P). On the other hand, cane burning can reduce the risk of soil acidification, due to the increase in pH associated with ash deposited on the soil.

Green harvesting leaves a trash layer, which includes the following benefits (e.g. Leal, 2006):

- an increase in soil organic matter and available N and P;
- enhanced soil diversity (in terms of microbial and earthworm populations);
- reduced risk of erosion;
- conservation of soil moisture; and
- reduction of diurnal fluctuation in soil temperature, as well as weed suppression.

Disadvantages of using a trash blanket on sugar cane fields include the increase in some pests, increased risk of soil acidification, increased harvesting costs, complications to irrigation and fertilizer application and slowing tiller emergence. The quantification of the effects on cane yield and economics results is not easy to determine (Leal, 2006). The South African Sugar Association (SASA) has comprehensive standards and guidelines on soil conservation within best practice for the crop under local conditions (SASA, 2006).

Air quality impacts

The environmental impacts on air are the product of two main activities: the pre-harvest burning of sugar cane fields; and emissions from either sugar or ethanol processing. Both of these have effects on public health and on ecosystems, besides contributing to GHG emissions.

Burning of cane to facilitate manual harvesting causes air pollution and increases soil erosion. Burning can be avoided by harvesting green cane, a practice that has spread from Cuba to Brazil and Australia. In most cases, green cane harvesting tends to be mechanized rather than manual, and cannot therefore be implemented by all industries owing to cost, suitability of terrain, etc. (IIED, 2004). Pollutants produced by burning are among those that better practices could limit; this would include local impacts from smoke and ash, and global impacts from CO_2, NO_x and CO. Smoke and ash are regarded as aggravating and nuisance pollutants respectively.

The use of bagasse as fuel in boilers produces particulate matter, and oxides of nitrogen and sulphur. If pollution control equipment is not installed, fly ash escapes to the atmosphere and can have negative health effects and can damage crops. Lime sludge and press mud are important solid waste generated by sugar mills (lime is used for purifying sugar cane juice; impurities from sugar cane juice are either vacuum filtered or press filtered and removed as press mud). Solid wastes are also generated from pollution control facilities (IIED, 2004).

The South African Sugar Association (SASA) standards also regulate burning of sugar cane prior to harvesting, which is restricted to certain hours a week or gazetted for magisterial districts. This is also regulated by local environment committees (LECs) (SASA, 2006).

Emissions from the soil under cultivation are essentially nitrogen compounds associated with fertilizer decomposition. The concern in terms of GHGs is nitrous oxide (N_2O), due to its high global warming potential. These emissions are highly dependent on soil conditions (moisture, nitrate content, etc.) and cultivation practices (Cheesman, 2005). A detailed study of the GHG balance of sugar cane in South Africa is provided in Chapter 13.

Biodiversity impacts

As a monoculture, sugar cane can have a negative impact on biodiversity in the local context, while the impact in a larger region may depend on the linkages and possible synergies and conflicts with agricultural and forest management strategies. The impact of an energy plantation on biodiversity in the local context depends on the system the energy crops are replacing. Biodiversity is generally higher in areas where a diverse range of habitats is found. Biodiversity is likely to remain unchanged or even increase if an energy plantation is replacing grasslands or annual agricultural crops, whereas it will decrease if the energy plantation is on land with high species diversity, such as unmanaged wetlands (Woods *et al.*, 2006). It is important to leave a buffer zone between the plantation and established woodland or hedgerows to preserve the crucial edge habitat necessary for a diversity of species. Energy plantations can also provide corridors between isolated habitats.

In some cases, the desire to avoid food conflicts may lead to interest in growing energy crops such as sugar cane in more fragile or marginal areas where other conventional crops are likely to fail or are too difficult to farm. Such areas include steep slopes or riparian areas or wetlands, where greater concentrations of biodiversity exist. Such marginal lands are often not good for conventional agriculture, as a higher level of inputs (lime and fertilizer) is required to sustain productivity on marginal soils. However, there are cases where farming on marginal land leads to a higher internal rate of return compared to farming on more typically suitable agricultural soils (WWF, 2005). The steeper slopes will make mechanization difficult and thus only manually harvested sugar cane will be feasible in such areas, and this has positive impacts for employment.

According to Von Maltitz and Brent (2008), if energy crops are planted in degraded lands, the net positive global climate change benefit may be higher with relatively limited biodiversity impacts. Conversely, if land is converted from indigenous vegetation to biofuel plantations, the biodiversity loss might be higher and the time taken to reclaim the carbon loss from land clearing will be longer.

Careful planning and management will be needed to mitigate negative biodiversity impacts. Walter and colleagues (2008) reported that loss of biodiversity is less for sugar cane in comparison to other energy and agricultural crops.

As with any large-scale use of land, it is important to avoid protected or vulnerable areas and to observe relevant national laws and regulations on conservation and nature protection. Many sustainability standards require that biofuels should not be produced using raw materials from high-carbon or high-biodiversity areas, including undisturbed forests, areas designated for nature protection and some types of grasslands (see Chapter 12). As mentioned above, there is a link between biodiversity conservation with protected areas and land use change. Land use change can be avoided by considering agro-ecological zoning and conservation areas (see Chapter 3).

Contribution to sustainability

If designed with the involvement of local communities, sensitivity toward local environmental constraints, and a clear objective of meeting the identified needs of the poor, bioenergy activities can contribute significantly to sustainable development of rural populations. Recent work (e.g. Amezaga et al., 2010) offers policy makers some additional tools to assess the sustainability of bioenergy projects in a developing country context. The discussion in this section is based largely on previous findings from CARENSA case studies in Mauritius and South Africa, two countries where comprehensive and high-quality data was available (Woods et al., 2008). Scenarios are based on different approaches to production and market development for sugar, ethanol and cogeneration, which depends in turn on supporting policies and regulations (see Chapter 9).

Description of methodology

To evaluate the sustainability of different sugar cane co-product scenarios, their respective potential environmental, social and economic impacts were examined.

An appropriate methodology of evaluation to organize the key factors or aspects of sustainable development into a systematic, consistent and transparent framework was required.

The Indicator-based Framework for Evaluation of Natural Resource Management Systems (acronym MESMIS, from the Spanish name) was selected. MESMIS was developed in Mexico in the mid-1990s as a methodological tool to evaluate the sustainability of natural resource management systems, with an emphasis on local contexts. MESMIS comprises a six-step cycle.

The first three steps focus on characterizing the systems, identifying critical points and selecting indicators for the environmental, social and economic dimensions of sustainability. The last three steps integrate the information obtained by means of the indicators using qualitative and quantitative techniques and multi-criteria analysis to obtain a value judgement about the resource management systems and to suggest possible improvements to their socio-environmental profile (López-Ridaura *et al.*, 2002).

The MESMIS approach was adapted to the requirements of the CARENSA country case studies to comprise the five following steps:

1. Definition of the object of evaluation.
2. Determination of sustainable development attributes.
3. Selection of indicators by first defining diagnostic criteria and then deriving strategic indicators.
4. Measurement of indicators, which includes designing appropriate analytic tools and methods for collecting data.
5. Integration of the results.

Scenarios assessment: South Africa and Mauritius case studies

The objects of evaluation in the CARENSA study were alternative co-product scenarios (ethanol and cogeneration) in the context of Mauritius and South Africa where the relevant data was available. Moreover, the two countries represent two contrasting geo-physiographic and agro-economic circumstances. Mauritius is a small country, with a high percentage of cultivated land under sugar cane compared to the other crops and to the total available land (sugar cane accounts for 73 per cent of total crops). Per capita primary energy consumption for Mauritius is low (12.84 MWh), as are the corresponding CO_2 emissions. On the contrary, South Africa is a large country with a small percentage of arable land cultivated under sugar cane (2 per cent), higher per capita primary energy consumptions (33.79 MWh) and much higher CO_2 emissions. From the economic point of view, Mauritius is a middle-income country, with a small economy in which sugar cane is the most important crop. On the contrary, South Africa has a large diversified economy in which sugar plays a relatively small role.

The CARENSA study *scenarios* took into account the following baselines and assumptions:

1. Timeline: 2000 was the base year; use of an earlier base year allowed for a stable and consistent dataset to be used; projections were reported up to 2030.
2. Scenarios are demand-driven, i.e. a percentage of national transport fuel and electricity, is met by locally produced ethanol or cogeneration.
3. Internal crystalline sugar demand (low and high projections).
4. Projected growth in sugar cane production – 2 per cent per year from the 2000 base.
5. Projected growth in electricity consumption – variable 2.1 per cent and 1.5 per cent for South Africa and 3.5 per cent and 2.5 per cent for the rest of SADC, respectively, for 2000–2020 and 2021–2030 (Yamba and Matsika, 2003).
6. Projected growth in transport energy consumption (petrol and diesel) – 2.1 per cent per year.

Ethanol scenarios

The ethanol scenarios were based on the following regional conditions and assumptions:

a. Baseline/current situation in 2000: fuel ethanol production occurring in Malawi only; four scenarios for production of ethanol to meet mandated shares of national forecast gasoline and diesel demand as per Table 11.5.
b. Internal crystalline sugar demand is met and produced from local/national cane resources, any surplus cane resources are used to produce ethanol.
c. Export markets for sugar or ethanol are not explicitly assessed.

The six different ethanol production scenarios analysed in this study were:

1. ethanol potentials after meeting lower national sugar demand (by population growth);
2. ethanol potentials after meeting higher national sugar demand (40 kg convergence);

TABLE 11.5 Projected targets for ethanol production

Scenario	2000 (baseline)	2005	2015	2030
	A	A1	B	D
% of national transport energy based on ethanol	0	0.1	0.79	18.4
% of national gasoline consumption displaced by ethanol	0	0.1	0.91	25.0
% of national diesel consumption displaced by ethanol	0	0.1	0.63	10.0

3. ethanol potentials after meeting lower national sugar demand and sugar exports declining to 0 in 2015;
4. ethanol potentials after meeting higher national sugar demand and sugar exports declining to 0 in 2015;
5. ethanol potentials after meeting lower national sugar demand and sugar exports stabilized at 2002 levels;
6. ethanol potentials after meeting higher national sugar demand and sugar exports stabilized at 2002 levels.

Cogeneration scenarios

The cogeneration scenarios assume gradual expansion of demand for surplus electricity from sugar mills in each country. Potential electricity production is estimated based on three cogeneration options (see Chapter 5 for relevant technical background):

a. 20 kWh per tonne of cane – optimization of current back pressure turbines, 2.0 megapascals (20 bar) and 350 °C;
b. 92 kWh/tc – condensing extraction steam turbine (CEST), 4.6 MPa (45 bar) and 440 °C;
c. 250 kWh/tc – biomass gasification and combined cycle.

Using Mauritius and South Africa case studies and the assumptions discussed earlier, sustainability evaluation was applied to the six ethanol and three cogeneration scenarios. The ethanol and cogeneration scenarios are independent; hence each of *six* ethanol scenarios was considered for the *three* cogeneration scenarios. Thus 18 sub-scenarios were considered for sustainability analysis.

The analysis then followed the five steps of the MESMIS process and the sustainability indicators in Table 11.6 were developed for step four. The indicators selected from the table and used in the sustainability assessment are the following: *employment (social), financial return (economic), water needs (environmental), hectares of land used (environmental) and tonnes of CO_2 equivalent avoided (environmental)*. The goal of the indicator selection process was to derive a small number of meaningful and clearly definable indicators for which data were available and which calculations could be performed with ease.[2]

Results

The main results of the sustainability assessment applied to the Mauritius and South Africa case studies are described below.

Number of jobs created (social)

The assessment shows how sugar cane-based bioenergy exploitation in Mauritius can prevent the decreasing trend of employment in the country's sugar cane sector.

TABLE 11.6 Sustainability indicators

	Attributes	Diagnostic criterion	Indicators of sustainability	Unit of measure
Economic	Sustainability of balance of payments	Contribution to foreign currency saving; contribution to money return	Importation of fossil fuel avoided – foreign currency saved or net energy import dependency; selling of ethanol and fossil fuel	US$
	Capacity development	Knowledge of people involved	Number of people trained	No. of persons
	Employment	Contribution to job creation	Numbers of jobs created	No. of jobs created
Social	Poverty alleviation	Increase in number of people living above income poverty line	Percentage of population living below poverty line	
	Social equity	Contribution to empowerment of excluded social groups and equitable wealth distribution	Energy use per capita	kWh
			Number and type of beneficiaries by gender, racial group	
		Income inequality	Gini index of income inequality	
	Employment (quality)	Improvement of job quality	Quality and length of job	
Environmental	Sustainable use of water	Water quality	Water needs	m^3
		Water quality	Concentrations of main pollutants (incl. BOD) and faecal coliform in fresh water	kg/m^3
	Land use change	Preserving natural forests	Area of natural forest preserved	ha
	Climate change mitigation	Net reduction of GHG emissions	Tonnes of CO_2 equivalent avoided	tonnes
	Contribution to air quality	Air quality	Ambient concentration of air pollutants (e.g. CO, SOx, NOx, HC, particulates)	Gg
	Land conservation	Soil erosion	Percentage of organic matter in the soil	

Source: adapted from Farioli, 2002.

On the contrary, in the case study of South Africa, an increase in cultivation is possible and thus increase in the number of jobs is relevant. The two case studies illustrate the potential social impact of sugar cane (cultivation, sugar processing, ethanol and energy production). In particular they show:

- a higher impact (i.e. considerable job creation) where there is potential for increase in sugar cane cultivation (South Africa)[3]; and
- where there is no increase in sugar cane cultivation (Mauritius), jobs created in energy production will halt the decrease in the number of sugar cane workers in the sector.

Therefore, energy production from sugar cane co-products can produce positive social and economic impacts by reducing unemployment and poverty levels.

Financial returns (economic)

The assessment shows significant potential economic impacts of sugar cane (cultivation, sugar processing, ethanol and energy production). In particular it shows how this impact is greater (higher financial returns) where sugar cane cultivation can be extended and where the current quantity of electricity produced does not meet total electricity demand (South Africa). This is because the financial returns from electricity production are higher than financial returns from ethanol production.

Tonnes of CO_2 equivalent avoided (environmental)

To calculate the total emissions avoided (CO_{2a}) through the use of bioenergy, the following CO_2 emissions were considered:

- avoided CO_2 emissions thanks to the use of bagasse in cogeneration (CO_{2kWh}); and
- avoided CO_2 emissions when ethanol is used for combustion engines (CO_{2eth}) instead of fossil fuels. CO_2 emissions produced[4] during ethanol production and distribution (CO_{2prod}) have to be subtracted from CO_{2eth} to get net CO_2 avoided.

  Thus: $CO_{2a} = CO_{2kWh} + CO_{2eth} - CO_{2prod}$

The two case studies show the following.

- It is possible to avoid all CO_2 emissions where sugar cane is the main crop and country's per capita primary energy consumptions are low (Mauritius);
- The amount of CO_2 emissions reduction is more dependent on cogeneration scenarios than on ethanol scenarios[5] in cases where the primary energy mix is based mainly on fossil fuels (in Mauritius 70 per cent of electricity is produced from diesel and coal, and in South Africa 90 per cent is produced from coal).

- Where a cogeneration plant is not efficient or there is no surplus electricity available for export (cogenerated electricity meets total electricity demand), emissions reductions are more dependent on ethanol scenarios.

Land use (environmental)

Land use was used as proxy for biodiversity conservation. However, high conservation-value areas have more relevance for sustainability considerations and should be used where more detailed data or methods can be applied (IUCN, 2011).

To evaluate the impact of the scenarios on land use, considering the 2 per cent per year projected growth in sugar cane production and 1 per cent yield increase, the following indicator has been used:

$$1 - \frac{SC}{AR \,\&\, PC}$$

where:
SC is area under new sugar cane cultivation (ha),
AR and PC are the arable[6] and permanent crops[7] land areas (ha).

Results from the assessment show that the percentage increase in land area cultivated under sugar cane is the same in the two countries, but the impact on total availability of land is particularly critical in Mauritius, where the land under sugar cane already amounts to a very high percentage (72 per cent of arable land); the land is being gradually liberated from cane cultivation to accommodate new areas of economic development in the country.

Water needs (environmental)

Water use impact analysis has been carried out, taking into consideration only water needs for cultivation, since these are an entire order of magnitude higher than those for sugar, ethanol and power production. The average value for irrigation and rainfall is 100–1000 m³/tonne, while the average value for water use in sugar/ethanol/power production is 1–20 m³/tonne). Impact on water use deriving from efficiency improvement at cultivation and process stage have not been considered since the focus was to derive the impact on availability of water due to increase of land under sugar cane.[8] Consequently, a mean value of water needs has been fixed at 220 m³/tonne. Water needs have been considered independent from the production process (sugar or ethanol) and dependent only on sugar cane yields and total sugar cane produced; therefore they vary only over time (and not among scenarios during the same period) to satisfy the increase of sugar cane production per year.

The results reported show water needs increasing over time. South Africa is, however, a water-stressed country (DWEA, 2010), and therefore such increases should be compensated or mitigated elsewhere.

Integration of the results

The results obtained from the sustainability assessment of the South Africa and Mauritius cases suggest positive net socio-economic and environmental impacts from sugar cane cultivation, despite the difference in importance at national level of sugar cane cultivation in the two countries. In particular they show that for South Africa there is the opportunity for considerable job creation due to potential increases in sugar cane cultivation. The Mauritius case shows that jobs created in energy production can halt the decrease of sugar cane workers on the agricultural side of the sector by creating expansion in cultivation. This is because the number of jobs related to sugar cane cultivation is greater than the number of jobs created in the production of sugar, ethanol and electricity.

Potential economic impacts (higher monetary returns) of bioenergy production are greater where there is extensive sugar cane cultivation (Mauritius) and where the quantity of electricity produced does not meet the total electricity demand (South Africa). This is because the monetary returns from electricity production were higher in this case than the monetary returns from ethanol production. This balance seems likely to change in the future since the returns on ethanol production will increase as oil prices increase and availability tightens.

Results of the two case studies also show that it is possible to reduce net CO_2 emissions where sugar cane is the main crop, and that the size of this reduction is more dependent on cogeneration scenarios than on ethanol scenarios. Where cogenerated electricity meets total electricity demand, emissions reduction is more dependent on ethanol scenarios.

In order to provide an overview of the sustainability evaluation for the proposed scenarios, the results obtained by measuring the indicators have been summarized and integrated using indices built for each indicator with reference to sustainability thresholds. In the following figures results are presented through amoeba diagrams[9] that enable a simple yet comprehensive graphical comparison of the advantages and limitations of the evaluated scenarios.

Generally, all scenarios have high social sustainability (i.e. all result in employment creation). Other indicators exhibit different impact levels by country. For instance, Plate 19a illustrates the normalized values for South Africa (years: 2005, 2010, 2015, 2030).[10] The analysed scenarios produce negative impacts on land use and water needs, but produce limited positive impacts on unemployment and significant positive impacts on CO_2 emissions and monetary returns, which vary by scenarios and over the years.

Integration of results from the sustainability assessment applied to the Mauritius case study is shown in Plate 19b.[11] This shows positive impacts on CO_2 emissions, unemployment and monetary returns (indicators near 1), and negative impacts on land use and water needs (low value of the indicators). On comparing the scenarios, it can be noted that all scenarios have high social sustainability (i.e. all result in employment creation). Scenarios B and C perform much better than scenario A with regards to environmental and economic sustainability (i.e. CO_2 reduction and monetary returns), mainly due to the improved efficiency of the cogeneration technologies considered.

In particular, in the best scenario for Mauritius in 2015 (B1), it is possible to satisfy electricity demand, generate a surplus, reduce CO_2 emissions to zero, increase GDP by 1.7 per cent and reduce the current unemployment rate by 50.5 per cent. This potential can be achieved through an 11 per cent increase in the amount of land under sugar cane, bringing the total amount of land under sugar cane to 83 per cent of total arable land. This means that in order for the scenario to be sustainable when the amount of land under sugar cane remains static, the yields should increase 2 per cent annually. In order to achieve a reduction from 72 per cent to 55 per cent of the land, the yields would need to increase by 3 per cent annually. Such yield increases are unlikely unless significant advances in breeding are considered (see Chapter 2). Furthermore, the significant share of land already devoted to sugar cane in Mauritius has led to pressures to decrease that share, which means that even higher yield increases would be necessary.

The best scenario for South Africa in 2015 (C1) shows a significant part of electricity demand satisfied through bagasse cogeneration (11 per cent), a CO_2 emissions reduction of 13 per cent, an increase in GDP of 0.9 per cent and reduction of current unemployment by about 1.4 per cent. Those results can be achieved with a modest increase in land and water use (a 10 per cent increase in the area under sugar cane out of total arable land and an 11 per cent increase in water consumption over total available water). Although modest, this increase in water consumption may imply an indirect impact on land use change that has to be considered in further analysis, due to the possible shift of sugar cane cultivation to rain-fed regions as a consequence of the water scarcity South Africa has already started to suffer.

The integration of the results also shows that in South Africa, the change of the cogeneration technology affects sustainability more than the change of ethanol production. In Mauritius (where electricity demand targets are reached in scenarios B and C) the change caused by cogeneration technology and the change relating to different ethanol production levels have a similar impact on sustainability. Therefore, as a general deduction, the decisive factor of sustainability for the scenarios is the cogeneration technology used.

Conclusions

Considering the level of poverty in Africa and the need to provide opportunities for income generation, there is a need to maximize the local benefits for small-scale stakeholders as well as addressing energy insecurity. Production of ethanol and bagasse for cogeneration would pay dividends in the region by addressing the need to cope with rising oil prices and the prevailing power shortages. There will also be benefits to the economies of the countries through import substitution.

The production and processing of sugar cane has significant sustainability impacts. The impacts are generally classified into water consumption, water pollution, biodiversity, soil impacts and air pollution. In comparison with many other commodity crops, pesticide use is relatively low, and chemical application is mainly restricted to herbicides. Other impacts are associated with expansion of land use to accommodate growth in cane production, which are highly location-specific.

The case studies showed high socio-economic sustainability based on the measure of employment creation. Other indicators exhibit rather different impact levels by country. The integration of the results also show that in both Mauritius and South Africa change in cogeneration technology affects sustainability more than change in ethanol production, due especially to the significant reduction in GHG emissions through cogeneration. Therefore, a decisive factor for sustainability in these scenarios is the cogeneration technology used.

With respect to the two case studies chosen, it is also important to note that, although the further expansion of bioenergy from sugar cane can support sustainable development, such expansion is best achieved at existing facilities through advances in agro-industrial efficiency. Further expansion of land under cane is constrained by soil and climatic conditions in both countries and especially by water scarcity in the case of South Africa. There are other African countries that do not face such limitations, and thus from a regional development perspective it would be preferable to concentrate elsewhere on expansion in the land under cane.

Notes

1 The economic viability of biogas options tends to be limited compared to other co-product scenarios.
2 For more detail about the construction of indicators, see Woods *et al.*, 2008.
3 This is because the number of jobs related to sugar cane cultivation is much greater than the number of jobs created in the production of sugar, ethanol and electricity.
4 The production processes consume fossil fuels and the entire life-cycle has to be considered in the calculation of net CO_2 avoided.
5 This is because (a) avoided CO_2 is more sensitive to cogeneration efficiency (and electricity production) than to changes in the quantity of ethanol production; and (b) given high cogeneration efficiency, avoided emissions from cogeneration of 1 tonne of sugar cane (bagasse) are higher than avoided emissions from ethanol production using 1 tonne of sugar cane (molasses).
6 *Arable land* is defined by FAOSTAT as land under temporary crops (double-cropped areas are counted only once), temporary meadows for mowing or pasture, land under market and kitchen gardens and land temporarily fallow (less than five years). Abandoned land resulting from shifting cultivation is not included in this category. Data for "arable land" are not meant to indicate the amount of land that is potentially cultivable.
7 *Permanent crops land* is defined by FAOSTAT as land cultivated with crops that occupy the land for long periods and need not be replanted after each harvest, such as cocoa, coffee and rubber; this category includes land under flowering shrubs, fruit trees, nut trees and vines, but excludes land under trees grown for wood or timber.
8 For impact on water needs related to use of diffusers, the recycling process and other practices, see Chapter 5.

9 The amoeba shows a radial diagram in which each axis represents one of the selected indicators with its normalized values, so the maximum value is always equal to 1. Each scenario is plotted in the diagram, by joining the different points (corresponding to each indicator) attained by the scenario in each axis, as well as the points attained in the ideal situation (represented by the outer pentagon).

10 Results are presented through groups of sub-scenarios that show similar values.

11 Results are presented through groups of sub-scenarios that give similar values (in Mauritius scenarios B and C are also grouped since they both reach electricity demand targets).

References

Agama (2003) "Employment Potential of Renewable Energy in South Africa", SECP.

Amezaga, J.M., Von Maltitz, G. and Boyes, S. (eds) (2010) *Assessing the Sustainability of Bioenergy Projects in Developing Countries: A Framework for Policy Evaluation*, Newcastle University.

Balaton Group (1998) "Indicators for Sustainable Development: Theory, Method, Applications", A Report to the Balaton Group, IISD, Canada.

Bekunda, M., Palm, C.A., de Fraiture, C., Leadley, P., Maene, L., Martinelli, L.A., McNeely, J., Otto, M., Ravindranath, N.H., Victoria, R.L., Watson, H. and Woods, J. (2009) "Biofuels in developing countries", in R.W. Howarth and S. Bringezu (eds) *Biofuels: Environmental Consequences and Interactions with Changing Land Use* (Proceedings of the Scientific Committee on Problems of the Environment (SCOPE) International Biofuels Project Rapid Assessment, 22–25 September, Gummersbach, Germany), Cornell University, Ithaca, NY, http://cip.cornell.edu/biofuels/.

BUN (2004) "Interviews of Illicit Ethanol Brewers in St Mary's and Epworth Townships, Harare", Internal Survey Report, Biomass Users Network (BUN), Harare.

Cheesman, O.D. (2005) *Environmental Impacts of Sugar Cane Production*, CABI Publishing, Wallingford.

Deepchand, K. (2001) "Bagasse-Based Cogeneration in Mauritius – A model for Eastern and Southern Africa", AFREPREN Occasional Paper No 2.

Cotula, L., Dyer, N. and Vermeulen, S. (2008) "Fuelling Exclusion? The Biofuels Boom and Poor People's Access to Land", FAO and IIED.

Cotula, L. and Leonard, R. (eds) (2010) *Alternatives to Land Acquisitions: Agricultural Investment and Collaborative Business Models*, IIED. London.

DWEA (2010) National Climate Change Response Green Paper. Department of Water and Environmental Affairs (DWEA), Government of South Africa, Pretoria, www.environment.gov.za/docs/DocumentHomepage.aspx?type=D&id=4108, accessed 28 February 2011.

Duku, M.H., Gua, S. and Hagan, E.B. (2010) "A comprehensive review of biomass resources and biofuels potential in Ghana", *Renewable and Sustainable Energy Review* (2010), doi:10.1016/j.rser.2010.09.033.

Eaton, C. and Shepherd, A.W. (2001) "Contract Farming. Partnerships for Growth", FAO.

Er, A.C. (2011) "A comparative analysis of the Brazilian bioethanol sector and the Malaysian palm biofuel sector", *Asian Social Science*, vol. 7, no. 2 (February).

Farioli, F. (2002) "The Implications of the Clean Development Mechanism for Sustainable Development: Prospects for Developing Countries", PhD thesis, University of Rome (December).

Farioli, F. and Portale, E. (CIRPS) (2010) "Diligent Tanzania Ltd. Good Practice Assessment for Bioenergy Projects", Report prepared for the COMPETE Project.

Gaye, A. (2007) "Access to Energy and Human Development", UNDP Human Development Report 2007–2008 Occasional Paper 2007/25, New York.

Glover, D. and Kusterer, K. (1990) *Small Farmers, Big Business: Contract Farming and Rural Development*, Macmillan Press, Basingstoke.

Goldemberg, J., Coelho, S.T. and Guardabassi, P. (2008) "The sustainability of ethanol production from sugarcane", *Energy Policy*, vol. 36, no. 6, pp. 2086–2097.

GTZ (2009) ProBEC Biofuel Newsletter, September, www.probec.org/precachesection. php?czacc=erika&zSelectedSectionID=sec1256029660_RelatedFiles, accessed on 12 August 2010.

Handley, G., Higgins, K. and Sharma, B. with Bird, K. and Cammack, D. (2009) "Sub-Saharan Africa: An Overview of Key Issues", Overseas Development Institute Working Paper 299, London.

IIED (2004) "Local Sustainable Development Effects of Forest Carbon Projects in Brazil and Bolivia", International Institute for Environment and Development, January.

IUCN (2011) "High Conservation Value", www.iucn.nl/our_themes/high_conservation_ value_hcv/, accessed on 3 March 2011.

Jaffee, S.M. and Little, P.D. (1994) "Living under contract: contract farming and agrarian transformation in sub-Saharan Africa", in P.D. Little and M.J. Watts (eds), *Contract Farming in the Shadow of Competitive Markets: The Experience of Kenyan Horticulture*, University of Wisconsin Press, Madison, WI.

Leal, M. (2006) "Environmental and Social Impacts", Technical Paper for IC.

Lloyd, P.J. and Visagie, V.M. (2007) "The Testing of Gel Stoves and their Comparison to Alternative Cooking Fuels", at the Proceedings of the International Conference on the Domestic Use of Energy, 11–12 April, Cape Town, pp. 59–64.

López-Ridaura, S., Masera, O. and Astier, M. (2002) *Evaluating the Sustainability of Complex Socio-Environmental Systems*, The MESMIS framework (Ecological Indicators 2), Elsevier, pp. 135–148.

Macedo, I.C. (2005) *Sugar Cane's Energy – Twelve Studies on Brazilian Sugar Cane Agribusiness and its Sustainability*, UNICA, Brazil.

Macedo, I.C., Leal, M.R.L.V. and Silva, J.E.A.R. (2004) "Assessment of Greenhouse Gas Emissions in the Production and Use of Fuel Ethanol in Brazil", Report to the Secretariat of the Environment of the State of São Paulo.

Neumayer, E. (2003) *Weak versus Strong Sustainability: Exploring the Limits of Two Opposing Paradigms*, Edward Elgar, Cheltenham.

OECD/IEA (2010) "Sustainable Production of Second Generation Biofuels, Potential and Perspectives in Major Economies and Developing Countries", Information Paper, February, www.iea.org/papers/2010.second_generation_biofuels.pdf.

PASASA (2011) Paraffin Matters newsletter (various), Paraffin Safety Association of South Africa (PASASA), www.paraffinsafety.org/paraffin-matters-newsletter-issue-7/, accessed 15 February 2011.

Porter, G. and Phillips-Howard, K. (1997) "Comparing contracts: an evaluation of contract farming schemes in Africa", *World Development*, vol. 25, no. 2 (February), pp. 227–238.

SADC (2005) "Feasibility Study for the Production and Use of Biofuel in the SADC Region", Gaborone.

SASA (2006) "Standards and Guidelines for Conservation and Environmental Management in the South African Sugar Industry", www.sasa.org.za/sasaservices/standards/contents. asp, accessed July 2006.

Sulle, E. and Nelson, F. (2009) *Biofuels, Land Access and Rural Livelihoods in Tanzania*, IIED, London.

UNDP (2010) Human Development Report 2010, UNDP, New York.

Von Maltitz, G.P. and Brent, A. (2008) "Assessing the biofuel options for Southern Africa", http://researchspace.csir.co.za/dspace/bitstream/10204/2579/1/Von per cent20Maltitz_ 2008.pdf, accessed 31 January 2011.

Walter, A., Dolzan, P., Quilodrán, O., Garcia, J., da Silva, C., Piacente, F. and Segerstedt, A. (2008) *A Sustainability Analysis of the Brazilian Bio-ethanol*. Universidad de Campinas and DEFRA, São Paulo.

WCED (1987) "Our Common Future (Brundtland Commission Report)", World Commission on Environment and Development (WCED), Oxford University Press, London.

Woods, J. (2006) "Science and technology options for harnessing bioenergy's potential", "Bioenergy and Agriculture: Promises and Challenges", IFPRI 2020 Focus 14.

Woods, J., Mapako, M., Farioli, F., Bocci, E., Zuccari, F., Diaz-Chavez, R. and Johnson, F.X. (2008) "The Impacts of Exploiting the Sugar Industry Bioenergy Potential in Southern Africa – Options for Sustainable Development", Thematic Report 4 on Social and Environmental Impacts, Cane Resources Network for Southern Africa (CARENSA).

Woods, J., Tipper, R., Brown, G., Diaz-Chavez, R., Lovell, J. and de Groot, P. (2006) "Evaluating the Sustainability of Co-firing in the UK", Report for the DTI.

WWF (2005) "Better Sugar Meeting Report", 23–24 June, UK, www.wwf.org.uk/filelibrary/pdf/ag_bsi_meeting.pdf, accessed 24 July 2006.

Yamba, F.D. and Matsika, E. (2003) "Assessment of the Potential of State of the Art Biomass Based Technologies in Contributing to a Sustainable SADC Regional Mitigation Energy Scenario", Paper presented at the Risoe Energy Conference, Copenhagen, Denmark, 19–21 May, http://130.226.56.153/rispubl/presentations/yamba.pdf, accessed on 2 December 2010.

12

CERTIFICATION AND STANDARDS FOR SUGAR CANE AND BIOENERGY

Experiences with development and application and their relevance for Africa

Rocio A. Diaz-Chavez and Anna Lerner

Introduction

The growing interest in biomass and its by-products for biofuel and bioenergy production has made its potential costs and benefits in terms of sustainable use more evident. The need for sustainability reporting and control mechanisms has become increasingly acknowledged. Sustainability reporting and control mechanisms, which frequently fall under the category of "sustainability standards", attempt to provide assurance that an activity or product is environmentally, socially and economically sustainable. With regard to bioenergy and biofuels, sustainability standards address several factors, including contribution to a reduction of greenhouse gases (GHG), impacts on the surrounding environment, impacts on economies and impacts on livelihoods (Woods *et al.*, 2007).

A *standard* refers to a set of principles and criteria to be used consistently as rules, guidelines or definitions of characteristics to ensure that materials, products, processes or services sufficiently address the metrics put forth by the standard. Standards also define indicators and methods to measure compliance with principles and criteria.

- *Standard:* set of principles and criteria to be used consistently as rules, guidelines or definitions of characteristics.
- *Principles:* general tenets of sustainable production.
- *Criteria:* conditions to be met in order to achieve these tenets, and which help define indicators.
- *Indicators:* individual proof or measurable unit that indicates if a farm, producer or company meets a particular criterion.

Standards have long existed to certify various crops and commodities (e.g. forestry, cotton, sugar). According to the International Organization for Standardization (ISO, 2011), standards are useful as they facilitate trade, and provide information on

health, environmental and technical issues to facilitate governments in creating legislation. An organization can be certified to show that it follows a quality standard.
The certification is issued by a "certification body", which has been accredited to
do so by a national "accreditation body".

In the case of biofuels, since the EU Directive on Biofuels (EC, 2003) came into
force the setting of targets to achieve a reduction in greenhouse gases in Europe has
promoted an increasing demand for biofuels imports from developing countries.
This increment was expected to come mainly from sugar cane, soya, palm oil, rape-
seed, wood products and other biofuel feedstock (see Walter *et al.*, 2008; Diaz-
Chavez and Woods, 2008).

One of the first calls to put an eco-certification system for biofuels in Europe
came from a report from WWF (WWF, 2006), not only for those biofuels produced internally but also for those imported. The European Commission (EC) also
acknowledged the concern, indicating that depending on the production process
and on the land used for this purpose biofuel production could be either an environmentally friendly process that contributes positively to climate change mitigation or the opposite (COM, 2006).

In April 2009, the European Parliament adopted the Directive on the Promotion of Energy from Renewable Sources (2009/28/EC). The Renewable Energy
Directive set targets to cut greenhouse gas emissions by 20 per cent, to establish a
20 per cent share for renewable energy and to improve energy efficiency by 20 per
cent, to be achieved by 2020 (known as the *20–20–20* targets). Additionally, a
specific target was announced to establish a 10 per cent share of renewables in transport by the same date. These commitments also included reporting obligations for
the Commission on the impact on social sustainability of increased demand for
biofuels in the EU and in third countries. In addition, the EU biofuel policy
required an assessment on the impacts on the availability of food at affordable prices,
in particular for people living in developing countries. Based on the results of these
reporting obligations on social sustainability, a revision of the Renewable Energy
Directive is foreseen possibly to include additional criteria ensuring the socio-
economic sustainability of biomass and biofuels (Diaz-Chavez, 2010).

Currently, most sugar cane-producing countries have local production standards
for emission reductions/allowance and apply the International Organization for
Standardization ISO 14000 environmental management standard. The series of ISO
14000 aim to establish procedures to control and improve environmental performance. Nevertheless, the development of better management practices can be seen as a
tool to understand and reduce environmental and social impacts, promoting sustainable development in producer countries. It is likely that producers who adopt better
management practices can expect more efficient production to increase their net
margins and improve social and environmental outcomes. Among the major agricultural crops that are of interest for bioenergy is sugar cane for ethanol production. For
bioethanol production on an industrial scale, the possibility of applying the better
management standards would benefit the producer countries by promoting a more
sustainable production of sugar cane. This chapter explores the available standards and

schemes to verify or certify sugar cane production worldwide and highlight some specifications in Africa. Lastly, an attempt is made at reviewing some of the implications from standards for the export markets in third countries.

Overview of sustainability issues and criteria in general

A number of issues need to be considered to ensure both a sustainable production and use of biomass oriented towards energy needs, while at the same time reducing GHG emissions. Amongst these are environmental and social concerns, which bring into consideration the area of land required from energy crops for producing electricity and biofuels for transport. Additionally, other concerns include the effects that the large-scale cultivation of energy crops and use of residues may have on biodiversity, soils, hydrology and landscape (Diaz-Chavez and Woods, 2008).

The main environmental criteria contained in a number of available and proposed standard systems include among others:

1. biodiversity (including GMO) and natural ecosystems;
2. water (efficient use, conservation and pollution);
3. soil conservation and maintenance;
4. crop management (use of fertilizers and pesticides);
5. waste management;
6. landscape impacts.

The social and economic impacts have focused on working conditions, land use rights, child labour, health and safety and forced labour.

In the UK, under the Renewable Transport Fuel Obligation (RTFO), a framework to verify the sustainability for imported biofuels was set (RFA, 2010). Although it applies to liquid biofuels to be used for transport, the social criteria selected from previous reports (e.g. Social Accountability in Sustainable Agriculture (SASA) (ISEAL, 2010)) are also applicable to biomass production and use in general, such as:

1. child labour;
2. freedom of association;
3. discrimination;
4. health and safety;
5. forced labour;
6. wages;
7. working hours (plus standard only);
8. contracts and subcontractors;
9. land rights.

Some of these criteria may be of greater relevance in developing countries (such as Brazil and a number of African countries). Nevertheless, some also apply to the EU member states, particularly new members from Eastern Europe (Diaz-Chavez and

Woods, 2008). Discussion of the socio-economic impacts of biofuel production must balance the significant potential for rural development and value added with the very real danger of environmental degradation and the social challenges associated with food security, access to land and conflict over resources such as water, both in developed and developing economies.

Standards and certification schemes. Purpose and criteria

The current efforts towards certification systems for biofuels and related commodities include the different international Roundtables on Sustainable Palm Oil (RSPO) and on Responsible Soy (RTRS), as well as the Roundtable on Sustainable Biofuels (RSB). Other national efforts in the European Union include the Cramer Report from the Netherlands (ETSF, 2006); the UK verification system for the Renewables Transport Fuel Obligation (RTFO) under the Renewable Fuels Agency (RFA) and the German initiative on biofuels (Meyer, 2008).

Other efforts outside the EU include those associated with the State of California, which has traditionally had some of the strictest air quality and environmental standards in the US, and thus included criteria within the Low Carbon Fuel Standard (LCFS) (Farrel and Sperling, 2007a, 2007b). International initiatives include the work of the United Nations Environment Programme (UNEP) and the Global Bioenergy Energy Partnership (GBEP), which has established a cooperation platform on bioenergy between the partner countries, several UN institutions and several international organizations (GBEP, 2010).

The following sections present some of the verification schemes and standards that are currently used in particular for sugar cane and ethanol production. Specific emphasis is on the standards and verification schemes that have been proposed for or used in various regions in Africa.

International initiatives

Bonsucro (previously Better Sugarcane Initiative)

Developed as an attempt to reduce the negative social and environmental impacts of sugar cane production, Bonsucro is a global multi-stakeholder non-profit initiative founded in 2008. Bonsucro has developed a production standard that is in line with the ISO 65 standard (ISO, 2011; on general requirements for bodies operating product certification systems), and is intended to constitute an auditable document serving to measure impacts and promote sustainable practices. In July 2010, the final version of the standard was presented after a multi-stakeholder consultative development process as outlined by the ISEAL Alliance and includes the following principles (ISEAL, 2010).

1.   Obey the law.
2.   Respect human rights and labour standards.

3.   Manage input, production and processing efficiencies to enhance sustainability.
4.   Actively manage biodiversity and ecosystem services.
5.   Continuously improve key areas of the business.

For entitlement to Bonsucro certification, members must adhere to principles as well as to their corresponding indicators. At least 80 per cent of the indicators must be satisfied, in addition to complying with a number of core criteria. The core criteria are:

1   To comply with relevant applicable laws.
2.1   To comply with the International Labour Organization (ILO) labour conventions governing child labour, forced labour, discrimination and freedom of association and the right to collective bargaining.
2.4   To provide employees and workers (including migrant, seasonal and other contract labour) with at least the national minimum wage.
4.1   To assess impacts of sugar cane enterprises on biodiversity and ecosystems services.
5.7   For greenfield expansion or new sugar cane projects, to ensure transparent, consultative and participatory processes that address cumulative and induced effects via an environmental and social impact assessment (ESIA).

It is expected that ethanol producers wishing to use the Bonsucro standard for the European market, will have to implement and demonstrate compliance with the above-specified standard. Furthermore, full compliance with the additional requirements listed under Section 6 of the production standard is mandatory. Section 6 covers the requirements for biofuels under the EU Renewable Energy Directive (RED) 2009/28/EC and the revised Fuel Quality Directive (FQD) 2009/30/EC (Bonsucro, 2011).

In order to facilitate trade for Bonsucro-certified biofuels in the EU, a warrant compiling the sustainability characteristics will be assigned to each certified biofuel consignment. The verification and audit requirements specified by the Bonsucro certification are outlined in the Certification Protocol that, together with the Chain of Custody Standard and its guidelines, is designed to ensure that compliance is honoured.

One of the main advantages of the Bonsucro standard is that it is the only one that addresses the three main products of sugar cane: sugar, ethanol and cogeneration through Principle 3, on "Manage input, production and processing efficiencies to enhance sustainability" (Bonsucro, 2011).

The Global Bioenergy Energy Partnership (GBEP)

The GBEP partners comprise the G8 countries plus five (Mexico, South Africa, China, India and Brazil). The GBEP Task Force on Sustainability, established under the leadership of the UK, is developing a set of global science-based criteria and indicators, with examples of experiences and best practices including benchmarks, regarding the sustainability of bioenergy (GBEP, 2010). These criteria are based on four themes: environmental, economic, social and energy security. The criteria and indicators developed have been under review by the GBEP members.

International Organization for Standardization (ISO)

The International Organization for Standardization (ISO) has been active since its founding in 1906 in establishing international standards to make the development, manufacture and supply of products and services more efficient, safer and cleaner (ISO, 2011). ISO standards are voluntary and are applied and monitored worldwide through 3,000 ISO technical groups. There are mainly two types of ISO standards, the ISO 9000, which is concerned with quality management, and the ISO 14000, which is concerned with environmental management (Diaz-Chavez, 2007).

Though the ISO system for production of biofuels is not yet specified it has some standards for agriculture and forestry that may well be considered for application (ISO, 2011). ISO also launched the new standard, ISO 14064, for greenhouse gas accounting and verification. This standard aims to provide government and industry with an integrated set of tools for programmes aimed at reducing greenhouse gas emissions, as well as for emissions trading (ISO, 2006). The ISO is currently working on a more specific sustainability standard for biofuels and bioenergy.

The International Social and Environmental Accreditation and Labelling Alliance (ISEAL)

ISEAL is an association of international organizations engaged in standard setting, certification and accreditation focused on social and environmental issues. ISEAL has a code of good practice (ISEAL, 2010) that provides a benchmark to assist standard-setting organizations in developing their social and environmental standards. The normative documents on which ISEAL based its Code of Practice include the ISO/IEC Guide 59 Code of Good Practice for Standardization, the ISO/IEC 14024 (environmental standards) and also the World Trade Organization Technical Barriers to Trade Agreement. Although ISEAL does not provide direct standards related to specific topics (e.g. agriculture, biofuels), the points marked in its Code of Practice are relevant to setting a standardization system with reference to biofuels from its production throughout the supply chain (e.g. 7. Effectiveness, relevance and international harmonization, ISEAL, 2010, p. 5)

The Roundtable on Sustainable Biofuels

The Roundtable on Sustainable Biofuels (RSB), created in 2007, is an international multi-stakeholder initiative that has brought together over 500 individuals from companies, NGOs, governments, along with experts from nearly 40 countries. Through a series of online consultations, teleconference discussions, and in-person stakeholder meetings in Brazil, China, South Africa and India held between June

2007 and July 2008, the Roundtable drafted a series of principles and criteria of a global sustainability standard called "Version Zero". Accompanied by some pilot testing of the draft standard, Version Zero has been reviewed and worked into "Version 1", and this is presently under consultation to establish "Version 2". The RSB initiative is considered as a key stakeholder in the sustainability standard arena, with potential to gather the various initiatives and somehow harmonize the biofuel sustainability schemes (RSB, 2010).

International Sustainability & Carbon Certification (ISCC)

The International Sustainability & Carbon Certification (ISCC) system is supported by the German Federal Ministry of Agriculture/Agency for Renewable Resources, and Méo Carbon Solutions GmbH is operating the system. The system was developed in 2006 to respond to the German Sustainability Regulation (BionachV) and the EU Directive on the Promotion of Renewable Energy Sources. After pilots in 2008, the first certifications are expected in 2011 (ISCC, 2011).

The objective of the system is to test an international, pragmatic certification system, with the lowest possible administrative burden, that reduces the risk of unsustainable production and can be used as a proof of GHG emissions of biofuels on a life-cycle basis. The standard includes ten principles, with corresponding criteria and indicators. Three of the ten principles are related to social matters, two are related to management, and five are related to the environment.

National and local schemes

The UK Renewable Transport Fuel Obligation

The UK formed the Renewable Fuels Agency (RFA), which introduced guidance on "Carbon and Sustainability" through the Renewable Transport Fuel Obligation (RFA, 2010). In the UK, the original RFA started to verify imported biofuels under the RTFO in April 2008, making it the first system of its kind to operate in the world.

Currently under the Department of Transport, the RTFO comprises seven sustainability principles: five environmental and two social. These seven principles have been used to define the RTFO sustainability meta-standard. A meta-standard approach enables the use of existing certification schemes to meet the standard. The suppliers of biofuels are required to report data on the volumes and source of biofuels introduced to the market in the UK (DfT, 2011).

The Cramer Report from the Netherlands

Along with the UK system, the Testing Framework for Sustainable Biomass, also known as the Cramer Report, was one of the first to be developed in Europe. The Cramer Report is the result of the Dutch "Sustainable Production of Biomass" report commissioned by the Energy Transition Task Force to formulate sustainability criteria for the production and processing of biomass for energy, fuels and chemistry (ETSF, 2006). The system looked at developing sustainability criteria to be used by 2007, and tightened criteria to be used by 2011. These criteria are arranged into six themes:

- greenhouse gas balance;
- competition with food, local energy supply, medicines and building materials;
- biodiversity;
- economic prosperity;
- social well-being;
- environment.

The certification in Brazil SMA (Secretary of Environment of São Paulo State)

The State of São Paulo in Brazil has a certification system that serves to avoid sugar field fires (*queimadas*) during the sugar cane harvesting season (SMA, 2007). This programme is called Green Fuel (from the Environmental Secretariat of São Paulo State) and rewards with a certificate those plantations or ethanol plants (*usinas*) that do not burn the cane fields and instead produce sugar cane in a sustainable form. This programme is implemented in collaboration with the National Union of Sugar Cane Producers (UNICA).

At the same time, there has been an initiative at the federal level coordinated by the National Institute of Metrology for the certification of biofuels, and this is currently working on the ISO bioenergy certification in collaboration with the USA.

Private initiatives

SEKAB

SEKAB has a "Verified Sustainable Ethanol Initiative", which uses a series of criteria and indicators currently under verification (SEKAB, 2008) for SEKAB's sources of ethanol on sale in Sweden. These criteria include the following.

- At least 85 per cent reduction in fossil carbon dioxide compared with petrol, from a well-to-wheel perspective.
- At least 30 per cent mechanization of the harvest now, plus a planned increase in the degree of mechanization to 100 per cent.
- Zero tolerance for felling of rainforests.
- Zero tolerance of child labour.
- Rights and safety measures for all employees in accordance with UN guidelines.
- Ecological consideration in accordance with UNICA's environmental initiative.
- Continuous monitoring to ensure that the criteria are being met.

SEKAB's verification system is used mainly for ethanol produced from sugar cane in Brazil.

Greenergy

According to the RFA (2008), Greenergy was the first UK company that reported fuel meeting the RTFO social sustainability meta-standard. The company receives supplies of Brazilian sugar cane ethanol. Greenergy produced a report interpreting the sustainability criteria from the RTFO especially for the Brazilian case (Greenergy, 2008). The report was completed during the same period that the Better Sugar Initiative was developing its standards. Greenergy continues to verify its ethanol with its own standard in compliance with the RTFO.

Current status in Europe

In July 2011, the European Commision (Energy Directorate) recognized seven voluntary sustainability schemes for biofuels to assure that the biofuels production comply with the Renewable Energy Directive sustainability issues. These comprise some of the previously mentioned standards (ISCC, Bonsucro, RSB, Greenergy) and other initiatives, including 2BSvs (Biomass Biofuels voluntary scheme), RTRS (Round Table on Responsible Soy EU RED) and the RBSA (Abengoa RED Bioenergy Sustainability Assurance) (EC, 2011).

Schemes in Africa

South African Sugar Association (SASA)

SASA published "Standards and Guidelines for Conservation and Environmental Management in the South African Sugar Industry" (SASA, 2008). The purpose of these guidelines is to set local priorities and standards for the achievement of good levels of environmental management in the mill area. They cover points related to legislation on environmental protection in the sugar cane industry considered for local and export markets. The issues covered are: field practices; water; air pollution; soil; traffic regulations and cane spillage; services; management and use of natural resources and cultural assets; and public recreational facilities and Environmental Management System (EMS) audits. The guidelines are supposed to be updated regularly (SASA, 2008).

West Africa initiatives

The African Biofuels and Renewable Energy Fund (ABREF) was set up through the Economic Community of West African States (ECOWAS) in Africa in order to facilitate investments for biofuels and renewable energy projects. This fund aims to promote development particularly in West Africa and will generate "certified emissions reductions" (CERs). It will be applicable to all Africa but the starting point is centred in West African countries. It includes all types of energy projects that generate CERs through the Clean Development Mechanism, including biofuels (Globalbiofuels, 2011).

The Union Economique et Monétaire Ouest Africaine (UEMOA) also has a biomass programme, which assesses biomass considering biochemical and thermo-chemical processes. The UEMOA and the United Nations Foundation also reported on the state and possibilities of bioenergy production in UEMOA member countries (UEMOA, 2008). Although there are currently no standard systems in West Africa some countries producing feedstocks (e.g. Ghana, which produces *Jatropha curcas* L.) are expected to consider sustainability requirements to access the EU market.

The Southern African Development Community (SADC) framework

As one of several recent attempts to improve regional collaboration and promote sustainable bioenergy development in the Southern African Development Community (SADC) region, the SADC Secretariat developed the SADC Framework for Sustainable Biofuel Production in 2009. The development of a sustainability framework came as a response to a call from SADC energy ministers in 2008, to accelerate the initiatives in developing biofuels as a source of alternative and cheap environmentally friendly fuel that could also contribute to improving rural development and reducing poverty (SADC, 2008). The framework was approved at the SADC energy ministers' meeting in Luanda in April 2010 (Lerner, 2009), and individual SADC member states were due to present national roadmaps on how to adopt and adjust the regional scheme to national conditions in early 2011.

As stated by Lerner and Stiles (2010), the SADC framework is designed as a set of broad general recommendations describing how regional biofuel production should adhere to environmental, economic and socially sustainable principles. These general regional guidelines should later be adapted by individual SADC member states to meet national requirements and preconditions recognizing, for example, different legislative regimes, national development priorities, land tenure issues and specific local conditions affecting biofuels crop selection. The final phase is for SADC member states to implement sustainable policies in line with the SADC framework and mainstream sustainable practices in biofuel relevant strategy documents (e.g. biofuel policy, poverty reduction strategy, green revolution strategy, food security strategy, budget planning). The policy document guiding the process in the member states is called the National Roadmap for the Implementation of SADC Sustainability Framework (Lerner, 2010).

The guiding principles of the SADC framework involve the:

> development of a biofuel industry that throughout the value chain promotes respect for, and inclusion of, SADC citizens in the biofuel production; the protection and sustainable management of biodiversity and natural resources, and a sustainable economic approach contributing to overall development and social well-being.
>
> (Lerner, 2010)

In August 2010, SADC member states gathered in Mozambique to discuss potential implementation strategies for the national adaptation of the regional sustainability framework. Given the differences in biofuel potentials, policy priorities and legislative frameworks in the various countries, a wide range of implementation roadmaps were proposed. South Africa, with its mandatory licensing requirement for biofuel operators, intends to include sustainability criteria within the existing licensing procedures. Swaziland and Botswana, both countries on track to finalize biofuel policies, suggest including the sustainability criteria in the general investment guidelines that will be elaborated. Mozambique, the member state that has come furthest in the implementation process, opts for a national definition of sustainability of biofuels and the utilization of existing procedures and regulations outlined for biofuel investments (Lerner, 2010).

Defining the meaning of sustainability on the national level – the case of Mozambique and its biofuel sustainability principles

Mozambique approved its National Biofuel Policy and Strategy in mid-2009. As one of several implementation pillars, the sustainability framework has been under development since the policy approval. Based on the outlined objectives and priorities of the government of Mozambique (GoM), the sustainability framework drafting process seeks to define the meaning of sustainability of biofuels in a Mozambican context, as expressed by Mozambican stakeholders. To ensure international compatibility, existing international sustainability schemes such as the EU Fuel Quality Directive, Renewable Transport Fuel Obligation, the Dutch Cramer criteria, the Roundtable on Sustainable Biofuels (RSB), the Better Sugarcane Initiative (BSI) and the Global Bioenergy Partnership (GBEP), as well as experience from social certification initiatives from Brazil, guide the development of the Mozambican framework (Schut et al., 2010).

The objective of the framework is to provide a set of minimum standards as to how sustainable biofuel production should be produced in Mozambique, in order to guarantee the availability of natural resources for future generations and to maximize the social and economic benefits of production for Mozambique (Lerner et al., 2010). Lerner and Schut (2010) further state that in order to facilitate the implementation of the sustainability framework, reduce the cost of compliance and increase the transparency of the system, the framework was designed to fit the existing Mozambican Biofuel Project Application and Land Acquisition Process. The guiding principles of the sustainability framework are:

- legalities;
- social responsibility;
- energy security;
- macro-economic benefits and financial viability;
- food security;
- agricultural productivity;
- environmental protection.

The first phase of the drafting process ended in October 2010 with stakeholder consultations on the proposed principles and criteria. It is worth noticing that the stakeholder feedback from the capital city Maputo and the provincial cities of Nampula and Beira were very different, strengthening the argument for extensive and diverse stakeholder consultations. Biofuel stakeholders, representing civil society, government, private investors and academia, discussed the first version of the framework, resulting in a "Versão 1 dos critérios de sustentabilidade de Mozambique". As discussed by Lerner and Schut (2010), the second phase that was initiated in late 2010 strives to outline indicators as well as an implementation strategy aiming for the most time and cost-efficient way of integrating the sustainability scheme into existing biofuel legislation and procedures.

BOX 12.1: ASSESSING POTENTIAL FOR SUSTAINABLE BIOFUEL PRODUCTION IN MOZAMBIQUE – THE HIGH CONSERVATION VALUE EXERCISE OF GTZ, PROFOREST AND ECOENERGIA

The high conservation value (HCV) concept was initiated by the Forest Stewardship Council in 1999, aiming to safeguard forest areas of critical biological or social value in the context of commercial forestry (HCV, 2009). However, the framework is now increasingly being used in commodity crop planning and management and can be found in several biofuel certification schemes, like the Roundtable on Sustainable Palm Oil (RSPO) principles and criteria, the Dutch Government's Cramer principles, and the UK's Renewable Transport Fuel Obligation (RTFO) meta-standard. While the HCV exercise is less of a certification scheme in the traditional sense and should be seen more as a strategic planning and development tool, it does provide both the site operator as well as other stakeholders with useful information to ensure compliance with traditional sustainability schemes for biofuel.

The HCV framework contains the following six types of high conservation value areas:

HCV1: Areas containing globally, regionally or nationally significant concentrations of biodiversity values (e.g. endemism, endangered species, refugia).

HCV2: Globally, regionally or nationally significant large landscape-level areas where viable populations of most if not all naturally occurring species exist in natural patterns of distribution and abundance.

HCV3: Areas that are in or contain rare, threatened or endangered ecosystems.

HCV4: Areas that provide basic ecosystem services in critical situations (e.g. watershed protection, erosion control).

HCV5: Areas fundamental to meeting the basic needs of local communities (e.g. subsistence, health).

HCV6: Areas critical to local communities' traditional cultural identity (areas of cultural, ecological, economic or religious significance identified in cooperation with such local communities).

The first step of the HCV process identifies which HCV areas are present on the assessment site (HCV, 2009). The next step identifies how the HCV areas must be managed in order to maintain or enhance the identified HCVs. The last step establishes an appropriate monitoring regime to ensure that the management practices are effective in their aim of maintaining or enhancing the HCVs.

Since 2007 the German Technical Cooperation (GTZ) has gathered extensive field experience from the Southern African Development Community (SADC) region, assessing the usefulness of social and ecological standards in relation to sustainable biofuel development (Meissner, 2009). In an attempt to assess potential for sustainable biofuel production in Mozambique, GTZ tested the applicability of various existing standards in the biofuels sector. Together with HCV specialist consultants ProForest, GTZ performed an HCV exercise on the proposed biofuel investment site of Swedish–Mozambican investors EcoEnergia, in northern Mozambique (Lerner, 2008). The investment focused on a mixed farming system consisting of sugar cane and sweet sorghum.

Findings from ProForest and local biodiversity and social experts indicated that attributes existed that could be considered high conservation values. One benefit for the biofuel operator of the Mozambican exercise was that EcoEnergia could incorporate information on areas where high conservation values were thought to exist at an early stage of project development. Integration of this information to a land use management plan enabled EcoEnergia to ensure that future land use change did not prevent the values from being maintained or enhanced within the project landscape.

(ProForest, 2009, p. 3)

As discussed by Lerner (2008), another useful result of the exercise was that it became clear to the government that existing land use zoning maps were insufficiently detailed to ensure a sustainable biofuel industry expansion while safeguarding important ecological and social values. According to Lerner and Schut (2010), a second phase of a land zoning exercise was initiated by the Government of Mozambique in 2009, striving to provide more detailed maps at both the national and provincial level and highlighting biodiversity hot spots and other conservation areas.

Implications of the use of standards in the African context

The review of the main standards and schemes that have or are expected to have an impact in Africa is presented in Table 12.1. This shows some of the main standards (e.g. Bonsucro) and main verification systems (e.g. the EU RED, Cramer and the

TABLE 12.1 Overview of the main standards and verification systems for sugar cane in Africa

Standard, scheme/ main criteria	EU (RED) sustainability topics	Bonsucro (BSI)	Cramer (NL)	RTFO (UK)	RSB	SADC
Operational (in 2010–2011)	✓ (defined in each country)	✓		✓		
GHG balance	✓	✓ climate change impacts	✓	✓	✓ climate change	✓ CC adaptation
Biodiversity	✓	✓	✓	✓	✓	✓ No forest degradation
Carbon	High carbon stock		Carbon sinks	High carbon stock		
Soil	Not specific, independent report	✓ improve status	Soil quality	Soil degradation	✓	
Water		✓ improve status	Water degradation	Water contamination	✓ (availability and quality)	✓ (availability and quality)
Air pollution		✓		✓	✓	✓ quality
Production		Monitor production and process efficiency				
Socio-economic	Voluntary report on • ILO conventions • food prices • food supply	ILO labour rights, human rights, land rights, stakeholders' participation, economic sustainability	No competition with food • local prosperity • social well-being	Workers' rights • community consultation • land rights	Food security • human and labour rights • socio-economic development consultation • land rights • legality	Food security • human and labour rights • rural development • land rights • legality • energy security • economic sustainability • local participation in value chain

RTFO). The main standards and verification systems that are actively in use in Africa for sugar cane production are Bonsucro, the RTFO and SADC. Bonsucro considers all the co-products of sugar and ethanol and their use in cogeneration. The RSB also considers them but is not yet operational.

The Competence Platform on Energy Crop and Agroforestry Systems for Arid and Semi-arid Ecosystems – Africa (COMPETE) explored different certification schemes. As was concluded in the COMPETE project, most countries in Africa lack national specific standards for biofuel production, with the exception of the ongoing development in the SADC region. Given this lack of implemented national schemes, COMPETE also reviewed the possibilities of Fair Trade as an entry level for standards systems for bioenergy production. There are many commodities grown in Africa that are exported to the international market and several of the commodities already show compliance to this standard. Some examples of implemented commodity certification schemes are Fair Trade, GlobalGap and Organic (e.g. for coffee, cotton, chocolate). Although the scale used for these commodities is different to the scale considered for bioenergy projects, the existence of implemented commodity schemes demonstrates the experience in Africa using standards. The COMPETE project, in response to what African partners and other stakeholders considered as main issues in bioenergy sustainability production, produced a guideline on good practices for bioenergy production in Africa with a set of principles and criteria (Diaz-Chavez, 2009, 2010a)

Most biofuel-relevant standards considered for African biofuel operations derive from the EU Directive. Given the recent development of the EU Directive and the relatively new emergence of the African biofuel industry it is hard to review impacts on the ground from outlined sustainability schemes. Nevertheless, it is expected that companies and countries that produce a non-verifiable or a non-sustainable biofuel will face restricted EU market access. The restricted market access is, however, a matter for trade disputes and disagreements in light of the non-discriminatory rules stipulated by the World Trade Organization (WTO).

The motivation and role of different organizations in sustainability certification and the expected impacts in the African sugar cane industry are briefly summarized in Box 12.2. Additionally, with the ISEAL Alliance, the "Code of Good Practice for Assessing the Impacts of Standards Systems" (ISEAL, 2010) is expected to create a requirement for all credible standards systems to measure and demonstrate their contributions to social and environmental impacts using consistent methodologies.

A positive side-effect that has been noted in the SADC region from implementing sustainability certification frameworks is the opportunity to reinforce national environmental legislation. Most countries in Southern Africa have comprehensive environmental laws but few of them have sufficient resources to enforce compliance. Biofuel certification can strengthen enforcement and monitoring of environmental regulations since most certification schemes lift the burden to demonstrate compliance from the government to the economic operator.

In the case of biofuel operations in Africa, most large-scale operations have at least one source of foreign investment funding. As a result of the increased foreign investment interest in the agricultural sector, SADC governments have had to revise their investment guidelines and capacitate their investment centres. Most governments have also decided to develop specific biofuel investment guidelines to make sure biofuel investments correspond to the national biofuel strategy and/ or policy objectives. The parallel drafting of sustainability frameworks has led to a number of biofuel investment guidelines being developed in close collaboration with sustainability framework drafting teams. In some cases, such as Botswana, the officials working on the investment guidelines are the same as those developing the national roadmap to implement the SADC sustainability framework. Amongst SADC countries the implementation of sustainability frameworks is hoped to be beneficial through its potential to attract sustainable investors ready to adhere to government biofuel policy objectives and overall poverty reduction strategies.

Another implication of the use of sustainability standards for biofuels in Africa is the realization that community consultation and land tenure processes still require more work. In the case of Mozambique, for example, the current land law is recognized on paper as being very community-sensitive, but the actual implementation of the land law has resulted in several inadequate land consultation processes.

BOX 12.2: SUGAR CANE EXPANSION AND SUSTAINABILITY STANDARDS (RICHARDSON, 2010)

During the reform of the EU sugar regime, the NGO Oxfam released an influential report calling for more meaningful market access for least developed countries in Southern Africa. The report also insisted that sugar companies improve the welfare of their workforce and extend investment in smallholder outgrower schemes (Oxfam, 2004). Such demands are likely to multiply as greater volumes of sugar enter the EU from the "new" LDC exporters and as exposure to media campaigns and consumer demands for fairer production systems increases.

In light of this scrutiny, many sugar cane producers have recognized the need to demonstrate their sustainability across a range of issues. The case of Brazil is instructive here. To mitigate mounting criticism of its cane industry, the federal government has enacted highly visible legislative measures to "zone" expansion away from sensitive ecosystems, and, with the support of the biggest sugar producer association, UNICA, negotiated a tripartite agreement with the trade unions. This agreement was designed to improve working practices beyond the legal minimum required, and support such initiatives as the retraining of manual cane cutters made redundant by mechanization. Alongside such national schemes, UNICA has also made clear its intent to work with multilateral certification bodies in order to demonstrate its compliance with various international sustainability requirements.

In the African context, efforts to address sustainability concerns are evident. In Southern Africa, for instance, Illovo has worked with the NGO WWF to reduce its effluent runoff and help restore the wetlands of the Kafue Flats in Zambia, while Tongaat Hulett continues to promote the economic empowerment of its black employees in South Africa. However, it remains difficult for such companies to attend to a broad scope and effective implementation of sustainability measures and to convey this internationally. This is due, first, to the commonplace reliance on self-regulation, which is perceived to be weak or prone to vested interests, and, second, to the fact that when technical external audits are used (such as the ISO standards) they address a minimum legal compliance rather than a more ambitious sustainability agenda appropriate for modern companies.

Meeting the growing need for a credible and progressive auditing and communication tool is the primary aim of the multilateral certification bodies. These bodies are comprised of different industry stakeholders – including farmers, millers and downstream buyers, as well as members of civil society – to develop an international standard that improves the economic, environmental and social impacts of sugar cane production. Crucially, these are "private" standards, so are not influenced by a particular state and do not interfere with international trade rules. They are instead a means to encourage prominent manufacturers and retailers to source sugar cane from certified producers, and also to enable producers to avert media attention and meet the necessary export regulations.

Conclusions and recommendations

Most of the schemes available for sugar cane certification worldwide are voluntary. Some have the objective of producing sugar cane in a more sustainable form and follow environmental local/national regulations. Others focus on providing best practice guidelines and consider the socio-economics of sugar cane production.

In Africa there have been increasing efforts in recent years in some sugar cane-producing countries to improve sustainability, efficiency and environmental and social regulation. A general conclusion of the SADC developments is that the SADC region is aware of the risks and potentials related to biofuel production and proactively engages to ensure that the developmental promises of biofuels are realized. As the Mozambican case shows, the region aims to use sustainability frameworks and investment guidelines to differentiate between investors ready to adhere to sustainability standards and those less prepared or able to do so. As a sign of ownership and awareness that country preconditions matter, the region initiated the development of a SADC-specific framework. Instead of joining up with international processes, the region chose to further regional collaboration by working with its member states towards a sustainable implementation strategy of biofuels, with the aim of providing a sufficiently flexible system to cater for the various agro-ecological preconditions and the biofuel policy objectives that guide biofuel development in the member states.

Other schemes (e.g. BSI and RSB) are now incorporating GHG assessments in order to respond to the EU Directive, which is the largest export market. Issues on land use change and HCV are also incorporated in some of the international certification schemes (e.g. RSB), reflecting their growing importance in the sustainability discourse. The possibilities of using a code like the one suggested by ISEAL to assess the impacts of certification and standards will aid in providing an idea of the particular implications of biofuels production in developing countries, with an emphasis on Africa.

One of the important results of the HCV assessment in Mozambique is to illustrate the usefulness of the HCV as a powerful land management tool, given that it takes into account global and regional data as well as the local site context. Additional benefit is provided by the global applicability of the HCV concept while it is still flexible enough to be adapted to local conditions. Meissner (2009) argues that the generic concept of High Conservation Values (HCV) can be applied globally across a range of ecosystems, but exactly what those values are, and how to maintain them, must be defined in the regional and local context.

As the HCV concept focuses on the values and not the characteristics of an area, certain land uses, which in other certification schemes would have been restricted, remain feasible under HCV characterization, provided that they do not destroy the value itself. Furthermore, the HCV concept promotes the involvement of the local population, as it is vital to the collection and interpretation of the necessary data. This is likely to strengthen further the important link between project developer and local communities.

Although there are few available standards and verification systems directly related to sugar cane production, it is expected that the export market will continue to grow and applicability of such systems will therefore be extended. Sustainability standards cannot fully ensure that sustainability is achieved in a country – regardless of its development status – but voluntary market standards may contribute to development and ensure that these commodities and co-products (including electricity generation) are produced in a more sustainable manner.

References

Bonsucro (2010) Bonsucro or Better Sugar Cane Initiative Production Standard (July), Better Sugarcane Initiative Ltd, www.bettersugarcane.org/, accessed December 2011.

COM (2006) "Biofuels Progress Report: Report on the Progress Made in the Use of Biofuels and other Renewable Fuels in the Member States of the European Union", accompanying document to the biofuel report, COM (2006) 845 final, SEC(2006) 1721/2, Brussels.

Diaz-Chavez, R. (2007) "Comparison of Draft Standards – Contribution to ECCM, 2006, Environmental Standards for Biofuels", a report commissioned by the Low Carbon Vehicle Partnership, www.lowcvp.org.uk/resources-library/reports-and-studies.asp?pg=11.

Diaz-Chavez, R. (2010a) "Good Practice Guidelines to Project Implementers", Deliverable D3.4/D1.4, Competence Platform on Energy Crop and Agroforestry Systems for Arid and Semi-Arid Ecosystems – Africa, www.compete-bioafrica.net/sustainability/Annex3-3-3-COMPETE-032448-3rdReport2009-D3.4-Good%20practices%20report-Final.pdf.

Diaz-Chavez, R. (2010b) "Sustainability Analysis of Biofuels Production and Use. The BEST Project Approach", Deliverable Report D9.28, Imperial College London, www.best-europe.org/upload/BEST_documents/info_documents/BEST_FinalReport_revfeb10.pdf.

Diaz-Chavez, R. and Woods, J. (2008) "Sustainability Assessment of Biofuels in Practice", Fifth International Biofuels Conference, Winrock International, New Delhi, 7–8 February, www.winrockindia.org/complete_publications_list.htm.

EC (2003) Directive 2003/30/EC of the European Parliament and the Council on the Promotion of the Use of Biofuels or Other Renewable Fuels for Transport, Official Journal of the European Union, Brussels.

EC (2009) Directive 2009/28/EC of the European Parliament and of the Council of 23 April 2009 on the Promotion of the Use of Energy from Renewable Sources and Amending and Subsequently Repealing Directives 2001/77/EC and 2003/30/EC.

EC (2011) Memo: Certification Schemes for biofuels, European Commision MEMO/11/522, http://europa.eu/rapid/pressReleasesAction.do?

ETSF (2006) "Criteria for Sustainable Biomass Production", Final report of the Project Group Sustainable Production of Biomass, Energy Transition Task Force, The Netherlands, www.lowcvp.org.uk/assets/reports/070427-cramer-finalreport_en.pdf.

Farrell, A.E. and Sperling, D. (2007a) "A Low-Carbon Fuel Standard for California – Part 1: Technical Analysis", Regents of the University of California.

Farrell, A.E. and Sperling, D. (2007b) "A Low-Carbon Fuel Standard for California – Part 2: Policy Analysis", Regents of the University of California.

FQD (2009) 2009/30/EC, EU Fuel Quality Directive, www.senternovem.nl/mmfiles/FQD_23042009_EN_tcm24–328226.pdf.

GBEP (2010) Task Force on Sustainability, www.globalbioenergy.org/fileadmin/user_upload/gbep/docs/2008_events/6th_Steering_Committee/TF_Sustainability-Report_to_6th_SC1.pdf, accessed September 2010.

Globalbiofuels (2011) "African Biofuels and Renewable Energy Fund (ABREF)", Globalbiofuels Ltd, http://globalbiofuelsltd.com/partners/partners16.html, accessed January 2011.

Greenergy (2008) "Full Sustainability Criteria", www.greenergy.co.uk/Biofuel_sustainability/PDFs/Standard_Brazilian_sugarcane.pdf, accessed June 2010.

HCV resource network (2009) "HCVs in Natural Resource Certification", High Conservation Value Resource Network briefing note, 3 October, www.hcvnetwork.org/site-info/The per cent20high-conservation-values-folder).

ISCC (2011) "Sustainability Requirements for the Production of Biomass", International Sustainability & Carbon Certification, www.iscc-system.org/e865/e890/e1491/e1496/ISCC202SustainabilityRequirements-RequirementsfortheProductionofBiomass_eng.pdf, accessed January 2011.

ISEAL (2006) "ISEAL Code of Good Practice for Setting Social and Environmental Standards", P005, Public version 4, January, www.isealalliance.org/documents/pdf/P005_PD4_Jan06.pdf, accessed September 2010.

ISEAL (2010) "Impacts Code of Good Practice", ISEAL Alliance, www.isealalliance.org/content/impacts-code, accessed September.

ISO (2006) "Standard 14064 on Greenhouse Gases", Parts 1–3, www.iso.org/iso/iso_catalogue/catalogue_tc/catalogue_detail.htm?csnumber=38381.

ISO (2011) International Organization for Standardization Catalogue of Available and In Process Standards, www.iso.org/iso/catalogue_detail.htm?csnumber=52528.

Lerner, A. (2008) "SADC – ProBEC Biofuel newsletter #8 – 2008", Programme for Basic Energy Conservation (GTZ-ProBEC), accessed through www.probec.org.

Lerner, A. (2009) "SADC – ProBEC Biofuel newsletter #22 – 2009", Programme for Basic Energy Conservation (GTZ-ProBEC), accessed through www.probec.org.

Lerner, A. (2010) "Biofuels, Poverty and Growth-Field Experiences from SADC", Power Point presentation at World Biofuel Markets, Amsterdam, March.

Lerner, A. and Schut, M. (2010) "Resultados do Workshop da consulta sobre a proposta do Quadro de Sustentabilidade dos Biocombustíveis, 18 Maio 2010", Programme for Basic Energy Conservation (GTZ – ProBEC).

Lerner, A. and Stiles, G. (2010) "SADC Framework for Sustainable Biofuel Production", Southern African Development Community Secretariat, Energy Unit.

Meissner, L. (2009) "Implementing Sustainability Standards, Experiences and Challenges From the Field. German Technical Cooperation", Power Point presentation from RSB Consultation, Nairobi, 23 March.

Meyer, S. (2008) "The German Biofuel Sustainability Ordinance (BSO)", Ecofys Germany GmbH.

Oxfam (2004) "A Sweeter Future? The Potential for EU Sugar Reform to Contribute to Poverty Reduction in Southern Africa", Oxfam Briefing Paper, Oxford.

ProForest (2009) "Identification of HCV areas in Cabo Delgado, Mozambique – Summary of key findings for EcoEnergia", Workshop report by ProForest.

Richardson, B. (2010) Personal communication, 20 September, University of Warwick.

RFA (2008) "Quarterly Report 2: 15 April–14 October 2008", Renewable Fuel Agency UK, www.renewablefuelsagency.org/reportsandpublications/rfforeports.cfm, accessed January 2009.

RFA (2010) "Carbon and Sustainability Guidance", Renewable Fuels Agency, www.dft.gov.uk/rfa/reportsandpublications/carbonandsustainabilityguidance.cfm, accessed September 2010.

RSB (2010) "The Roundtable on Sustainable Biofuels", Version 1. http://energycenter.epfl.ch/.

SADC (2008) "Protocol on Energy", Southern Africa Development Community, www.sadc.int/index/browse/page/147#preamble, accessed September 2011.

SASA (2008) "Standards and Guidelines for Conservation and Environmental Management in the South African Sugar Cane Industry", South African Sugar Cane Association, South African Sugar Cane Research Institute.

Schut, M., Slingerland, M. and Locke, A. (2010) "Biofuel developments in Mozambique, update and analysis of policy, potential and reality", *Energy Policy*, vol. 38, pp. 5151–5165.

SEKAB (2008) "Verified Sustainable Ethanol Initiative", url:www.sustainableethanolinitiative.com/default.asp?id=1062, accessed September.

SMA (2007) "Queima de Cana de Açucar", www.ambiente.sp.gov.br/cana/default.asp, accessed December 2007.

Subgroup Sustainability Criteria and Development Models (2010) "Versao 1 – Quadro de Sustentabilidade do Biocombustiveis em Mozambique", March, Governo de Mozambique, Ministerio de Energia (DNER).

UEMOA (2008) "Sustainable Bioenergy Development in UEMOA Member Countries", UEMOA, UN Foundation and the Hub, www.unfoundation.org/press-center/publications/sustainable-bioenergy-report.html, accessed December 2010.

Walter, A., Rosillo-Calle, F., Dolian, P., Piacente, E. and Borges Da Cunha, K. (2008) "Perspectives on fuel ethanol consumption and trade", *Biomass and Bioenergy*, vol. 32, no. 8, pp. 730–748.

WWF (2006) "Agriculture. WWF Asks for Mandatory Eco-Certification for Biofuels", http://panda.org/about_wwf/what-we-do/policy/agriculture_environment/index, accessed 28 February 2006.

13

SUGAR CANE ENERGY PRODUCTION AND CLIMATE CHANGE MITIGATION

A case study for South Africa

Joaquim Seabra, Isaias Macedo, Tarryn Eustice and Rianto van Antwerpen

Introduction

Among the current feedstock options for bioenergy applications, sugar cane in Brazil attracts special attention due to its significant contribution to mitigation of greenhouse gases (GHG) emissions through the production and use of ethanol and electricity (Macedo *et al.*, 2008). Such environmental benefit could also be explored in other countries where sugar cane is cultivated. In this chapter, a case study of such application to the sugar industry in South Africa is presented.

In South Africa, biomass is commercially used in the sugar industry and the pulp and paper mills by burning bagasse, logs and black liquor to generate process heat. The sugar industry, which has been (in most cases) self-sufficient in energy requirements could, however, generate surplus electricity for export to the grid upon modifications of its energy production configuration, but such development will be driven by a number of factors among which the electricity price and environmental regulations will be key determinants. The White Paper on Renewable Energy (2004), for instance, has set a medium-term target of 10,000 GWh renewable energy contribution to final energy consumption by 2013 (DoE, 2009) while the national Department of Energy (DoE) has developed a Biofuel Strategy focused on the production of renewable energy to reduce local dependence on imported crude oil. About 1.4 per cent of the national arable land would be utilized to achieve a market penetration of 2 per cent of liquid fuels used for road transport by 2013 (DoE, 2009). Besides providing energy security, the introduction of renewable energy sources would significantly contribute to mitigation of GHG emissions.

This chapter presents an assessment of GHG emissions associated with sugar cane production in South Africa, distinguishing rain-fed from irrigated areas. Projected ethanol emissions have been estimated for hypothetical scenarios involving ethanol production either from molasses in distilleries adjacent to sugar mills or directly from cane juice in autonomous distilleries. Possibilities to reduce ethanol life-cycle emissions were also investigated, taking into consideration the production in autonomous distilleries. Potential emissions derived from land use change due to cane expansion in Africa are also discussed. Finally, comparisons with life-cycle emissions for other biofuels are reviewed, followed by a discussion on the effects of climate change on biomass yields for different photosynthetic cycles.

The sugar industry in South Africa

Overview

South Africa produces approximately 20 million tonnes of sugar cane from which around 2 million tonnes of sugar is produced annually. The bulk of cane is produced by the country's 1,600 large-scale sugar cane planters. Miller-planters contribute about 6 per cent of the total cane production. South Africa also has more than 33,000 registered small-scale sugar cane planters and approximately half of these deliver cane to the mills in any given year, which accounts for around 10 per cent of the country's total sugar cane production (SASA, 2010). Sugar cane is cultivated on approximately 400,000 ha extending from Northern Pondoland in the Eastern Cape to the Mpumalanga Lowveld (Plate 20). Each year, sugar cane from approximately 330,000 ha is harvested and sent to the country's 14 mills.

In some areas, particularly in the northern part of the country, irrigation is essential for sugar cane production, while the remaining area relies entirely on rainfall (SASA, 2010). Table 13.1 gives the mill supply areas, number of hectares of sugar cane grown under irrigated and rain fed conditions as recorded for 2007, and average long-term rainfall availability. The rain-fed areas typically produce approximately 60 t/ha per season, while the irrigated areas produce about 100 t/ha per season.

Climate

In comparison to most other sugar cane-producing regions in the world, South Africa experiences a very mild climate with an average annual temperature of approximately 18 °C. The mean maximum and minimum temperatures are approximately 24 °C and 13 °C respectively. During winter, temperatures in some of the inland areas (>600 m above sea level) are often so low that frost can occur (Schulze and Maharaj, 2008a).

The areas under sugar cane receive most of their rain during summer, with March as the wettest month (peak rainfall occurring from December to March). The driest part of the year extends from June to August, with July being generally the driest month. The average long-term annual rainfall is approximately 833 mm (SASRI Weather Database, 2011). Rainfall across those areas occupied by the sugar industry ranges from approximately 600 to 1,200 mm/year. Table 13.2 shows the average rainfall ranges for January, April, July and October (Schulze and Maharaj, 2008b). In certain areas, rainfall is insufficient to provide for the water needs of the crop (Beater, 1936), and under these circumstances the fields are irrigated.

TABLE 13.1 List of mill supply areas and number of hectares of land under irrigated and rain-fed conditions for 2007 for large-scale and small-scale growers

Areas	Large-scale growers (ha)			Small-scale growers (ha)			Long-term average rainfall (mm)
	Rain-fed	Irrigated	Total	Rain-fed	Irrigated	Total	
South Coast	71,706	0	71,706	12,012	0	12,012	1,031
North Coast	94,611	5,594	100,205	19,146	0	19,146	992
Midlands	82,833	1,288	84,121	5,356	0	5,356	812
Zululand	59,082	12,452	71,534	39,065	0	39,065	884
Irrigated North	0	51,145	51,145	0	11,226	11,226	604

Source: Adapted from SASRI, 2007; SASRI Weather Database, 2011.

TABLE 13.2 Average rainfall ranges for January, April, July and October for the South African sugar industry

Month	Average (mm)	Rainfall range (mm)[a]
January	130	120–160
April	40	20–80
July	18	<5–40
October	70	40–160

Note
a Schulze and Maharaj, 2008b.

Production practices

The following section briefly describes the sugar cane cultivation practices in South Africa (see Chapters 2 and 4 for related analyses on sugar cane agriculture). It is discussed in four parts: fallow crop operations, planting operations and crop management, soil management and harvesting operations.

Fallow operations

It is recommended that at the end of each season about 10 per cent of all fields are ploughed for replanting (SASRI, 2008). Once harvested, all stools must be completely eradicated from the soil to prevent the spread of diseases to the next crop, and depending on region (and thus climate, topography, etc.) and/or management choice a grower will choose one of the following means of stool eradication.

- Chemical eradication – makes use of herbicides to kill the sugar cane stools. It is generally used in areas that are too steep for mechanical stool eradication.
- Mechanical eradication – makes use of a mouldboard plough, fitted with a depth wheel or skid, and a disc harrow to remove the stools.
- Combination methods – makes use of herbicides to kill the stools after which they are mechanically ploughed out two weeks later.

Small-scale growers remove the old stools by hand hoeing, using chipping hoes to dig them up. The choice of method is based on regional management practices, including the type of tillage that is used (e.g. low tillage).

Generally, producers are encouraged to allow for a four to six-month fallow, preferably with a green manure crop, which helps to suppress weeds and allows the soil to replenish some of the nutrients and encourage microbial activity. Green manures are usually planted by discing the soil directly after stool eradication and spreading the seeds over the field (except for soybeans, which must be planted in rows). At the end of the green manure fallow, the crop can be mowed down and the soil can be prepared for planting (ripping, ridging, ploughing, etc.).

Planting operations and crop management

The time for planting varies from region to region, based on the specific conditions of the cane-growing regions in South Africa. In the Northern Irrigated region, for example, where water is not limiting and temperatures are not too low, cane can be planted in winter (exceptions are June and July). Conversely, the Midlands region has a very short window for planting and it is thus recommended that planting be completed by October, after which any further plantation makes the crop vulnerable to the low winter temperature of the Midlands as well as to mosaic.[1] The general recommendation for the South African industry is therefore that planting be completed by early summer, preferably early August to the end of October.

Sugar cane is planted in furrows which are drawn with a tractor-drawn ridger to a depth of approximately 100 mm. Space between furrows is dependent on variety, climate, region, slope, soils, irrigation and harvesting technique. There are four categories of row spacing, as given in Table 13.3.

Prior to laying cane in the furrow, fertilizers are applied at a predetermined rate based on soil analysis. In some regions organic fertilizers are used and supplemented with inorganic fertilizers to fulfil the nutrient requirements. The seed cane is commonly laid in the furrow by hand and then covered with soil using hoes. Mechanized covering is practised by certain growers, however (a practice which is gaining popularity when used in combination with minimum tillage).

Managing and maintaining the crop in subsequent years (ratoons) is very important because in rain-fed regions cane will be ratooned six to eight times, and in irrigated areas five to seven times. Approximately 90 per cent of growers in the South African sugar industry burn their crop at harvest to remove trash. The green tops that are left behind after burning are spread evenly over the soil surface. Herbicides are then applied to kill existing weeds and prevent further weed growth, after which an appropriate dosage of fertilizers is applied. In some places where the soils are very compacted growers may shallow rip the interrows from 50 mm to 200 mm, and this is a recommended practice each season for soils that tend to

TABLE 13.3 Row spacing categories and distances used in South Africa

Row spacing category	Row spacing (meters)	Description and use
Close	1.0 to 1.1	Used in cooler climates where water is limited and slopes are steep.
Intermediate	1.2 to 1.3	Used in warmer climates where water is limited.
Wide	1.4 to 1.5	Used in warmer climates where water is unlimited, soils are shallow and land is flat.
Tramline	1.8 (1.2 followed by 0.6)	Used to favour mechanization and to reduce compaction. Design: two cane rows situated close together (0.4 to 0.6 m); on either side of the double cane row is a wider interrow (1.2 m). The distance from one interrow to the next is 1.8 to 1.9 m.

develop crusts. In South Africa, sugar cane producers are encouraged to treat/ alleviate compaction issues prior to planting and they are also advised to keep the wheels of equipment off the cane rows in order to limit stool damage.

Harvesting and transport

In irrigated areas, cane is harvested each season at a cutting age of roughly 12 months, and at higher altitude (>600 m) rain-fed areas the cutting age is 24 months. In other rain-fed areas, the cutting age is somewhere between these extremes. Manual harvesting is performed on standing cane by cane-cutters who cut the stalks at ground level and strip it of any leaves before it is topped. The stalks are then loaded into bundles either by means of grab loaders or manually. With mechanical cutting, the cane can be harvested burnt or unburnt ("green") with the leaves stripped from the stalk in both cases and left in the field. The harvested cane is finally transported to the mill yard by alternative transport means such as tractor-drawn trailers, heavy road vehicles or by rail.

Cane processing

In South African mills all cane is processed into sugar, with molasses and bagasse obtained as by-products. The first process involves cane juice extraction using diffusers. The juice, after treatment, is evaporated to produce syrup, which is boiled in vacuum pans for crystal growth. The crystals are then separated from the surrounding mother liquor in centrifuges, and dried to produce sugar (see Chapter 5 for more details on cane processing). In South Africa, most of the molasses produced is sent to distilleries for the production of potable ethanol. The cane fibre (bagasse) separated in the extraction process is used for generating the internal energy required for sugar recovery. Coal or wood is sometimes used as a complementary fuel whenever there is a deficit in bagasse for energy production.

Methodology for GHG analysis

Sugar cane production and processing parameters

A national database comprising technical parameters of sugar cane production in South Africa is not currently available. However, the national Cane Growers Association regularly surveys cane productivity and expenditure data by sugar cane producers, which can be used as technical parameters if proper information on costs is available. The technical parameters for cane production considered in this study are given in Table 13.4, which distinguishes between rain-fed and irrigated areas. Information on cane productivity was provided by the Cane Growers Association, while diesel consumption for farming was estimated from the cost figures provided by farmers. Data on the fertilizers application rate were provided by SASRI (South African Sugarcane Research Institute). Other parameters, such as electricity consumption for irrigation, are estimations provided by local specialists.

TABLE 13.4 Basic parameters for sugar cane production[a]

Parameter	Units	Rain fed	Irrigated
Cane productivity	t/ha	65	89
Farming diesel consumption	litres/t cane	2.5	2.5
Transportation distance: tractors	km	8	8
Tractor fuel consumption	litres/100 km	60	60
Transportation distance: trucks	km	20	20
Truck fuel consumption	litres/100 km	65–80	65–80
Share of trucks for transportation	%	80	80
Burnt area	%	90.5	99.5
Ratio of cane : trash[b]	t/t$_{dry}$	9.3	9.3
Effective burning[c]	%	50	50
Unburnt trash remaining on the ground	%	80	80
Electricity consumption (irrigation)	kWh/ha	0	3,000
Agrochemicals			
N (plant)	kg/ha	120	120
N (ratoon)	kg/ha	140	160
P_2O_5 (plant)	kg/ha	69	69
P_2O_5 (ratoon)	kg/ha	46	69
K_2O (plant)	kg/ha	144	144
K_2O (ratoon)	kg/ha	144	180
Dolomite[d]	kg/ha	3,000	0
Herbicides	kg/ha	7	7
Insecticides[e]	kg/ha	2	2

Notes
a   Parameters refer to the whole SA cane industry.
b   Not including tops, which produces additional $2\,t_{dry}$/ha (Donaldson *et al.*, 2008).
c   Represents the amount of trash that is actually burnt in the burnt areas.
d   On average, 3 t/ha is applied every ten years in rain-fed areas.
e   Include other defensives.

For cane processing, SMRI (Sugar Milling Research Institute) keeps comprehensive chemical control data for all 14 national mills and other associates from neighbouring countries. Table 13.5 gives the average cane processing data over the 2005–2008 period. In order to estimate the potential ethanol yields for the case studies, estimations regarding fermentation and distillation yields were included based on the Brazilian industry performance. For ethanol transport and distribution, it was arbitrarily assumed that ethanol would be transported by trucks over an average distance of 300 km.

GHG emissions evaluation

The analysis of GHG emissions in the sugar cane production chain comprised a field-to-factory gate analysis, which included emissions from cane farming and transport to the mills. For the GHG analysis on ethanol production option, a

TABLE 13.5 Basic parameters for sugar cane processing[a]

	%
Sucrose % cane	13.44
Fibre % cane	14.86
Bagasse % cane	30.24
Pol % fibre in bagasse	2.01
Pol % bagasse	0.97
Moisture % bagasse	49.84
Sucrose lost % sucrose in cane	
Bagasse	2.18
Filter cake	0.18
Final molasses	9.69
Undetermined	2.04
Boiling house	11.90
Total losses	14.07
Additional fuel per 1,000 tonnes of cane	
Tonnes of coal	9.36
Tonnes of wood	0.06

Source: Based on Davis *et al.*, 2009.

Note
a Parameters refer to all 14 SA mills.

field-to-wheels approach was adopted. Two levels of energy flows were considered in the energy balance and GHG emissions evaluation: the direct consumption of external fuels and electricity (direct energy inputs); and the energy required for the production of chemicals inputs in agricultural and industrial processes.

The emissions associated with the production of fertilizers and fossil fuels life-cycle were estimated based on the GREET 1.8c.0 model[2] parameters, using available South African data. The energy required to produce dolomite was assumed to be similar to limestone requirements, for which values used in the Brazilian cane industry were adopted. For chemical inputs to the mill, average data from the Brazilian chemical industry association was used (Abiquim, 2008). These emissions were estimated from the use of all chemicals in the mill and no distinction was made between the different types of chemicals used. Such assumptions do not compromise the results since the impacts of these emissions on final ethanol emissions are very low.

In addition to fossil fuel utilization, emissions from cane trash burning and from the field due to fertilizers/dolomite application and crop residues that are returned to soil were included. Residues returned to soil include only unburnt cane trash that is left in the field, for which it was assumed that a fraction of the nitrogen available would be emitted as N_2O (IPCC, 2006). Concentrated molasses stillage (CMS), produced by concentrating vinasse at the ethanol distilleries, is a relevant option for vinasse disposal in South Africa, given that it can be applied as part of the nitrogen fertilizer recommendations for cane. Molasses from several mills is transported to a central distillery where the vinasse is produced. The current evaporators in South Africa can only concentrate about half of all the available vinasse, and the rest is still pumped out to sea (van Antwerpen, 2009).

A low application rate of approximately 3 tonnes CMS/ha (around 150 kg potassium per hectare) is applied to sugar cane fields due to limited availability. As a result, only about 10 per cent of all cane fields can be fertilized to receive the required potassium from CMS, and consequently CMS application was neglected in this study. However, it must be noted that if cane mills were producing large quantities of fuel ethanol, larger amounts of vinasse would be available, and it would then be reasonable to assume that at least part of this effluent is applied to cane fields. GHG emissions derived from such practice should thus be considered, but due to the high uncertainties involved this option was not considered in this study. Other residues are also applied to soil (molasses, for instance) but since they represent low amounts, for which precise data is not available (and is difficult to estimate), these emission sources were neglected.

Carbon emissions from direct and indirect land use change (LUC and iLUC) were not included either, since the area under cane in South Africa has changed little in recent years. However, for future scenarios possibly involving the expansion of cane area related to sugar/ethanol production, it is important to consider the impacts of land use change on carbon emissions given that these could be substantial. Indirect effects of land use change may be of concern, but there is no scientific consensus to date on a methodology to evaluate such carbon emissions properly. A brief discussion regarding land use change aspects is presented later in this chapter.

In the South African cane industry, molasses is one of the main co-products derived from the recovery of sugar from cane. Hence, a proper method must be adopted to allocate the emissions associated with each product during cane production. This will consequently impact on ethanol emissions produced from molasses in the case study. According to the ISO standards, where allocation cannot be avoided, the inputs and outputs of the system should be partitioned between its different products or functions in a way that reflects the underlying physical relationships between them (ISO, 2006). In this study, the emissions from sugar cane production and processing were allocated between sugar and molasses based on the sucrose mass balance related to these output streams. For the autonomous distillery scenario such a procedure is not necessary since all cane juice is used for ethanol production. For other scenarios involving the co-production of electricity, the displacement method was used, whereby emission credits were assigned to ethanol considering that the bagasse-derived electricity would displace the electricity derived from the national average mix.

The GHG emissions were estimated for rain-fed and irrigated conditions considering two hypothetical industrial configurations: an adjacent distillery and an autonomous distillery. These represent the *reference cases*, which were evaluated assuming the average conditions given in Tables 13.4 and 13.5. Three alternative scenarios were proposed for an autonomous distillery in order to investigate their impacts on ethanol environmental performance, as follows.

- *Without coal* scenario: it is assumed that bagasse would be the only source of energy of the distillery (i.e. the additional use of coal or wood as an energy source would be unnecessary). This is the current practice in Brazilian mills and almost all other mills in the world.
- *Electricity surplus* scenario: bagasse would be the only source of energy and 60 kWh/t cane of surplus electricity would be exported to the grid. Such value represents a potential benchmark estimated for Brazilian mills equipped with high-pressure CHP systems, and a process steam consumption of 500 kg/t cane (NAE, 2005). Actually, some mills in Brazil are already exporting such an amount, and mills in other countries (e.g. Mauritius) can export more than twice this amount depending on their boiler configuration and biomass availability.
- *Unburnt cane* scenario: this scenario assumes the conditions for the *Electricity surplus* scenario and the elimination of cane trash burning for harvesting.

In all scenarios, GHG emissions were estimated for rain-fed and irrigated conditions, although no impacts of these cultivation practices on the cane processing phase were considered.

Energy supply in South Africa

Liquid fuels

Energy production in South Africa is characterized by its high dependence on cheap and abundantly available coal, which contributes to almost two-thirds of the total primary energy supply. As of 2006, about 61 per cent of the coal consumed in South Africa was used by Eskom, the country's electricity utility, in its power stations. Sasol consumed around 25 per cent for its refinery activities, while the rest was consumed by the industry and small consumers. In terms of supply use, total coal production in 2006 was utilized in electricity generation (44 per cent), export (28 per cent), conversion into synthetic fuels for transport (18 per cent) and final consumption (11 per cent) (DoE, 2009).

South Africa consumed more than 25 billion litres of liquid fuel products in 2003. Of this demand, 36 per cent was met by synthetic fuels (synfuels) produced locally, largely from coal and a small amount from natural gas (GCIS, 2007). The rest was met by products refined locally from imported crude oil; the country imports about 60 per cent of its crude oil requirements, mainly from the Middle East (DoE, 2009).

From a GHG emissions perspective, such a supply system is not favourable. The coal-to-liquids (CTL) products emit more CO_2 than conventional diesel from oil on a life-cycle basis. Van Vliet and colleagues (2009), for instance, estimated that CTL diesel without carbon capture and storage would emit 341 g CO_2eq/km, compared to values lower than 200 g CO_2eq/km for the conventional diesel technology (these values vary depending on diesel use). The South African diesel life-cycle emissions were estimated using the GREET 1.8c.0 model, taking into consideration that 36 per cent of the domestic diesel would be produced via CTL and the remainder from oil.

Electricity

South Africa produces two-thirds of Africa's electricity and is one of the four cheapest electricity producers in the world. Almost 90 per cent of South Africa's electricity is generated in coal-fired power stations. Koeberg, a large nuclear station near Cape Town, provides about 5 per cent of the capacity. An additional 5 per cent is provided by hydroelectric and pumped storage schemes. Generation is currently dominated by Eskom, which also owns and operates the national electricity grid. Eskom currently supplies about 95 per cent of South Africa's electricity (GCIS, 2007).

The national statistics show that coal is used to generate about 92 per cent of the country's electricity, but the government is committed to reducing this to 78 per cent by 2012 and 70 per cent by 2025. Nuclear energy accounts for 4.2 per cent of the electricity generation, and the remaining comprises small contributions of hydro and pumped-storage generation (DoE, 2009). These shares were used as input parameters in the GREET model to assess the average GHG emission related to electricity generation in South Africa.

Uncertainty analysis of ethanol life-cycle emissions

Evaluation of cane production and ethanol life-cycle emissions depend on the inventory information needed to map the energy and emissions flows into and out of the selected systems. Uncertainties in inventory data may arise from errors in measurement and/or use of inconsistent or heterogeneous data sets due to time, space or financial limitations. Since the dispersion of data is relevant in agricultural activities, uncertainties are usually high. Due to the differences in soil and climatic conditions and farming techniques, there is dispersion of the data regarding productivity and the specific use of agrochemicals and energy from farm to farm.

To deal with the uncertain parameters in the evaluation model for ethanol life-cycle emissions, the Monte Carlo method was applied. The Monte Carlo method is based on the assumption that all the uncertain model parameters and input and output variables are random variables and the probability distributions of these variables are known. Uncertain variables in the model are generated using their respective probability distributions and are subsequently used in model simulations to produce desired predictions. This process is repeated many times to provide enough information to construct a probability distribution of the model output (Yu and Tao, 2009). In this study uncertainty distributions were assigned to 11 parameters for rain-fed and irrigated areas (Table 13.6). Distributional characteristics (minimum and maximum for triangularly distributed parameters, mean and standard deviation for normally distributed parameters) are set to reflect reasonable bounds of national average conditions.

GHG emissions analysis

The GHG emissions from sugar cane production in South Africa are given in Table 13.7, which includes values for irrigated and rain-fed areas. The results differ substantially between these two cultivation practices, mainly due to the high consumption of carbon-intensive electricity for irrigation. Emissions for the rain-fed areas were estimated at 56.3 kg CO_2eq/t, whereas cane from irrigated areas is produced with emissions of 86.2 kg CO_2eq/t. Despite the higher cane yields, the coal-based electricity results in irrigated cane cultivation emitting approximately 50 per cent more GHG than the rain-fed system.

For the rain-fed areas, emissions related to agrochemicals production represent the main share of total emissions, while cane transport and trash burning are the

TABLE 13.6 Parameters for the Monte Carlo analysis

Parameter	Units	Assigned distribution	Values[a]	
			Rain fed	Irrigated
N appl.	kg/ha	Normal	(138, 28)	(154, 31)
Dolomite appl.	kg/ha	Normal	(398, 80)	–
Diesel (farming)	litres/tc	Normal	(2.5, 0.5)	(2.5, 0.5)
Electricity[b]	kWh/ha	Normal	–	(3,000, 170)
Burning area	%	Triangular	(80%, 100%)	(90%, 100%)
Diesel (transport)	litres/tc	Normal	(0.89, 0.18)	(0.89, 0.18)
Productivity	tc/ha	Normal	(65, 10)	(89, 13)
Ethanol yield	litres/tc	Normal	(80, 4)	(80, 4)
Coal	kg/tc	Uniform	(0, 32)	(0, 32)
Ethanol T&D	km	Normal	(300, 60)	(300, 60)
N_2O EF min[c]	%	Triangular	(0.4%, 4.0%)	(0.4%, 4.0%)
N_2O EF org[d]	%	Triangular	(0.4%, 4.0%)	(0.4%, 4.0%)

Notes
a Normal distribution (mean, SD); triangular (min., max.); uniform (min., max.).
b Electricity consumption for irrigation.
c N_2O emission factor for N mineral fertilizer.
d N_2O emission factor for residues that are returned to soil.

TABLE 13.7 GHG emissions from sugar cane production in South Africa (kg CO_2eq/t cane)

	Rain-fed areas	Irrigated areas
Farming	12.1	50.2
Burning	4.7	5.1
Soil emissions	17.6	12.2
Agrochemicals inputs	17.6	14.3
Cane transport	4.3	4.3
Total	56.3	86.2

minor sources of emissions (for both cultivation practices). The high use of nitrogen for cane fertilization has significant impact on total emissions. The production of mineral nitrogen fertilizer accounts for almost 70 per cent of the emissions from agrochemicals production, and at the same time substantial emissions arise from the soil due to nitrogen application, even though uncertainties related to the N_2O emission factors exist.

The results in Table 13.7 comprise the production of cane to delivery at the mill gate. When these values are added to the emissions from cane processing, fuel transport and distribution and combustion, the ethanol field-to-wheels emissions are obtained. Figure 13.1 gives the GHG emissions and the projected range of values for the uncertainty analysis for the different ethanol production scenarios considered in this study; namely, the production of ethanol from cane juice in an autonomous distillery or from molasses in an adjacent distillery. In Figure 13.1, the reference cases represent the results obtained for the average conditions given in Tables 13.5 and 13.6.

Ethanol emissions were estimated at between 49 g CO_2eq/MJ and 72 g CO_2eq/MJ for the reference cases, which would lead to emissions savings between 22 per cent and 47 per cent when compared to fossil gasoline (considering gasoline life-cycle emissions as 92.5 g CO_2eq/MJ, based on GREET parameters). Results of the Monte Carlo analysis indicate a significant uncertainty in the estimates. For instance, the 90 per cent confidence interval is between 43 and 98 g CO_2eq/MJ for an autonomous distillery processing rain-fed cane. The feedstock production represents the major part of ethanol life-cycle emissions, but the direct use of coal in the mill's boilers is a very significant source of emissions, as depicted in Figure 13.2 for an autonomous distillery. These values differ substantially from the Brazilian sugar cane ethanol performance, for which emissions have been estimated at around 22 g CO_2eq/MJ. If credits from co-products are accounted for, the net life-cycle emissions are as low as 12 g CO_2eq/MJ in Brazil (Macedo and Seabra, 2008). However,

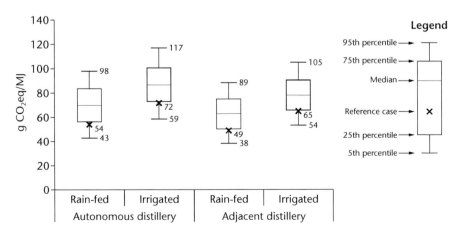

FIGURE 13.1 Range of values for ethanol life-cycle emissions

if bagasse were the only source of energy in South African mills, the ethanol emissions could be significantly reduced, as shown in Figure 13.2 (*without coal* scenario).

Emissions could be further reduced if mills export excess electricity to the grid (*electricity surplus* scenario), and emissions credits assigned to ethanol (note that emissions for the conversion step would be negative). There are actually no significant technological challenges for bagasse cogeneration, since commercial options for advanced CHP systems and alternative processes with reduced energy demand are available and are currently adopted in sugar cane mills in different countries (see Chapter 5).

Finally, a minor contribution to emissions reduction could be obtained by eliminating sugar cane burning (*unburnt cane* scenario). The N_2O emissions from trash that would be left on the soil were estimated, but the additional diesel consumption due to (possibly) increased mechanization is uncertain and was not considered. The combined effects could even offset the benefits of the elimination of trash burning in terms of GHG emissions.

GHG emissions associated to land use change (direct and indirect) for ethanol production from sugar cane in Africa

Direct land use change (LUC) emissions depend essentially on the change in soil carbon stocks, in the aerial biomass and underground (root system) carbon stocks. Indirect land use change (iLUC) emissions are still in a preliminary stage of discussion (as a result of different concepts, inadequate methodologies and lack of sufficient information on the drivers for change). It is possible that for some time the carbon stock values for some regions in Africa will be estimated using default data

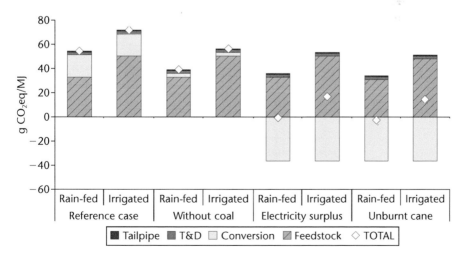

FIGURE 13.2 Alternatives to reduce ethanol life-cycle emissions: analysis for an autonomous distillery

(mostly from IPCC). However, it is important to present some available information, as discussed below.

A recent survey (Watson, 2011) evaluated the potential for sustainable sugar cane production in six selected countries (from the areas previously identified as potentially interesting) in Southern Africa: Angola, Malawi, Mozambique, Tanzania, Zambia and Zimbabwe (see Chapter 4). It looked for protected areas, land cover, climate, elevation and soil data. An agro-ecological zoning was prepared; protected areas, closed-canopy forests and wetlands were excluded, as well as all areas under food and/or cash crop production. Areas unsuitable for sugar cane because of climate, terrain slopes and soil limitations were also excluded.

This preliminary assessment (using GIS, $1\,km^2$ resolution) could be fine-tuned with ground information; but the main results indicate a significant potential. The assessment found nearly 6 million hectares of suitable land available for sugar cane production in these countries. By recognizing that this might not be entirely adequate (the protection areas needed may be larger than those legally protected today; some of the small, family-based agricultural areas may not be well estimated, and they include low-productivity grazing and cultivation) it appears that land availability is not, for the current decade, a limiting factor for implementing a biofuel production policy in this region. Mozambique is particularly suitable, with an available area of 2.3 million hectares.

Taking into consideration the agro-ecological and social guidelines for expansion in cane production, the areas to be used will be covered by grasslands and savannah (low density) as well as some degraded areas. The map of soil carbon distribution used until recently (FAO-UNESCO Soil Map of the World) was updated by Winrock (Harris *et al.*, 2009), using a new data set (Harmonized World Soil Database, released in March 2009) to estimate soil carbon stocks. This data set still relies on the FAO-UNESCO Soil Map of the World for some regions, but incorporates improved datasets where they exist (e.g. China, South America). The new soil dataset was produced at $1\,km^2$ resolution, which is a significant improvement over the FAO-UNESCO map. These estimates were used for the EPA analyses of biofuels (EPA, 2010). The land to be used in Southern Africa is classified as "grasslands" and "savannah" (does not include "woody savannah") in the IGBP land cover classes, and in the Winrock estimates. The basic data for carbon stocks in the six countries is given in Table 13.8.

It must be noted that the actual carbon stock in soil is still evaluated applying to the SOC_{ref} the same factors (for each new crop, or perennial) recommended by IPCC (IPCC, 2006). It is also important to note that sugar cane is considered as perennial for the purpose of carbon stocks (soil and biomass) (Winrock, 2009). This has also been verified by extensive data from Brazil (Donzelli, 2010).

So the change in soil organic carbon when changing from grasslands or savannah to sugar cane is zero; and the changes in biomass carbon stocks are close to zero, when changing from savannah to sugar cane, and negative (increase in carbon stocks) when changing from grasslands to sugar cane. There would be a loss of carbon stocks in biomass if shrublands were used, but the value is relatively small. As a concluding remark, the overall direct emissions from LUC are zero or negative.

TABLE 13.8 Carbon stocks in soil (SOC_{ref}), and in biomass (above and below ground) for the six countries ($t\,CO_2$/ha)

	Soil, SOC_{ref}	Biomass (above and below ground)			
		Grasslands	Savannah	Shrubland	Perennial
Angola	105	24	43	81	44
Malawi	158	23	42	79	44
Mozambique	129	20	36	68	44
Tanzania	213	19	35	66	44
Zambia	153	24	42	80	44
Zimbabwe	87	15	27	52	44

Source: Winrock, 2009.

Finally, regarding the iLUC emissions, it is considered that (whatever the methodology, or concepts under discussion used) for the conditions proposed for the analysis (limitations imposed on land use, preservation of all food production/cattle-raising areas, as well as the large land availability compared to expected use in the decade) there will be no iLUC involved in the expansion of sugar cane production in Southern Africa.

GHG emissions of biofuels

GHG emissions and energy requirements for biofuel production vary widely, depending on the feedstock, technology considered and boundary conditions assumed. Along the production chain, the processes contribute differently to the overall performance. The GHG balance of biomass production depends on the direct use of fossil energy, the inputs needed for cultivation (e.g. machinery, fertilizers, pesticides, etc.) and the amount of fossil fuels replaced by the biofuels produced. It varies significantly for different plants, because of their specific yields and input requirements (Bringezu *et al.*, 2009). Selected allocation procedures, reference land and soil N_2O emissions are also key factors for the analyses (Hoefnagels *et al.*, 2010). In contrast to agricultural production, the provision of waste materials and residuals does not require significant energy input and usually leads to lower GHG emissions.

The production process of the fuels contributes, on average, much lower GHG emissions than the agricultural cultivation. Emissions from the fermentation of bioethanol are variable: emissions are high when fossil energy carriers are used (as is often the case with bioethanol from corn); they are low when wastes or residual materials are used to generate process energy (e.g. use of bagasse for sugar cane processing) (Bringezu *et al.*, 2009).

According to Hoefnagels and colleagues (2010), life-cycle assessments (LCAs) generate wide-ranging conclusions, which is mainly due to differences in (data) quality, the setting in which production is assumed to take place, the method used to account for co-products and the assumptions on changes of above- and below-ground biomass, soil organic carbon, litter and dead wood due to direct/indirect land use change. The authors have conducted a comprehensive life-cycle analysis on the energy and GHG performance of different biofuel production systems, aimed at providing comparisons on a consistent basis. The results for the systems assessed are given in Figure 13.3.

Even though there is no consensus on the basic principles about the sustainability of biofuels, the need to lead to GHG emissions reduction (compared to the equivalent fossil option) is one of the most relevant aspects. Because of this goal, international initiatives aimed at promoting the use of biofuels have been established in different countries, paying attention to other sustainability criteria as well. One example is the Regulatory Impact Analysis (RIA) performed by the US Environmental Protection Agency (EPA), which assessed the impacts of an increase in production, distribution and use of the renewable fuels sufficient to meet the Renewable Fuels Standard volumes established by Congress in the Energy Independence and Security Act of 2007 (EISA). The life-cycle emissions of different biofuels were investigated, including the potential emissions derived from the direct and indirect land use change. Emission reductions of the Brazilian sugar cane ethanol, for instance, were evaluated at between 59 per cent and 91 per cent for the four scenarios considered (EPA, 2010).

Climate change and impacts on productivity

In general, photosynthesis may be considered as the process that stores light energy of the sun as carbohydrates by assimilating CO_2 and H_2O. Mineral nutrients are also required for the functioning of the photosynthetic system (Janssens *et al.*, 2009). In higher plants, virtually all of the carbon is assimilated through the reductive pentose phosphate cycle, often referred to as the Calvin cycle (DaMatta *et al.*, 2009). In C_3 species, which comprise the great majority of plant species, including important food crops such as wheat, rice, soybeans and potatoes, CO_2 is exclusively assimilated through the Calvin cycle. In contrast, C_4 plants, including the most productive crop species such as maize, sorghum and sugar cane, use a series of enzymes that initially combine CO_2 (HCO_3) with a three-carbon molecule (phosphoenolpyruvate, PEP), producing oxaloacetate, a four-carbon compound (DaMatta *et al.*, 2009).

The application of the advances in biology and biotechnology to the design of plants and organisms that are more efficient energy-conversion machines has been identified as one of the major solar energy research goals for the coming decades. Zhu *et al.* (2009) report that the highest solar energy conversion efficiency reported for C_3 crops is about 2.4 per cent and about 3.7 per cent for C_4 crops across a full growing season based on solar radiation intercepted by the leaf canopy. Higher short-term conversion efficiencies are observed for brief periods during the life of a crop reaching 3.5 per cent for C_3 and 4.3 per cent for C_4 (Zhu *et al.*, 2008). The theoretical maximal photosynthetic energy conversion efficiency was calculated as 4.6 per cent for C_3 and 6 per cent C_4 plants (Zhu *et al.*, 2008), based on the total

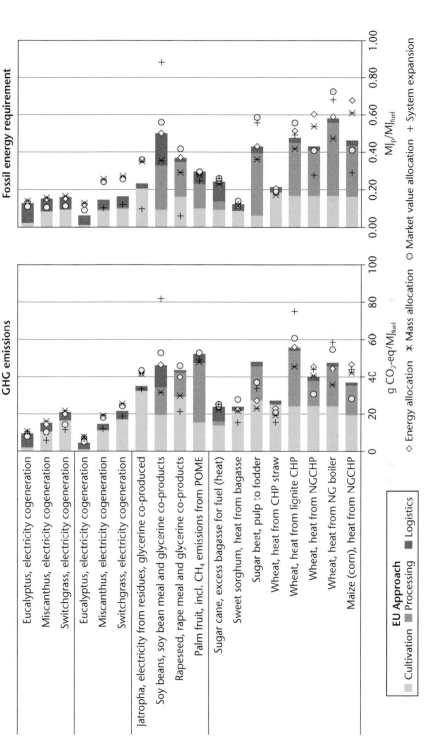

FIGURE 13.3 Fossil energy requirement and GHG emissions from biofuels production excluding land use change (reprinted from Hoefnagels *et al.*, with permission from Elsevier)

initial solar energy and the final energy stored in biomass. These calculations assumed a leaf temperature of 30 °C and an atmospheric CO_2 of 380 parts per million (ppm). The authors comment that the simulations show a very strong advantage for C_4 photosynthesis at the atmospheric CO_2 that prevailed for most of the past 25 million years, but the recent rises in CO_2 may affect these advantages.

Over the past 800,000 years, atmospheric CO_2 changed between 180 ppm (glacial periods) and 280 ppm (interglacial periods) as Earth moved between ice ages. From pre-industrial levels of 280 ppm, CO_2 has increased steadily to 384 ppm in current days, and projections to the end of this century suggest that it may surpass 700 ppm (DaMatta *et al.*, 2009; IPCC, 2007). An increase in atmospheric CO_2 and other greenhouse gases would increase the mean global temperature as well as alter the precipitation patterns in many areas of the globe. The rise in atmospheric CO_2, along with potential global warming and changes in precipitation, will undoubtedly have significant economical and ecological impacts on many agricultural crop plants (Vu and Allen Jr, 2009).

According to DaMatta *et al.*, 2009, there is growing evidence suggesting that many crops, notably C_3 crops, may respond positively to increased atmospheric [CO_2] in the absence of other stressful conditions, but the beneficial direct impact of elevated [CO_2] can be offset by other effects of climate change, such as elevated temperatures, higher tropospheric ozone concentrations and altered patterns of precipitation. In contrast, Vu and Allen Jr (2009) remark that the photosynthetic mechanisms of C_4 plants in response to future elevated atmospheric [CO_2] and elevated [CO_2] interacting with unfavourable climate change components is still uncertain. Plants of the C_4 photosynthetic category, which have developed a CO_2-concentrating mechanism to overcome the limitations of low atmospheric [CO_2] and photorespiration, possess a near-saturating photosynthesis at current atmospheric [CO_2]. A rise in atmospheric [CO_2], therefore, would theoretically have little direct effect on C_4 photosynthesis according to authors. Nevertheless, they report that studies on a number of C_4 crop plants at elevated [CO_2] reveal a positive response in biomass, although to a smaller extent as compared with C_3 plants. This unexpected increase in C_4 biomass has been widely explained by an indirect increase in the efficient use of water under elevated CO_2 (Vu and Allen Jr, 2009).

Souza and colleagues (2008) studied the effects of increased [CO_2] on photosynthesis, development and carbohydrate metabolism in sugar cane. The plants grown under elevated [CO_2] showed an increase of about 30 per cent in photosynthesis and 17 per cent in height, and accumulated 40 per cent more biomass in comparison with the plants grown at ambient [CO_2]. Industrial productivity analysis showed an increase of about 29 per cent in sucrose content. These data suggest that sugar cane crops increase productivity in higher [CO_2], and that this might be related, as previously observed for maize and sorghum, to transient drought stress. The authors comment that, for sugar cane, productivity is not gauged by the results of reproductive effort but the direct allocation of sucrose to the stems. In other words, sugar cane plants probably spend proportionally less energy in storing carbon. From this point of view, their data (from experiments not performed under field conditions) suggest that sugar cane productivity might increase in a scenario of future increase in atmospheric [CO_2].

Overall climate change mitigation potential of cane bioenergy

In this study the GHG emissions from cane production in South Africa were estimated to be 56 kg CO_2eq/t and 86 kg CO_2eq/t respectively for rain-fed and irrigated conditions. Comparatively, significantly lower emissions are verified for the Brazilian conditions (the major cane producer in the world), which can be essentially explained by the lower cane yield, the higher use of mineral nitrogen and the consumption of carbon-intensive electricity for irrigation (for irrigated areas) in the South African industry. Still, the climate change mitigation potential associated with cane bioenergy in South Africa is substantial. It was shown that the production of ethanol would lead to considerable emissions mitigation due to the displacement of oil-derived gasoline. In addition, the improvement of mills' CHP systems could enable the generation of elevated electricity surpluses, leading to the displacement of carbon-intensive electricity from coal generation.

In Table 13.9 the potential emissions avoidance associated with cane bioenergy in South Africa is given for the different scenarios explored in this study. For the reference cases, one tonne of rain-fed cane could avoid the emission of 65 kg CO_2eq through the production of ethanol in an autonomous distillery. Considering all the improvements proposed in the *unburnt cane* scenario, such net avoidance could be greater than 160 kg CO_2eq/t of cane, considering the production of ethanol and electricity surplus.

Concluding remarks

Sugar cane is an important commercial crop in South Africa and has a good potential for expansion of its cultivation in other African countries for bioenergy purposes. It has been found that irrigation has a significant impact on GHG emissions of cane production in South Africa due to the consumption of carbon-intensive electricity. Based on different scenarios for ethanol production in South African mills, estimations suggest that ethanol could mitigate emissions in comparison to gasoline in South Africa. For the *reference cases*, the use of coal in addition to bagasse in the mills' boilers would represent a significant emission source in the ethanol life-cycle. Such practice, however, reflects the current conditions for sugar production in the South African mills, which could easily be altered by using commercially available technologies. Therefore, ethanol emissions could be significantly lowered by the simple elimination of the direct coal consumption. Additional reductions could be reached through the export of excess electricity generated from bagasse, which would contribute in diversifying the national electricity mix thereby favouring exploitation of renewable energy sources in the country. The direct emissions from LUC are expected to be zero or negative while for the conditions proposed there will be no iLUC involved in the expansion of sugar cane in Southern Africa.

TABLE 13.9 Potential GHG emissions balance for the South African sugar cane industry (kgCO_2eq/t cane)

Scenario[a]	Reference case		Without coal		Electricity surplus		Unburnt cane	
	Rain_fed	Irrigated	Rain_fed	Irrigated	Rainfed	Irrigated	Rain_fed	Irrigated
Emissions	**93**	**123**	**67**	**97**	**67**	**97**	**63**	**93**
Cane production	56	86	56	86	56	86	53	82
Cane processing	31	31	5	5	5	5	5	5
Ethanol T&D	4	4	4	4	4	4	4	4
Tailpipe	1	1	1	1	1	1	1	1
Avoided emissions	**-158**	**-158**	**-158**	**-158**	**-226**	**-226**	**-226**	**-226**
Gasoline[b] displaced	-158	-158	-158	-158	-158	-158	-158	-158
Electricity[c] displaced					-68	-68	-68	-68
Net avoided emissions	**-65**	**-35**	**-92**	**-62**	**-160**	**-130**	**-163**	**-133**

Notes

a Estimated for a hypothetical autonomous distillery.

b Gasoline life-cycle emissions may vary considerably from country to country. In this study we have assumed a life-cycle emission of 92.5 gCO_2eq/MJ based on GREET parameters.

c Based on the SA fuel mix for electricity generation.

Acknowledgements

The authors thankfully acknowledge the specialists of the SMRI (Gavin Smith), SASRI (Peter Lyne, Peter Tweddle, Neil Miles, Geoff Maher, Ruth Rhodes, Sanesh Ramburan, Michiel Smit, Matthew Jones, Abraham Singels, Peta Campbell, Shaun Berry, Ashiel Jumman, Neil Lecler, Graeme Leslie and Poovie Govender) and CANEGROWERS (Olivia Finnamore, Stuart Ferrer, Minette Landman and Kathy Hurley) for their valuable contributions to this work, and Manoel R.L.V. Leal and Brian Purchase for their support. Joaquim Seabra also acknowledges financial support from the Swedish International Development Cooperation Agency (SIDA) through the Stockholm Environment Institute; however, SIDA had no part in the design of the research or in the choice of the case study, and does not necessarily agree with any conclusions reached.

Notes

1 This is caused by the sugar cane mosaic virus (SCMV) and is more common, but not restricted, to the cooler, higher-altitude sugar cane-growing regions (e.g. the Midlands). It can cause severe yield losses.
2 The Greenhouse Gases, Regulated Emissions and Energy Use in Transportation Model, developed by the Argonne National Laboratory (USA). It was developed as a multidimensional spreadsheet model in Microsoft Excel. This public domain model is available free of charge at the Argonne website (www.anl.gov).

References

Abiquim (Associação Brasileira da Indústria Química) (2008) "Indústria Química Reduziu Emissões de Gás Carbônico e Consumo de Água", Comunicação Abiquim, São Paulo.

Beater, B.E. (1936) "Climatic conditions in the Natal Sugar Belt as indicated by records taken at Mount Edgecombe", *Proceedings of the South African Sugar Technologists Association*, vol. 10, pp. 95–106.

Beater, B.E. (1970) "Soil Series of the Natal Sugar Belt", South African Sugar Association, Durban.

Bringezu, S., Schütz, H., O´Brien, M., Kauppi, L., Howarth, R.W., McNeely, J. (key authors) (2009) "Towards Sustainable Production and Use of Resources: Assessing Biofuels", United Nations Environment Programme (UNEP).

DaMatta, F.M., Grandis, A., Arenque, B.C. and Buckeridge, M.S. (2009) "Impacts of climate changes on crop physiology and food quality", *Food Research International*, doi:10.1016/j.foodres.2009.11.001.

Davis, S.B., Smith, G. and Achary, M. (2009) "Eighty-fourth annual review of the milling season in South Africa (2008–2009)", *Proceedings South African Sugar Technologists' Association*, vol. 82, pp. 1–29.

DoE (2009) "Digest of South African Energy Statistics 2009", Department of Energy, South Africa.

Donaldson, R.A., Redshaw, K.A. and van Antwerpen, R. (2008) "Season effects on productivity of some commercial South African sugarcane cultivars, II: trash production", *Proceedings South African Sugar Technologists' Association*, vol. 81, pp. 528–538.

Donzelli, J.L. (2010) "Carbono Orgânico em Solos Cultivados com Cana-de-Açúcar no Centro Sul do Brasil", Relatório Técnico, Centro de Tecnologia Canavieira (CTC), São Paulo.

EPA (2010) "Renewable Fuel Standard Program (RFS2) Regulatory Impact Analysis", Assessment and Standards Division, Office of Transportation and Air Quality, US Environmental Protection Agency, EPA-420-R-10–006, February.

GCIS (2007) "South Africa Yearbook 2006/07", Government Communication and Information System, South Africa.

GREET (2009) "The Greenhouse Gases, Regulated Emissions, and Energy Use in Transportation (GREET) Model", Version 1.8c.0, Argonne National Laboratory.

Harris, N., Grimland, S. and Brown, S. (2009) "Land Use Change and Emission Factors: updates since the RFS Proposed Rule", Report to EPA, Winrock International.

Hoefnagels, R., Smeets, E. and Faaij, A. (2010) "Greenhouse gas footprints of different biofuel production systems", *Renewable and Sustainable Energy Reviews*, vol. 14, no. 7, pp. 1661–1694.

IPCC (2006) "2006 IPCC Guidelines for National Greenhouse Gas Inventories", prepared by the National Greenhouse Gas Inventories Programme (eds: H.S. Eggleston, L. Buendia, K. Miwa, T. Ngara and K. Tanabe), IGES, Japan.

IPCC (2007) "Summary for policymakers", in *Climate Change 2007: The Physical Science Basis, Contribution of Working Group I to the Fourth Assessment Report of the Intergovernmental Panel on Climate Change* (eds: S. Solomon, D. Qin, M. Manning, Z. Chen, M. Marquis, K.B. Averyt, M. Tignor and H.L. Miller), Cambridge University Press, Cambridge and New York.

ISO (2006) "ISO 14044:2006, Environmental Management – Life Cycle Assessment – Requirements and Guidelines".

Janssens, M.J.J., Keutgen, N. and Pohlan, J. (2009) "The role of bio-productivity on bio-energy yields", *Journal of Agriculture and Rural Development in the Tropics and Subtropics*, vol. 110, pp. 39–47.

Macedo, I.C. and Seabra, J.E.A. (2008) "Mitigation of GHG emissions using sugarcane bioethanol", in P. Zuurbier, and J. van de Vooren (eds), *Sugarcane Ethanol: Contributions to Climate Change Mitigation and the Environment*, Wageningen Academic Publishers, Netherlands, pp. 95–111.

Macedo, I.C., Seabra, J.E.A. and Silva, J.E.A.R. (2008) "Greenhouse gases emissions in the production and use of ethanol from sugarcane in Brazil: The 2005/2006 averages and a prediction for 2020", *Biomass and Bioenergy*, vol. 32, pp. 582–595.

NAE (2005) "Biocombustíveis", Cadernos NAE n° 2, Núcleo de Assuntos Estratégicos da Presidência da República, Brasília, Secretaria de Comunicação de Governo e Gestão Estratégica.

SASA (2010) "South African Sugar Industry Directory: 2010/11", South African Sugar Association, Durban.

SASRI (2007) "List of Mill Supply Areas and Number of Hectares of Land under Irrigated and Rain-Fed Conditions for 2007 for Large-Scale and Small-Scale Growers (Extension Survey)", Durban.

SASRI (2008) "Land Use Planning: Soil Conservation", Senior Certificate Course in Agriculture (Course notes), Durban.

SASRI (2011) "South African Sugar Industry Indicating Sugar Mills and Rain-Fed and Irrigated Sugarcane Production Areas (Map)", Durban.

SASRI Weather Database (2011) SASRI Weather Web, http://portal.sasa.org.za/weatherweb.

Schulze, R.E. and Maharaj, M. (2008a) "Temperature over South Africa: Mean annual and monthly means of temperature and their variables", in R.E. Schulze, P.J. Hull and C.N. Bezuidenhout (2008) *South African Sugarcane Atlas*, University of KwaZulu-Natal, School of Bioresources Engineering and Environmental Hydrology, Pietermaritzburg, RSA (ACRUcons Report, 57, Section 3.2), pp. 37–48.

Schulze, R. E. and Maharaj, M. (2008b) "Rainfall over South Africa: Mean annual and median monthly rainfalls and their variabilities", in R.E. Schulze, P.J. Hull and C.N. Bezuidenhout (2008) *South African Sugarcane Atlas*, University of KwaZulu-Natal, School of Bioresources Engineering and Environmental Hydrology, Pietermaritzburg, RSA (ACRUcons Report, 57, Section 3.1), pp. 29–36.

Souza, A.P., Gaspar, M., da Silva, E.M., Ulian, E.C., Waclawosvsky, A.J., Nishiyama Jr, M.Y., dos Santos, R.V., Teixeira, M.M., Souza, G.M. and Buckeridge, M.S. (2008) "Elevated CO_2 increases photosynthesis, biomass and productivity, and modifies gene expression in sugarcane", *Plant, Cell and Environment*, vol. 31, pp. 1116–1127.

van Antwerpen (2009) "The use of CMS and related products as potassium fertilisers", *The Link*, vol. 18, no. 1, pp. 8–9.

van Vliet, O.P.R., Faaij, A.P.C. and Turkenburg, W.C. (2009) "Fischer Tropsch diesel production in a well-to-wheel perspective: A carbon, energy flow and cost analysis", *Energy Conversion and Management*, vol. 50, pp. 855–876.

Vu, J.C.V. and Allen Jr, L.H. (2009) "Growth at elevated CO_2 delays the adverse effects of drought stress on leaf photosynthesis of the C4 sugarcane", *Journal of Plant Physiology*, vol. 166, pp. 107–116.

Watson, H.K. (2011) "Potential to expand sustainable bioenergy from sugar cane in Southern Africa", *Energy Policy*, vol. 39, no. 10, pp. 5746–5750.

Winrock (2009) "Emissions Factors Docket", Report to EPA, Winrock International.

Yu, S. and Tao, J. (2009) "Energy efficiency assessment by life cycle simulation of cassava based fuel ethanol for automotive use in Chinese Guangxi context", *Energy*, vol. 34, pp. 22–31.

Zhu, X.G., Long, S.L. and Ort, D.R. (2008) "What is the maximum efficiency with which photosynthesis can convert solar energy into biomass?", *Current Opinion in Biotechnology*, vol. 19, pp. 153–159.

PART V

Strategic issues and comparisons

14

THE DEVELOPMENT OF BIOFUEL CAPACITIES

Strengthening the position of African countries through increased energy security

Henrique Pacini and Bothwell Batidzirai

Introduction

Among the key challenges to sustainable development in sub-Saharan Africa (SSA) are the issues of energy access and energy security. Fears of oil supply disruptions, high prices, power blackouts and fuelwood shortages are a major concern, especially for poor oil-importing countries in SSA (Batidzirai and Wamukonya, 2009; UNECE, 2007). The difficulties are compounded in landlocked countries, where transportation costs of fuel are high and supply lines are vulnerable to disruptions in the case of civil unrest, natural disasters, geopolitical instability or barriers to trade (Habitat, 1993; Scurlock *et al.*, 1991).

For most of SSA, the ability to meet growing demand for energy is among the top national priorities. The availability of reliable and affordable energy in sufficient quantities is essential for meeting basic needs and driving economic development. In seeking to improve their energy security, SSA countries must tackle a number of challenges; including diversification of supply, securing capital and financing for energy infrastructure, and developing technical expertise and technical solutions tailored to specific national needs.

Where conditions are conducive, biofuels have the potential to substitute a significant amount of energy used in the transport, electricity and household sectors. Since biofuels do not require complex technologies compared to other energy options, there are attractive opportunities for developing countries (Mathews, 2006). Furthermore, many SSA countries possess abundant natural resources and favourable climatic and soil conditions, which are prerequisites for the development of successful biofuels industries. Recent studies (Smeets *et al.*, 2007; Batidzirai *et al.*, 2006; Johnson and Matsika, 2006) have shown that the potential for biomass energy is very high in SSA. The sugar industry is one of the key sectors, with potential to contribute significantly to the supply of bioethanol fuel for transportation as well as to the supply of bagasse-fired electricity.

This chapter evaluates the potential contribution of biofuels in improving energy security in SSA, focusing on the role of sugar cane-based bioenergy. The next section provides some background on the development of biofuels. Section 2 explores some key energy security issues and their linkages to economic development. Section 3 gives an overview of the role that biofuels can play in improving energy security in SSA, using specific references to three African countries (Mozambique, Tanzania and Zambia) to support the analysis and discussion.

Section 1: background

Traditional biomass energy still accounts for up to 90 per cent of national energy supply in most SSA countries, and limited access to affordable and reliable modern energy services continues to stifle economic growth and income generation, especially in rural areas (Johnson and Rosillo-Calle, 2007). This continued dependence on traditional biomass puts natural resources under ever increasing pressure, and brings with it health risks and a vicious poverty circle. The International Energy Agency (IEA) estimates that the share of traditional biomass in total final energy supply will not change in the medium term, unless drastic measures are implemented to improve access to modern energy (IEA, 2002). A promising approach to addressing some of these concerns is the modernization of biomass energy in the region.

Modern bioenergy can play an important role in the future energy mix; current analysis suggests considerable potential within the SSA region to produce biofuels and thereby contribute to improved energy security (Batidzirai et al., 2006; Johnson and Matsika, 2006; Goldemberg, 2007; Azoumah et al., 2011). Because modern biomass energy (considering in this case mainly first-generation biofuels and cogeneration of heat and electricity) can be implemented as modular and decentralized systems using existing technologies, it is possible to provide energy to communities that are not accessed by existing conventional energy supply systems.

On the other hand, modern bioenergy has also courted controversy due to potential conflicts with other agro-based human needs as well as environmental impacts. In addition, there are mixed perceptions on whether activities based on natural resources are an effective path to achieve progress in developing countries (Sachs and Warner, 1995).[1] For example, Hall and colleagues (2009), and Reddy (2008), argue that biofuels represent a development trap, where the purported gains are outweighed by the serious risks they pose. However, other analysts (e.g. Matthews, 2007; Goldemberg, 2007) maintain that given the technological maturity and political momentum, utilizing biofuels to improve energy security is a viable path for development. As a dynamic industry that is undergoing rapid development, a considerable learning curve has to be transcended for the risks and benefits to become clearer. Nevertheless, modern bioenergy clearly offers alternatives, which, if the risks are carefully managed, can contribute to improved energy access and energy security in developing countries.

Section 2: energy security in sub-Saharan Africa

Energy security can be categorized in two main forms: direct and economic. Direct energy security relates to safe and stable access to energy sources, which offers both reliability and resiliency for national energy systems. Reliability means users are able to access energy services on demand and in proper form and quality. Resilience is the ability of the system to cope with shocks and change; for example, due to an interruption of imports.

Economic energy security entails the capacity to afford energy at higher prices without significant welfare losses; or to diversify economic activity in order to reduce the burden of energy imports in the overall economy. It can also be seen as the ability to supply energy to meet demand at a price that protects economic growth, a common objective of energy policy of many governments. Both direct and economic energy security are complementary to each other and can contribute towards the same goals.

SSA countries can potentially improve energy security through the production, supply and utilization of biofuels, by reducing their vulnerability to supply and price shocks associated with imported energy. First, physical energy security can be improved by substituting imported energy with indigenous resources (i.e. biofuels and by-products). Second, economic energy security can be enhanced through earning revenue from the biofuels trade, thus improving SSA's ability to pay for imported energy. By maximizing income via trade-offs between internal and foreign markets, the additional revenue expands the options available for SSA countries to supply their energy needs at the lowest cost.

A brief assessment of energy security is provided here from the perspective of three SSA countries: Zambia, Mozambique and Tanzania. The energy dependency[2] rates for the three countries are evaluated against international aid and economic growth rates. The three countries were selected based on the following criteria:

1. They are developing countries with pressing socio-economic demands (according to the Human Development Index, HDI <0.6) and deficits between aid inflow and energy import payments (see Table 14.1).
2. They possess limited or no hydrocarbon reserves.
3. These are countries where according to Jumbe and colleagues (2009), Van Gent and colleagues (2007) and Bekunda and colleagues (2009) concrete biofuels potential has been identified.

All three countries are dependent on official development assistance (ODA), which amounts to more than 10 per cent of their annual gross domestic product (GDP). Their local economies often rely on bartered transactions, with locally produced goods being cheap in their own currency, while imported goods are expensive and unaffordable to many. A clear indication of this dichotomy is the higher transport fuel prices in the three countries compared to the world average or even

TABLE 14.1 Basic national statistics for Zambia, Mozambique and Tanzania

Country	Population (million)	Area (km²)	HDI (2007)	Average GDP growth (since 1975) (%)	Unemployment rate
Zambia	12.9	752,618	0.481	−2.27	14% (2006)
Mozambique	22.8	801,590	0.402	2	18.7% (2005)
Tanzania	43.7	945,200	0.530	5.91	12.7% (2001)

Sources: Zambia Statistics Agency (www.zamstats.gov.zm), Tanzania Statistics Agency (www.nbs.go.tz) and UNSTATS.

compared to the highly taxed equivalent in European countries such as Sweden (Pacini and Silveira, 2011).

Table 14.2 shows the energy (i.e. oil and electricity) dependencies for the three countries as of 2009.

According to the data, reliance on imported energy is more accentuated for oil than for electricity. However, household access to electricity is low in all three countries; 18.8 per cent in Zambia, 11.7 per cent in Mozambique and 13.9 per cent in Tanzania (IEA, 2010). These low access rates and the strong interest in obtaining electricity access in these countries indicate a huge unmet demand that could over-load the current capacities if access rates were to improve.

Oil dependency is high for all three countries and requires shifting substantial resources towards energy imports. Figure 14.1 compares the costs of oil depend-ency with aid inflows and the average GDP growth between 2005 and 2007.

Among many costly trade-offs, these countries effectively channel part of the aid inflows received towards energy import bills, depriving other needy sectors of the economy and consequently affecting growth and development.

Section 3: improving energy security: a role for sugar cane

The SSA region has significant experience with bioethanol production from sugar cane since this crop is widely grown in many African countries and the technology is mature. According to Woods and colleagues (2008), the region is home to some of the most efficient sugar industries in the world. However, most of the ethanol produced in the region has been for industrial and beverage usages and most of it has been exported to the world market.

TABLE 14.2 Energy dependency in selected SSA countries

Country	Oil dependency (%)	Electricity dependency (%)
Zambia	92	0
Mozambique	92	−35
Tanzania	98	6

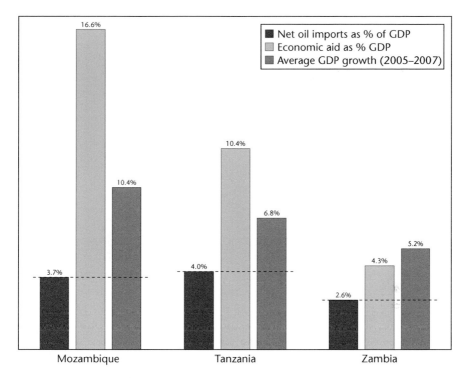

FIGURE 14.1 Costs of oil dependency compared to aid inflows (ODA) and GDP growth (sources: Based on OECD Development Co-operation Directorate (www.oecd.org/dac/stats), national statistics agencies and IEA (2010))

Historical context

Bioethanol fuel programmes in Africa were initiated in the 1970s in Zimbabwe, Malawi and Kenya to reduce dependency on oil imports, save foreign exchange and develop domestic industries. Zimbabwe pioneered ethanol production from sugar cane molasses and blending with petrol in 1980, followed by Malawi in 1982 and Kenya in 1983 (Habitat, 1993). Ethanol blending is still practised in Malawi, it was done in Zimbabwe until 1992 and it is likely to be introduced in South Africa in the near future (Takavarasha *et al.*, 2005). New ethanol blending programmes have been announced in several African countries including Nigeria, Ethiopia, Sudan and Mozambique as well as Zambia, Swaziland and Mauritius (see also Chapters 8 and 10).

Plans for a fuel ethanol plant in Zimbabwe began during the 1970s when the country was under international sanctions. Zimbabwe's landlocked position, the political vulnerability of supply routes and foreign exchange limitations also influenced the development of the fuel-blending programme. Even though the Triangle ethanol plant started operating in 1980, when sanctions had been lifted, disruptions of oil supplies through Mozambique (due to civil war) meant an alternative fuel source was still vital (Scurlock *et al.*, 1991). The ethanol programme in Malawi was motivated by the high oil prices: oil was imported by road via South Africa due to the war in Mozambique and this longer route coupled with currency devaluation raised the landed costs of oil by 33 per cent (Habitat, 1993). These experiences in improving energy security through biofuels form a good basis for replication and expansion in the region.

Estimated sugar cane potential

Based on data from various reports from the COMPETE[3] network and Watson (2011), Table 14.3 provides estimates of the potential area suitable for the cultivation of sugar cane ethanol in Mozambique, Tanzania and Zambia. Figures presented exclude protected areas.

The analysis demonstrates a large potential for all three countries to produce bioethanol, with significant contribution towards improving their physical and economic energy securities. The production potential for sugar cane-based ethanol presented in Table 14.3 is based on the assumption that the estimated amount of suitable land will be developed for this purpose by 2030, at yield levels comparable to those achieved currently in Tanzania and Zambia (103–106 tonnes/ha). The full conversion of sugar cane into ethanol would be constrained by developments in sugar markets (see Chapter 8). However, even with a much lower utilization (i.e. greater sugar production) it is clear that all three countries could cover their domestic oil demand with indigenous biofuel resources and still have considerable excess for export.

Bioethanol fuel for transportation

Alternative energy options appear to be most limited in the transport sector, especially in the heavy vehicle, aviation and water transport modes where viable alternatives or complements to oil are proving difficult to find. Biofuels, in this context, present perhaps the only feasible option (in the short to medium term) that can benefit from local production and substitute for a share of the – often imported – oil derivates currently used in SSA fleets. Only a few SSA countries have proven oil resources and imports are a heavy burden on national accounts. The volatility of oil prices remains a major cause for concern for heavily indebted oil-importing countries in SSA. Crude oil prices have been volatile since 2004, rising from less than US$40 a barrel to a record high of US$147 in July 2008, boosted by a jump in global demand for oil, and exacerbated by worries about supply from key producers as well as political instability in the Middle East (UNECE, 2007; EIA, 2009). Global oil resources are also on a downward trend with new reserves facing higher extraction costs, besides the fact that many are located in politically risky areas. The effects of this tightening supply base will worsen energy security further in the oil-importing least developed countries of SSA.

TABLE 14.3 Theoretical production potentials, physical offset of oil and potential revenues which could be secured via the uptake of biofuels in selected countries

| | Suitable farmland for sugar cane expansion (1,000'ha)[b] | Potentials by 2030 | | |
		Ethanol production (billion litres)	Ethanol production as % of estimated domestic gasoline consumption by 2030 (physical energy security)[a]	Ethanol revenue as % of net payments for oil imports by 2030 (economic energy security)[a]
Mozambique	2,338	18.06	10,113%	1,929%
Tanzania	467	3.71	526%	153%
Zambia	1,178	9.37	2,663%	1,151%

Sources: a – 2030 consumption estimated from average growth in energy consumption between 1990 and 2008 (IEA, 2010). b – Watson, 2011. Price ratios between ethanol and gasoline are assumed to remain constant.

Notes

Ethanol prices assumed to be US$0.6/litre (CEPEA/ESALQ and ethanolmarket.com). National energy balances extracted from IEA (2010). Sugar cane yields are assumed to be: for Mozambique, 103t/ha; Tanzania 106t/ha; Zambia 106t/ha (Watson, 2011). Also considered is an ethanol productivity of 75 litres/tonne of sugar cane processed (Watson, 2011). For this scenario the imports of oil products are considered constant as a share of total energy consumption by 2030.

Countries in SSA could produce and supply biofuels to international and/or domestic energy markets, depending on national biofuel strategies and other trade-off criteria. As shown in Figure 14.2 such a strategy is dependent on the scale of production, prices of fossil fuels and progress in technological learning. International markets can in some cases afford to pay the price premiums necessary for the commercialization of biofuels, through policy mechanisms such as blending mandates and sustainability requirements. As the cumulative production increases, biofuel prices may fall below the fossil fuel threshold and then become also attractive for internal markets. However, political and environmental concerns could outweigh the economic dimension, depending on differences in countries' strategic priorities.

The entry of African producers into the biofuels market takes place in a different commodity market context: biofuel technologies have matured since the 1970s, when some early biofuel programmes such as *Proalcool* in Brazil were initiated. This can provide advantages to SSA countries with potential for bioethanol production from high-yielding energy crops such as sugar cane.

Lessons from biofuel experiences in Brazil

Brazil offers some unique lessons in the national development of renewable energy resources in a developing country setting. The Brazilian experience with biofuels helped to change the negative perception of natural resources as a tool for development. After a learning phase of 30 years, bioethanol as a transport fuel in Brazil has matured and now contributes to:

- reduced demand for fossil energy in the transport sector;
- generation of renewable electricity via combined heat-and-power plants;
- economic growth, by providing employment and innovation interlinkages (production of higher value-added commodities);

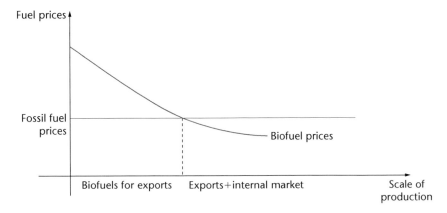

FIGURE 14.2 Biofuel market supply strategy curve for developing countries (source: Adapted from Peters and Thielmann (2008))

- lower pressure on balance of payments and locally financed biofuel activities which help in macro-economic stability;
- an improved geopolitical position due to reduced dependence on foreign energy and aid.

Brazil represents an interesting case of how a developing country can reduce its dependence on oil, despite enjoying only limited bargaining power in world markets. The national ethanol programme (PROALCOOL), primarily motivated by the oil crisis of 1973, was a major success which made optimal usage of existing geographical advantages along with the industrial knowledge on bioenergy accumulated during decades of sugar cane activity.

Coupled with other parallel initiatives in the electricity and the state-run oil sector, PROALCOOL helped to alleviate Brazil's vulnerability to oil shocks and its associated impacts on the economy. The successful Brazilian bioethanol story cannot be fully replicated in other developing countries due to the unique circumstances in Brazil (GTZ, 2008). However, important lessons can still be learnt from the implementation of PROALCOOL. Furthermore, even though other countries may not be able to replicate the Brazilian example, the increasing economic integration in Africa could make it possible to take a regional rather than national approach to such programmes (Johnson and Matsika, 2006).

Before the 1973 oil crisis, Brazil's economy was growing rapidly with the energy-intensive steel industry at its core, favoured by a cheap energy supply. The sudden increase in oil prices in 1973 struck a blow to Brazil's terms of trade and its foreign debt worsened (Siggel, 2005). Figure 14.3 depicts the historical trends in oil imports to Brazil, showing the degree of oil dependency in 1973.

Three major initiatives were undertaken to improve energy security in Brazil: the establishment of the PROALCOOL programme; investments in renewable electricity (hydroelectric power); and restructuring of Petrobrás (the state-owned oil company). Given the availability of production capacity for ethanol and a long-established sugar industry, Brazil saw potential in this local resource and national coordination began between farmers, sugar cane mills, financiers and the automobile industry.[4] Overcoming technical scepticism about the feasibility of large-scale blending of ethanol in gasoline, the initial blend was 10 per cent, but this rose rapidly to more than 20 per cent, including a large number of pure ethanol-driven vehicles. The introduction of flex-fuel technology in 2003 loosened another constraint on the expansion of fuel ethanol markets, by allowing consumers to choose between ethanol or gasoline blends[5] based on price and other attributes (Hira and de Oliveira, 2009). Goldemberg (2007) estimated the subsidies to ethanol at US$20 billion in 30 years, but the programme saved more than US$50 billion in petroleum imports.

Bioenergy and electricity production in Africa

The availability of affordable and reliable energy services, especially electricity, is key to unlocking development potential in SSA. However, the electricity sector

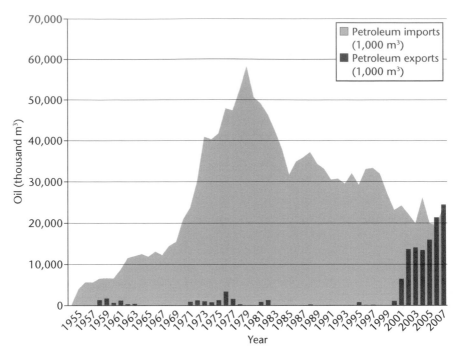

FIGURE 14.3 Historical trends in Brazilian oil imports and exports, in thousands of cubic metres (1955–2007) (source: IPEADATA Statistical database, 2010)

remains one of the key challenging areas in Africa, largely lacking in necessary investment, which limits grid access to a minority of the population. The sector is characterized by poor infrastructure, limited investments and overdependence on expensive emergency generation capacity.

Table 14.4 demonstrates the slow rate of electrification in the region. Only 25 per cent of SSA's population has access to electricity and electrification is as low as 5 per cent in some countries, while per capita electricity consumption is below 50 kWh in parts of the region (IEA, 2010). SSA has the lowest per capita electricity consumption in the world, with an average of 178 kWh (excluding South Africa). This is equivalent to only 2.4 per cent that of developed regions (Karekezi et al., 2008).

The lack of access to electricity is considered a major impediment to economic development (Gomez and Silveira, 2010). While diesel generators provide most of the off-grid and stand-alone electricity generation in SSA, bioenergy has a large potential to improve electric output both directly via off-grid systems and indirectly via electricity cogeneration of heat and power.

The capacity utilization and availability of electricity generation plants in SSA (excluding South Africa and Mauritius) is poor, typically in the range of 30–40 per cent of the installed capacities. Consequently, supplies are intermittent, with frequent power cuts, load shedding and sometimes outright grid collapse.[6] An

TABLE 14.4 Electricity access in selected SSA countries (2005)

Country	Electrification rate (%)	Per capita electricity consumption (kWh)*
Angola	26.2	124
Botswana	45.4	1,325
DR of Congo	11.1	93
Malawi	9.0	72**
Mauritius	99.4	1,382
Mozambique	11.7	367
Namibia	34.0	1,389
South Africa	75.0	4,976
Tanzania	13.9	53
Zambia	18.8	692
Zimbabwe	41.5	795

Source: * IEA, 2010; ** Mhango, 2005.

increasingly common response to the crisis has been short-term leases for emergency power generation by a handful of global operators. Though this capacity can be put in place within a few weeks, it is expensive. The costs of small-scale diesel units, for example, are typically about 35 US cents per kWh. In Eastern and Western Africa, about one-third of installed capacity consists of diesel-based generators. An estimated 700 MW of emergency generation is currently operating in SSA; this represents more than 20 per cent of the total installed capacity. As shown in Table 14.5, the total price tag ranges from 0.5 per cent of the gross domestic product (GDP) in Gabon to 4.3 per cent in Sierra Leone (World Bank, 2008).

The large share of emergency generation capacity in the electricity system of SSA presents a heavy cost burden in the region; the usage of diesel generators damages the balance of payments in the country and drives the cost of electricity

TABLE 14.5 Emergency power generation in selected SSA countries

Country	Year	Emergency capacity (MW)	% of total installed capacity	Estimated annual cost as % of GDP
Angola	2006	150	18	1.04
Gabon	N/A	14	3.4	0.45
Ghana	2007	80	5.4	1.90
Kenya	2006	100	8.3	1.45
Madagascar	2004	50	35.7	2.79
Rwanda	2005	15	48.4	1.84
Senegal	2005	40	16.5	1.37
Sierra Leone	2007	20	133.3	4.25
Tanzania	2006	180	20.4	0.96
Uganda	2006	100	41	3.29

Source: World Bank, 2008.

up. However, there are opportunities to use biofuels in off-grid electricity generation systems. For example, small-scale biodiesel-powered electricity generation has potentially significant economic and environmental benefits at the local level (Azoumah *et al.*, 2011). In addition, cogeneration of heat and electricity in sugar mills is an established procedure, which is being explored with a view to possible scaling up in the region. Apart from sugar cane bagasse, several other agro-based industries in the continent have potential for heat and power production. These include the wood, palm, rice and pulp and paper industries.

Cogeneration holds great potential for Africa. For example, Mauritius is an island state with no local fossil fuel resources, but has still made strides in reducing its dependency on imported coal and petroleum by using sugar cane bagasse to generate electricity for sale on the public grid. Already, Mauritius provides a highly successful example of the use of bagasse-based cogeneration (of process heat and electricity) which now meets over 20 per cent of the country's national electricity generation (Karekezi and Kithyoma, 2005). There are currently plans in other countries, such as Malawi, Swaziland, Zimbabwe and South Africa, to modernize sugar mills and generate excess electricity to feed the public grid. This also provides opportunities for electrifying rural areas and small towns in the vicinity without costly grid extensions. Considering the scale of production in most African countries (given in Table 14.5), large industrial-scale bagasse cogeneration plants can contribute significantly to increasing the renewable share of power generation. Batidzirai (2002) illustrates the economic and technical benefits of embedded generation of electricity in a sugar mill at Hippo Valley Estates in Zimbabwe. Costs of forced outages and unmet demand to the regional economies are difficult to quantify but acknowledged as being quite significant. In addition, meeting peak demand using inefficient coal thermal plants, network reinforcement and building new power plants is also costly to the economy. Distributed generators embedded in the national grid can play a significant role in stabilizing power in local networks and reducing overall system losses. In addition, the construction period for such plants is shorter; they require smaller capital investment, enjoy lower running costs and achieve quicker returns on investment compared to an equivalent fossil fuel-based power plant.

Bioethanol and household cooking

Traditional biomass, especially firewood, will continue to dominate cooking energy in the region for many years to come under the current trends. For many SSA countries, traditional biomass accounts for 50–90 per cent of national energy supply, with the household sector being the largest single consumer. However, because of unsustainable biomass production, inefficient transformation practices and the high costs of distribution and end use, the overall cost burden of traditional biomass is increasing considerably (Johnson and Rosillo-Calle, 2007). Depletion of fuelwood inevitably leads to fuel poverty through the use of poor-quality fuels such as animal dung or, for those that can afford it, purchased fuels. This situation hinders the development of rural economies and undermines efforts to enhance income-generating activities and alleviate poverty.

Among the currently available cooking options, LPG, electric and ethanol stoves offer cleaner alternatives, through high efficiency and reductions in harmful emissions. The uptake of clean cooking options is hindered in large parts of SSA by high prices, limited infrastructure and distribution systems as well as lack of end-use equipment (e.g. ethanol stoves). Use of traditional biomass such as firewood, peat and cow dung results in continuing health problems due to indoor air pollution, deforestation and an overburden on women, who must forage for firewood.

Promotion and use of ethanol as a cooking fuel is an option that is worth pursuing in SSA, given the household characteristics and cooking practices. This alternative fuel requires significant industrial capability to mass-produce ethanol stoves and improve affordability to the mostly poor target group. Initiatives such as Project Gaia[7] promote the use of ethanol stoves, which offer a number of health and environmental benefits (SEI, 2009; Takama *et al.*, 2011). There are, however, socio-economic constraints that prevent the broader adoption of clean cooking options such as ethanol in SSA. According to Tsephel and colleagues (2010), low-income groups in SSA cannot afford the higher up-front cost, although they do place value on some non-price impacts such as health and safety; some means of compensating for the higher up-front cost that also incorporates consumer preferences would support this market.

Section 4: discussion

The analysis suggests a useful role for biofuels in contributing to energy security and economic development in sub-Saharan Africa, in the sectors of transport, heat and power production, as well as cooking energy. Sugar cane-derived fuel ethanol is already a proven, albeit limited, success in the region, as is industrial-scale cogeneration using bagasse and other agro-residues. Developing, producing and supplying biofuels in the region contribute to improved energy security: the example of sugar cane is useful because it is already established in the region and can be harnessed on the basis of additional or complementary investments rather than having to start from the beginning.

Despite the enormous opportunities apparent in modern bioenergy as a potential source of energy in SSA, large-scale biofuel production and cogeneration are yet to be widely developed in the region. Apart from bioethanol production and use in Malawi, there has been no sustained biofuel programme in Africa. Currently there are numerous biofuel initiatives in many SSA countries but there is limited experience with large-scale production and use of biofuels in the region. Few commercial-scale plants have been established, although a number of projects and programmes are under development.

While SSA is seen as a key "missing link" towards a higher commoditization of bioethanol (Jumbe *et al.*, 2009; UNCTAD, 2009a; Johnson and Matsika, 2006), the development of the biofuel potential in sub-Saharan Africa faces multiple challenges and risks. These risks include lack of credible and consistent investment information, political scepticism and scientific uncertainties, lack of access to affordable financing and investment, lack of capacity and technological know-how, and in some cases inadequate policy and regulatory frameworks.

Political enthusiasm for tapping the potential of biofuels has been increasing in response to the sustained increases in oil prices starting in 2004. But the controversies around land and the food–fuel debate have confused many stakeholders and policy makers, resulting in some uncertainties in the sector. Despite this, many countries continue to support pilot projects to build experience and evaluate the performance of new schemes. Sugar cane-based bioethanol and cogeneration offer the advantage of being mature technologies that will only require replication and scaling up within the region. However, the investment costs can be high by African standards, although the initial costs are often much lower than similar-capacity investments in hydropower fossil fuel plants.

The development of biofuels should be accompanied by a strong policy and regulatory framework that encourages sustainable energy options, if the goal is to create long-term stable energy capacity in the region. For example, energy pricing distortions that favour fossil fuels put bioenergy at a disadvantage. Government policies are important because of their ability to create an enabling environment for mobilizing resources and encouraging investment in renewables. Most SSA governments do not have a clear-cut policy or strategy on bioenergy (with the exception of a few countries). As a result, bioenergy development follows an ad hoc path, with little recourse to national energy plans, which are sometimes out of date or inadequate. In addition, countries do not have data collection systems that can capture vital information to support the development and implementation of sound policies in the bioenergy subsector.

Successful introduction of bioenergy technologies hinges on the development of requisite capacities and competencies. The importance of technical know-how in the increased utilization of bioenergy has been recognized in the region, but there remains a continuing shortage of qualified personnel. Technical knowledge is needed to build a critical mass of policy analysts, economic managers and engineers who will be able to manage all aspects of bioenergy development.

Capacity building in Africa can be viewed as a twofold process: capacity building for new biomass energy technologies, which are not well disseminated in the region – this requires significant investment in training and capacity building over a long period of time; and capacity mobilization for mature bioenergy technologies, which are already well developed in the region – this would require more modest investment aimed at facilitating the utilization of an existing skills base.

In contrast to conventional energy technologies that are mature and have evolved into highly sophisticated industries, most bioenergy technologies are relatively new, with a relatively short learning curve behind them. They are also relatively less complicated in technical terms, meaning that a significant industry could be developed in Africa, even where technical expertise is limited. This is especially true for sugar industry-based bioenergy options that are mature and have been implemented in the region over a long period of time.

The absence of low-cost, long-term financing is a key barrier to realization of bioenergy. The financing problem is complicated by unfavourable macro-economic conditions and high investment risks in many SSA countries. Greater emphasis is often put on the traditional petroleum and power sectors, which, although important for the national economy, supply energy to only a small share of the population, whereas modern biomass energy can support a broader segment of the population.

Section 5: conclusions

This chapter explored the potential to improve energy security in the countries of sub-Saharan Africa (SSA) through biofuel production. This is on the back of a continued lack of access to modern energy services and lack of opportunities for income generation, issues which are affecting development and contributing to perpetuating poverty in the region. The continued dependency on traditional biomass energy remains one of the biggest challenges to African economies. But given the natural endowment of the region, development of a viable bioenergy industry is an interesting prospect that could potentially assist many SSA countries to solve many of their pressing needs.

As a natural resource-based activity, biomass energy has courted some controversy among analysts and various stakeholders; however, through careful implementation and use of sustainability criteria, modern bioenergy can play an important role in reducing energy poverty and decreasing energy import bills, while promoting health, economic growth and environment protection.

While agro-based activities have been long been labelled as delivering unfair terms of trade for those on the supply side, biofuels have the potential to add value to agricultural activities, thereby avoiding the tendency in Africa to export raw materials. Biofuels could boost energy security via direct reduction in oil imports (*physical energy security*) or through increased economic opportunities, meaning a higher ability to pay for energy (*economic energy security*).

Important lessons can be learnt from the bioenergy experiences of Mauritius and Brazil, countries that have demonstrated that political commitment to the exploitation of indigenous resources can achieve far-reaching economic benefits with the necessary policy framework. The widespread use of ethanol in Brazil has proven that biofuels are technically feasible on a large scale, while the adoption of bagasse cogeneration in Mauritius has shown how economic restructuring and improved competitiveness can be based on expanding the development of domestic biomass resources.

Indicators for three countries in SSA (Mozambique, Tanzania and Zambia) show strong dependency on foreign energy and very high fuel prices in their internal markets compared to average global markets. The high dependence on imported oil and the low reliability and availability of modern energy services (especially electricity) impose considerable costs on their economies and constrain economic development.

Countries in SSA could, in principle, cover their domestic energy demand with indigenous biofuel resources and still export surplus production to the world market. Whether countries will opt to export some of the produced biofuels will depend on national policy priorities, favouring physical or economic energy security. However, in order to realize their biofuels production potential, SSA countries will need to address a number of policy, technological and financial obstacles.

Acknowledgements

The authors would like to extend their appreciation for the helpful inputs provided by Mr Armin Wagner (GTZ), Mr Flavio Freitas (Universidade de São Paulo), Ms Anna Åberg and Professors Per Högselius, Arne Kajser and Semida Silveira from the Royal Institute of Technology, Sweden.

Notes

1 Although it is widely believed that natural endowments give a competitive hedge for the primary sector in developing countries, there is concern about the "paradox of the plenty", a reference to the apparent underperformance of countries focusing on large-scale agriculture and mineral extraction (Sachs and Warner, 1995). See also UN (2006) for the views of Raul Prebish and Celso Furtado of the Economic Commission for Latin America and the Caribbean.
2 Energy dependency is a measure of the net reliance of a country on foreign energy inputs for its energy needs. It is represented by the ratio of net energy imports to gross inland consumption (Sözen, 2009).
3 Only the theoretical biofuel potential is discussed here, based on studies available at www.compete-africa.net. See also Watson, 2007 and 2009.
4 Spearheaded by Banco do Brasil; see Krugmann, 1994.
5 A minimum blend of ethanol in gasoline is specified; the consumer with a flex-fuel car can effectively choose any level at or above this amount.
6 Foley (1990) evaluates the costs of supply outages according to studies made in different developing countries. These range from US$1.00–7.00 per kWh for industrial users, US$1.50–1.65 per kWh for commercial and US$0.05–1.50 per kWh for domestic users.
7 See www.projectgaia.com/.

References

AfDB (African Development Bank) (2006) "High Oil Prices and the African Economy", Concept paper prepared for the 2006 African Development Bank Annual Meetings, Ouagadougou, Burkina Faso.
Azoumah, Y., Yamegueu, D., Ginies, P., Coulibaly, Y. and Girard, P. (2011) "Sustainable electricity generation for rural and peri-urban populations of sub-Saharan Africa: the 'flex-energy' concept", Energy Policy, vol. 39, no. 1, pp. 131–141.
Batidzirai, B. (2002) "Cogeneration in Zimbabwe: A Utility Perspective", AFREPREN Occasional Paper No. 19. AFREPREN/FWD, Nairobi.
Batidzirai, B., Faaij, A. and Smeets, E. (2006) "Biomass and bioenergy supply from Mozambique," Energy for Sustainable Development, vol. 10, no. 1 (March), pp. 54–81.
Batidzirai, B. and Wamukonya, N. (2009) "Biofuels and their implications for Africa's development", FEMA background paper.

Bekunda, M., Palm, C.A., de Fraiture, C., Leadley, P., Maene, L., Martinelli, L.A., McNeely, J., Otto, M., Ravindranath, N.H., Victoria, R.L., Watson, H. and Woods, J. (2009) "Biofuels in developing countries", in R.W. Howarth and S. Bringezu (eds), *Biofuels: Environmental Consequences and Interactions with Changing Land Use*, Proceedings of the Scientific Committee on Problems of the Environment (SCOPE) International Biofuels Project Rapid Assessment, 22–25 September 2008, Gummersbach, Germany and Cornell University, Ithaca, NY. Avaliable at: http://cip.cornell.edu/biofuels/.

BNDES (National Development Bank of Brazil) and CGEE (Center for Strategic Studies and Management) with CEPAL (UN Economic Commission for Latin America and the Caribbean) and FAO (Food and Agriculture Organization) (2008), *Sugarcane-Based Bioethanol: Energy for Sustainable Development*, November, BNDES, CGEE, Rio de Janeiro.

Davidson, O. (2008) "Scaling Up Markets for Renewable Energy in Africa", Thematic Background paper for International Conference on Renewable Energy in Africa, 16–18 April, Dakar.

EIA (2009) "Weekly United States Spot Price FOB Weighted by Estimated Import Volume", www.eia.doe.gov/.

European Commission (2009) Directive on the Promotion of the Use of Energy from Renewable Sources, Directive 2009/28/EC.

Foley, G. (1990) "Electricity for Rural People", Panos Institute.

Goldemberg, J. (2007) "Ethanol for a sustainable energy future", *Science*, 315, pp. 808–810.

Gomez, M.F. and Silveira, S. (2010) "Rural electrification of the Brazilian Amazon achievements and lessons", *Energy Policy*, vol. 38, no. 10, pp. 6251–6260.

GTZ (2008) International Fuel Prices Data, November, pp. 4–5. Available at: www.gtz.de/de/dokumente/en-international-fuel-prices-data-preview-2009.pdf.

Habitat (1993) "Application of Biomass-Energy Technologies", United Nations Centre for Human Settlements (Habitat), Nairobi.

Hall, J., Matos, S., Severino, L. and Beltrao, N. (2009) "Brazilian biofuels and social exclusion: established and concentrated ethanol versus emerging and dispersed biodiesel", *Journal of Cleaner Production*, vol. 17, no. 1, pp. S77–S85.

Hira, A. and de Oliveira, L.G. (2009) "No substitute for oil? – How Brazil developed its ethanol industry", *Energy Policy*, vol. 37, no. 6, pp. 2450–2456.

IEA (2002) *World Energy Outlook*, OECD/IEA, Paris.

IEA (2004) *World Energy Outlook*, OECD/IEA, Paris.

IEA (2008) "End User Petroleum Product Prices and Average Crude Oil Import Costs", International Energy Agency, Paris.

IEA (2010) World Energy Outlook: the Electricity Access Database.

IMF (International Monetary Fund) (2005) "Zimbabwe: Selected Issues and Statistical Appendix", IMF Country Report No. 05/359, October, Washington, DC.

IMF (2006a) "Botswana: Statistical Appendix", IMF Country Report No. 06/65, February, Washington, DC.

IMF (2006b) "Mauritius: Selected Issues and Statistical Appendix", IMF Country Report No. 06/224, June, Washington, DC.

IMF (2006c) "Namibia: Selected Issues and Statistical Appendix", IMF Country Report No. 06/153. April, Washington, DC.

IMF (2006d) "The Kingdom of Swaziland: Statistical Appendix", IMF Country Report No. 06/109, March, Washington, DC.

Jank, M., Kutas, G., Do Amaral, L.F. and Nassar, A.M. (2007) "EU and US Policies on Biofuels: Potential Impacts on Developing Countries", The German Marshall Fund of the United States.

Johnson, F.X. and Matsika, E. (2006) "Bio-energy trade and regional development: the case study of bio-ethanol in Southern Africa," *Energy for Sustainable Development*, vol. 10, no. 1 (March), pp. 42–53.

Johnson, F.X. and Rosillo-Calle, F. (2007) "Biomass, Livelihoods and International Trade: Challenges and Opportunities for the EU and Southern Africa", SEI Climate and Energy Report 2007–01.

Jumbe, C.B.L., Msiska, F.B.M. and Madjera, M. (2009) "Biofuels development in Sub-Saharan Africa: are the policies conducive?" *Energy Policy*, 37, pp. 4980–4986.

Karekezi, S. and Kithyoma, W. (2005) "Sustainable Energy in Africa: Cogeneration and Geothermal in the East and Horn of Africa – Status and Potential", AFREPREN/FWD and HBF.

Karekezi, S., Kityoma, W. and Oruta, A. (2008) "Renewable Energy for Africa: Potential, Markets and Strategies", Report REN21/GTZ/BMZ.

Krugmann, H. (1994) "National Alcohol Program: Impacts and Role in a New Energy Context (Brazil)", The International Development Research Centre of Canada.

Lee, T.S.G. and Bressan, E.A. (2007) "The potential of ethanol production from sugarcane in Brazil", *Sugar Tech*, 8, pp. 195–198.

Lusaka Times (2008) "Zambians should brace for high fuel prices", 28 June. Available at www.lusakatimes.com/?p=3108.

Mathews, J. (2006) "A Biofuels Manifesto: Why Biofuels Industry Creation should be 'Priority Number One' for the World Bank and for Developing Countries", Macquarie University, Sydney, 2 September.

Matthews, J. (2007) "Can renewable energies be turned to a source of advantage by developing countries?", *Revue d'Énergie*, 576, pp. 1–10.

Mhango, L.B. (2005) "Pre-Feasibility Study for the Cogeneration Project: Malawi Country Report", Lilongwe.

Osava, M. (2008) "Brazil: Biodiesel to Bring Electricity to Amazon Villages", Galdu. Available at: www.galdu.org/web/index.php?odas=3050&giella1=eng.

Pacini, H. and Silveira, S. (2010) "Ethanol or gasoline? Consumer choice in face of different fuel pricing systems in Brazil and Sweden", *Biofuels*, vol. 1, no. 5, pp. 685–695.

Pacini, H. and Silveira, S. (2011) "Consumer choice between ethanol and gasoline: Lessons from Brazil and Sweden," *Energy Policy*, vol. 39, no. 11, pp. 6936–6942.

Peters, J. and Thielmann, S. (2008) "Promoting biofuels: implications for developing countries", *Energy Policy*, 36, p. 4.

Reddy, T. (2008) "Africa's Green Gold Rush or Another Resource Curse?", Institute for Security Studies – Corruption and Governance Programme.

Sachs, J.D. and Warner, A.M. (1995) "Natural Resource Abundance and Economic Growth", National Bureau of Economic Research, NBER working paper no. 5398.

Saghir, J. (2006) World Bank Perspective on Global Energy Security, presentation to G8 Energy Ministerial Meeting, Moscow, 16 March.

Scurlock, J., Rosenschein, A. and Hall, D.O. (1991) *Fuelling the Future: Power Alcohol in Zimbabwe*, Acts Press/BUN.

Sierra, K. (2006) "Meeting Energy Security and Environment Challenges: The Crucial Role of Renewable Energy Policy", Keynote Speech at the International Grid-Connected Renewable Energy Policy Forum, Mexico City, 1–3 February 2006.

Siggel, E. (2005) "Development Economics: A Policy Analysis Approach, Case Study 8 – the Brazilian Debt Crisis", Ashgate Innovative Economics Textbooks.

Smeets, E.M.W., Faaij, A.P.C., Lewandowski, I.M. and Turkenburg, W.C. (2007) "A bottom-up assessment and review of global bio-energy potentials to 2050", *Progress in Energy and Combustion Science*, vol. 33, no. 1, pp. 56–106.

SEI (2009) "Household Energy in Developing Countries: A Burning Issue", Stockholm Environment Institute, Policy Brief. Available at: www.sei.se.

Sözen, Adnan (2009) "Future projection of energy dependency of Turkey using artificial neural network", *Energy Policy*, vol. 37, no. 11, pp. 4827–4833.

Takama, T., Lambe, F., Johnson, F.X., Arvidson, A., Atanassov, B., Debebe, M., Nilsson, L., Tella, P. and Tsephel, S. (2011) "Will African Consumers Buy Cleaner Fuels and Stoves? – A Household Energy Economic Analysis Model for the Market Introduction of Bio-Ethanol Cooking Stoves in Ethiopia, Tanzania, and Mozambique", SEI Research Report, Stockholm Environment Institute, Stockholm, ISBN 9789186125257.

Takavarasha, T., Uppal, J. and Hongo, H. (2005) "Feasibility Study for the Production and Use of Biofuel in the SADC Region", SADC, Gaborone.

Tsephel, S., Takeshi, T., Lambe, F. and Johnson, F.X. (2010) "Why perfect stoves are not always chosen: A new approach for understanding stove and fuel choice at the household level", *Boiling Point*, 57, pp. 6–8.

UN (2006) CEPAL Review 88, United Nations Economic Commission for Latin America and the Caribbean, Santiago, Chile.

UNCTAD (2009a) "The Biofuels Market: Current Situation and Alternative Scenarios", Chapter IV – "Trade opportunities for developing countries", United Nations Publications.

UNCTAD (2009b) "The Biofuels Market: Current Situation and Alternative Scenarios", Chapter 1 – "The role and implications of biofuel blending targets", United Nations Publications.

UNECE (2007) "Secure and Sustainable Energy Supplies", United Nations Economic Commission for Europe Annual Report 2007, New York and Geneva, www.unece.org/pub_cat/topics/annual_report_2007.pdf.

Van Gent, C.F., Kalff, B. and Van Peperstraten, J. (2007) "A Theory on the Effects of Biofuels Production on Countries in Sub-Saharan Africa: A Regional Perspective", Rotterdam School of Management.

Watson, H. (2007) "Potential to Expand Sugarcane Cultivation in Southern Africa – An Assessment of Suitable and Available Land", Proceedings of COMPETE Workshop on Improved Energy Crop and Agroforestry Systems for Sustainable Development in Africa, 22 June, Le Reduit, Mauritius. Available at: www.compete-bioafrica.net.

Watson, H. (2009) "COMPETE Second Periodic Activity Report – Annex 1–2–1: Second Task Report on WP1Activities – Current Land Use Patterns and Impacts", University of KwaZulu-Natal, Durban.

Watson, H. (2011) "Potential to expand sustainable energy from sugarcane in southern Africa", *Energy Policy*, in press.

Woods, J., Mapako, M., Farioli, F., Bocci, E., Zuccari, F., Diaz-Chavez, R. and Johnson, F.X. (2008) "The Impacts of Exploiting the Sugar Industry Bioenergy Potential in Southern Africa," CARENSA Thematic Report No. 4, Cane Resources Network for Southern Africa (CARENSA)/Stockholm Environment Institute.

World Bank (2008) "Africa's Power Supply Crisis: Unravelling the Paradoxes", in IMF, *Regional Economic Outlook: Sub-Saharan Africa*, IMF, Washington, DC.

Zuzarte, F. (2007) "Ethanol for Cooking – Feasibility of Small-Scale Ethanol Supply and its Demand as a Cooking Fuel: Tanzania Case Study", KTH School of Energy and Environmental Technology, Heat and Power Technology, Stockholm.

15

THE ROLE OF ETHANOL FROM SUGAR CANE IN MITIGATING CLIMATE CHANGE AND PROMOTING SUSTAINABLE DEVELOPMENT IN LDCs

The case of Nepal

Semida Silveira and Dilip Khatiwada

Introduction

Global concerns related to the sustainability of energy systems and climate change, as well as increasing concerns about energy security have led to increased interest and support for biofuels around the world. In this context, ethanol has emerged as a concrete alternative transport fuel, particularly in the past decade. Although ethanol as transport fuel is not a novelty, the use of ethanol on a large scale and the formation of international ethanol commodity markets definitely mark a new stage in the realization of its potential. As part of discussions about the merits of ethanol, a large debate has taken shape that puts the production of biofuels and the production of food as conflicting objectives, and as competitors for land and water. In this context, the least developed countries (LDCs) are sometimes seen as the most negatively affected due to potential increases in food prices and reduced availability of land. However, this debate often ignores the established economic realities in various developing countries, the potential to create synergies between fuel production and existing industries, and the broad environmental and economic benefits that such synergies could generate.

The large interest of many industrialized countries to exploit renewable energy sources as part of environmental or national security policies coupled with rising oil prices and the development of a global climate agenda have served as major drivers to promote biofuels such as ethanol. Poor countries, on the other hand, have a pressing development agenda. Thus economic, social and environmental benefits at the national level provide the strongest motivation for choosing biofuels. Each country needs to find its appropriate entry point to realize bioenergy potential (Silveira, 2005a). But despite the progress achieved in this direction, many countries have not yet taken advantage of their great bioenergy potential. In this chapter, we use the case of Nepal to illustrate how a country's potential for ethanol production and fuel substitution can be used to promote sustainable development at national and global levels simultaneously. What is observed in Nepal has relevance for the LDCs of Africa that are considering similar development paths linking ethanol production with climate and development policies.

Sugar cane and bioethanol in LDCs

Many developing countries that produce sugar cane already have the necessary basic resources from which they can develop an ethanol industry. Following recent diversification strategies, sugar industries in many developing countries have integrated their sugar production with ethanol and electricity production (through bagasse cogeneration), including Argentina, Colombia, India, Mexico, Nicaragua, Thailand and Uruguay. Bagasse, a residue of sugar and ethanol production, when used efficiently, generates surplus electricity that can be exported to the grid, thereby improving the total economy of the sugar-ethanol industry significantly. While the US and Brazil are still by far the largest ethanol producers, an increasing number of countries all over the world are producing the fuel, including Australia, Canada, China, Colombia, Costa Rica, Cuba, the Dominican Republic, Zambia, Thailand and many EU countries, for example Sweden.

When it comes to the LDCs, many opportunities remain untapped. Although many of these countries produce sugar cane and sugar today, they have not yet moved into ethanol production and use. There are a variety of reasons for LDCs' reluctance to initiate ethanol programmes. In this chapter, we show that there are economic, political, environmental and health issues motivating the search for alternative transport fuels in LDCs, as illustrated with the case of Nepal, where bioethanol can be produced from sugar cane molasses. The study shows that fossil fuel costs have become a major drain on the national economy, compromising political stability and development. Consequently, immediate societal and environmental gains can be achieved by implementing bioethanol production and fuel substitution in the country. In order to initiate fuel ethanol production and use in Nepal, some bottlenecks need to be addressed, which are reviewed in this chapter.

Global motivation also exists for the realization of the biofuel potential in a poor country such as Nepal. In this context, the case study provides a contrasting view to generic arguments against biofuels. The case of Nepal shows that ethanol production and use can play an important role in development beyond just the transport fuel substitution itself and, at the same time, contribute to climate change mitigation. The lessons drawn are highly relevant for many poor countries in Africa that are also sugar producers and are reviewing their options and potential for ethanol production.

Our analysis draws on a life-cycle assessment (LCA) methodology to assess the energy and greenhouse balance, and direct economic and environmental benefits of molasses-based bioethanol production and use in Nepal. In the application of LCA, economic allocation is used to divide the resource consumption (primary energy) and environmental burdens (GHG emissions) in upstream operations when there are co-products (e.g. molasses). An economic allocation method is preferred over physical allocation since the relative prices drive the decisions about the type and quantity of co-products associated within the sugar market (see Chapter 8). Thus, the market prices of sugar and molasses determine the division of energy consumption and greenhouse gas (GHG) emissions between these two products.

Alternative fuels for transport in Nepal

Nepal is among the poorest and least industrialized countries in the world, with an uneven development distribution of wealth. According to the Asian Development Bank, Nepal presents the highest Gini coefficient level in Asia, at 47.3 per cent in 2003.[1] It is a landlocked country with an area of 147,200 km^2 and a population of 26.9 million inhabitants (CBS, 2008). The population is growing rapidly despite policies to slow down population growth. Globally, the country is best known for Mount Everest, the highest mountain peak in the world. Nepal is basically an agrarian country with more than 80 per cent of the population living in rural areas. At the same time, it is a country in transition both in political terms and in its pattern of population distribution.

Nepal does not have fossil fuel reserves. Being a poor landlocked country, energy security for transport is a serious problem. A number of African countries share similar energy security concerns: 42 African countries are net importers of oil, 15 of which are landlocked (e.g. Malawi, Zimbabwe, Ethiopia), and 33 of which (e.g. Malawi, Mozambique, Ethiopia, Zambia) are among the LDCs. Traditional biomass (e.g. firewood) consumption contributes 78 per cent of the total energy in LDCs (UNDP, 2006). Commercial energy (coal, oil and electricity) consumption in LDCs was only 67 kgoe (kg of oil equivalent) per capita in 2004, while other developing countries consumed 718 kgoe per capita (UNCTAD, 2008). At the same time, the average value of domestic resources available for financing governance and investment was only 41 US cents per capita in LDCs compared to US$3.2 and US$36.4 in lower-middle- and high-income countries respectively (UNCTAD, 2009). Thus, exploring alternative energy sources, particularly domestic ones, is an urgent issue for all LDCs.

In Nepal, the transport sector is the second-largest consumer of petroleum products. As many as 56 per cent of the total number of vehicles registered in the country run in the Kathmandu Valley, home to the capital. This is also where 30 per cent of the total urban population of the country live (KVEO, 2007).[2] The topography of the region is bowl shaped, posing serious local pollution problems. Vehicle emissions are major sources of air pollution in the Kathmandu Valley. A number of regulations to improve the local environment have been initiated; however, air pollution has not improved significantly. A report by the Asian Development Bank (ADB) points to Kathmandu as one of the most polluted cities in the world (ADB, 2006).

The state-owned company, Nepal Oil Corporation (NOC) is responsible for the supply of petroleum products for transport in Nepal. In the past, this company has accumulated increasing debts related to the rising cost of oil imports. Meanwhile, the Kathmandu Valley has suffered from frequent shortages of transport fuels, as well as public unrest due to price increases for petroleum products and air pollution reaching alarming rates. There are strong environmental, social, political and economic reasons, therefore, to explore and deploy alternative energy sources for automobiles in the Kathmandu Valley.

Two major new fuel options are at hand to achieve improvements in the transport sector of Nepal. One option is based on the huge potential to generate electricity from hydropower in Nepal, and the other option is based on ethanol from molasses (Silveira and Khatiwada, 2010). Battery-operated electric vehicles (three-wheelers) are already in use in the Kathmandu Valley and will most probably continue to play an important role. In this chapter, we focus on the ethanol alternative, since Nepal is a sugar producer and thus has the potential to derive ethanol from sugar cane molasses.

Nepal has nine operational sugar plants with a total capacity to process 17,050 tonnes of cane per day, indicating a significant potential to produce ethanol. Molasses is a by-product of the sugar industry, from which no further sugar can be economically recovered but which can be used for ethanol and alco-chemicals production as well as other uses. This molasses is also known as C molasses or final molasses, typically a low-value by-product in terms of sucrose content, but still having 40–50 per cent (w/w) fermentable sugars (including simple sugars like glucose and fructose) that are suitable as raw material for the fermentation process. At this moment, the country still lacks necessary production of sugar for domestic consumption. Therefore, it would be too early in the agro-industrial development process to opt for intermediate or higher-grade feedstocks such as B molasses. However, it is attractive to convert low-grade molasses (in terms of energy and market value) into a commercial energy product, bioethanol.

Sri Ram Sugar Mills Pvt. Ltd (SRSM) has installed an ethanol plant with a production capacity of $30 \, m^3/day$ ($1 \, m^3 = 1$ kilolitre (KL)) to produce molasses-based bioethanol for transport (Plate 15). The distillery can produce $4,500 \, m^3$ ethanol during 150 crop days and when molasses is purchased from other sugar mills to operate the ethanol plant the whole year, production reaches $8,760 \, m^3$ at 80 per cent of plant utilization factor. The national bioethanol production potential was $18,045 \, m^3$ in 2006/2007, considering 4.3 per cent molasses recovery. The government of Nepal has already decided, in principle, to blend 10 per cent ethanol in petrol. Yet, this has not been implemented due to technical, economic and institutional problems. How can Nepal explore its ethanol potential? Could ethanol production and use be explored as a way to improve rural conditions and promote development? Is there a risk that ethanol and food production will collide?

How to capitalize on the bioethanol potential in Nepal

The need to break oil dependency

Nepal's per capita energy consumption is very low even when compared with other developing countries. Total primary energy supply in Nepal was 14.28 GJ/capita in

2007 (IEA, 2007).[3] This means that a Nepalese citizen consumes only 5 per cent of the energy consumed by the average Swedish citizen, for example. The residential sector accounts for 90 per cent of the energy consumption, followed by the transport sector with 3.8 per cent, and industrial activities with 3.5 per cent. The rest goes for commercial activities and agriculture (WECS, 2006).

Kathmandu Valley has around 56 per cent of the total vehicle registration in Nepal. Almost 95 per cent of the Kathmandu Valley vehicle fleet is used for passenger transport and the other 5 per cent for goods transport. Figure 15.1 shows the registered vehicles in the local fleet, which includes mainly motorcycles (70 per cent) and car/jeeps/vans (21 per cent), but also a small number of buses (0.9 per cent), mini-buses (0.9 per cent), three-wheelers/tempos (1.9 per cent) and micro-buses (0.5 per cent). Figure 15.2 illustrates the growth of vehicle use in the Kathmandu Valley by vehicle type between 1990 and 2005. The annual average growth rate of motorized vehicles in the region was 13.5 per cent in this period.

When it comes to total energy use in transport, diesel contributes the major share, at 63.4 per cent, followed by gasoline (18.2 per cent), air turbine fuel (17.4 per cent), LPG (0.8 per cent) and electricity (0.2 per cent) (WECS, 2006). Electricity consumption in the transport sector amounted to 6.6 GWh in 2007 (NEA, 2007). There has been a relative reduction in the use of gasoline and a comparable increase in diesel and LPG compared to percentage shares of transport fuel consumption by energy type in 2000. The change in electricity use has been marginal. Biofuels have not been introduced but there is an immediate opportunity to substitute gasoline in 80 per cent of the individual car/jeep/van fleet, and 60 per cent of the total public transport taxis, which all run on gasoline (Silveira and Khatiwada, 2010).

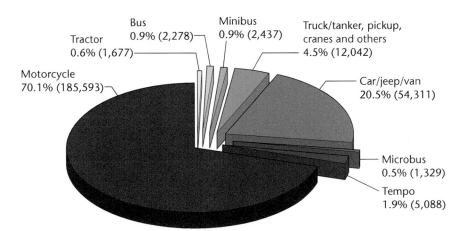

FIGURE 15.1 Motor vehicles in the Kathmandu Valley, 2005 (percentage of total and absolute number in parentheses) (source: Department of Transport Management (DOTM), Government of Nepal)

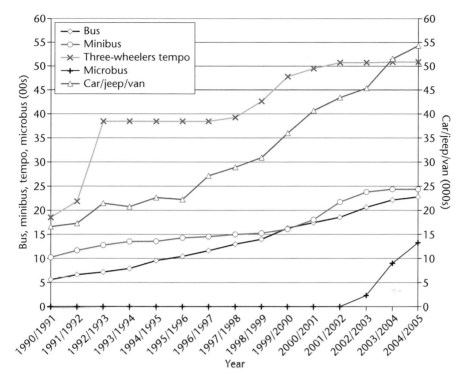

FIGURE 15.2 Growth of vehicle types in the Kathmandu Valley, 1990 to 2005 (source: Department of Transport Management (DOTM), Government of Nepal)

Gasoline prices have more than doubled over the last ten years in Nepal. Diesel prices have increased by a factor of 3.5 in the same period. Despite significant price increases, fuel prices do not reflect the actual increase in crude oil prices during this period. Oil prices soared up to US$130 a barrel in 2008, showing more than a tenfold increase as compared with the 1999 price level. The price of oil products is fixed by the board of Nepal Oil Corporation (NOC) upon approval by the government. The government of Nepal has deliberately refrained from passing the full costs of oil imports to users for fear of potential political unrest. As a result, NOC is selling the petroleum products for less than it pays the Indian Oil Corporation for the imported oil.

Thus dependency on oil imports has placed a huge burden on the Nepalese economy. The Nepal Oil Corporation imported 752,446 KL (kilo-litres; 1 KL = 1 m³) of oil products in the form of diesel (40 per cent), gasoline (13 per cent) and kerosene (26.5 per cent) from India in 2006/2007. The Kathmandu Valley consumes about 21 per cent of the diesel and 70 per cent of the gasoline imported.[4] While gasoline is used only in the transport sector, diesel is also used in the industrial and commercial sectors. The national sales of diesel have remained almost constant over the past ten years, but the sales of gasoline and LPG (liquefied petroleum gas) have increased by about twofold and fourfold respectively between 1997/1998 and 2006/2007 (see Figure 15.3). Only 7.2 per cent of the total LPG sales in the Kathmandu Valley is used in transport. The rest is mainly used as cooking fuel, since LPG is gradually substituting cooking fuels in the urban centres, with a consumption increase of 11 per cent per year accompanied by decreasing sales of kerosene, as shown in Figure 15.3.

Addressing environmental problems in the Kathmandu Valley

The average growth rate of vehicles in the Valley is approximately 13.5 per cent per year. Meanwhile, this growth has not been matched with proper expansion and upgrading of the road infrastructure. Traffic congestion, air pollution and acute shortage of transport fuels lead to serious consequences in terms of economics, politics, health and the environment, and are the cause of increasing concern in the Valley.

Air pollution and energy security are pressing problems related to fossil-based transport in the Kathmandu Valley. Six monitoring stations have been installed in different locations to check air quality. The monthly average air quality of Kathmandu Valley is far worse than the WHO recommended standards ($70\,\mu g/m^3$ for

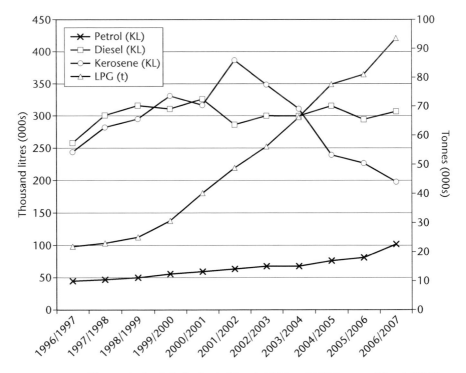

FIGURE 15.3   Change in fossil fuel sales in Nepal, 1996 to 2007 (source: Nepal Oil Corporation Ltd (NOC, 2008))

PM_{10}, $120\,\mu g/m^3$ for TSP for a 24-hour average) with direct implications for the health of Kathmandu's residents. Most PM_{10} (dust particles/particulate matters with a diameter smaller or equal to 10 micrometres) values in urban and traffic-congested areas are above the national ambient air quality standard ($120\,\mu g/m^3$). For example, the annual average concentration of particulate dust particles (PM_{10}) was $274\,\mu g/m^3$ (2003) and $261\,\mu g/m^3$ (2005) at Putalisadak in the Kathmandu Valley (CAI-Asia, 2005). According to WHO assessments, if concentrations of PM_{10} in the Kathmandu Valley could be reduced below $50\,\mu g/m^3$, 1,600 deaths could be avoided each year (MOEST, 2005).

Kathmandu Valley consumes 70 per cent of the gasoline imported to Nepal, or $71{,}338\,m^3$ annually. Most light vehicles (cars, jeeps and vans) use gasoline, and a huge fleet of two-wheeler motorbikes also consumes gasoline. Gasoline vehicles can use E5–E10 fuel in current engine and equipment fuel systems with only minor adjustments. By using E10 in the Kathmandu Valley, Nepal could save $4{,}860\,m^3$ of gasoline per year, which is a reduction of 6.8 per cent of the import (Silveira and Khatiwada, 2010). As much as 14 per cent of import reduction is possible if the vehicles go for E20. The annual requirements for ethanol for a blend of up to 20 per cent with gasoline will be lower than the total ethanol production potential from all molasses available in the country.

A study on environmental life-cycle analysis of E10 and pure gasoline cars in Nepal shows that E10 fuel offers improvements over pure gasoline with regard to greenhouse gas emissions, release of carcinogens and ozone layer destruction, although the quantity of substances causing acidification/eutrophication increases with E10 (Khatiwada, 2007). Research conducted by the AEPC (Alternative Energy Promotion Centre) and CRE (Centre for Renewable Energy) on the performance of E10 and E20 fuels in automobiles in the Kathmandu Valley indicates that there is a significant reduction in CO_2 (19 per cent with E10 and 86 per cent with E20) and hydrocarbon (20 per cent with E10 and 46 per cent with E20) pollutants when using ethanol blends as compared with pure gasoline in light vehicle cars (AEPC, 2008). These amounts are indeed very significant. The same study claims that there is also a decrease in CO_2 emissions using both E10 and E20 for motorbikes (two-wheelers); emissions of HC (hydrocarbon) increase when E10 is used, but are reduced when the blend is increased to E20.

The opportunity to modernize sugar cane agriculture

Of a total land area of 14.72 million ha in Nepal, 21 per cent is occupied by cultivated land, 7 per cent by non-cultivated land, 39.6 per cent by forest (including shrubs), 12 per cent grassland and 20.6 per cent by other uses (AICC, 2006). Cultivated land is used for cereal/food crops (paddy, maize, millet, wheat and barley), cash crops (oilseed, potato, tobacco, sugar cane and jute), pulses (also locally called *dal*) and fruits/vegetables production. Of the cultivated land, 13.2 per cent is used for cash crops, of which sugar cane occupies 14.5 per cent. For the purpose of comparison, Table 15.1 provides information on land allocated for major cash crops

TABLE 15.1 Area and production of cash crops in Nepal, 2004/2005 (hectares and tonnes)

Crops	Area (ha)	Production (tonnes)
Oilseed	187,823	141,989
Potato	146,789	1,738,840
Tobacco	3,003	3,016
Sugar cane	59,082	2,376,103
Jute	11,159	16,207
Total	407,856	4,276,155

Source: AICC, 2006.

and respective production in 2004/2005. It is worth noting that oilseeds (e.g. soybean, rapeseed/mustard, sesame) are primarily used to produce edible oils for cooking purposes in Nepal; they are not available for biodiesel production at present. As can be seen, the area occupied by sugar cane is significantly smaller than the area dedicated to oilseeds and potatoes.

According to the three-year interim plan prepared by the government of Nepal for 2007–2009, agriculture accounted for 33 per cent of the nation's GDP in 2007, compared with approximately 40 per cent in the mid-1990s. The agricultural sector employs 74 per cent of the national labour force (CBS, 2008). Nepal produced 2.6 million tonnes of sugar cane in 2006/2007 using 64,000 hectares of land. The average yield was 40.6 tonnes per hectare (AICC, 2007), as compared to the average yield of 68.2 tonnes/ha in India during 2001/2002. Nepal's adjoining Indian states had lower comparative yields: Bihar (48.2 tonnes/ha), Assam (37.2 tonnes/ha) and Uttar Pradesh (58.0 tonnes/ha) in the same year (Purohit and Michaelowa, 2007). In Brazil, sugar cane production reached around 74 tonnes/ha in 2004 (FAO, 2006).

The trends of production, cultivation area and productivity are shown in Figure 15.4. The increase in sugar cane production in recent decades was mainly from expansion of planted area, while improvement in yields has been only marginal since the 1990s.

Not all Nepalese sugar cane is used for sugar production. Part of the production is used for two other purposes – chewing and production of sweeteners (called *chaku* and *shakhar*). *Chaku* and *shakhar* are unrefined sugar (dark brown in colour) that is obtained after boiling juice extracted from sugar cane stalks. These products are consumed in rural areas and traditional occasions in the Kathmandu Valley. Production of *chaku* and *shakhar* does not require sophisticated technology; so-called non-centrifugal sugar cane processing is a very low-cost process. The actual amount of sugar cane used in the production of these sweeteners has not been estimated in Nepal, but neighbouring figures from India and Pakistan show that about 20–30 per cent of sugar cane is used for these local products (called *gur* and *khandsari* in India and Pakistan respectively) (FAO, 1999). Thus around 70 per cent of the total amount of sugar cane produced in the country is used for sugar manufacturing, which is equivalent to around 1.8 million tonnes per year. Given that there are nine

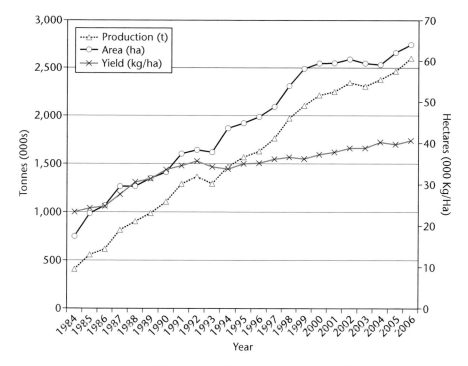

FIGURE 15.4 Sugar cane in Nepal: planted area (hectares), yields (kg/ha) and total pro-
duction (tonnes), 1984 to 2006 (source: Agriculture Information and
Communication Centre (AICC, 2007), Ministry of Agriculture and
Cooperative (MOAC), Government of Nepal)

operating sugar mills in Nepal at the moment with total capacity for 17,050 tonnes
per day, this gives a total plant utilization factor of 71 per cent for a period of 150
crushing days per year.

NARC, the Nepal Agriculture Research Council, is involved in the research of
agricultural crops in Nepal. Some large sugar industries have their own cane research
division working for the improvement of agricultural practices to increase sugar
cane yields. NARC and SRSM are conducting tests for sugar cane varieties. NARC
released two varieties of sugar cane in a crops research workshop in 2004, claiming
that one variety would give a yield of 79.3 t/ha of sugar cane in rain-fed natural
conditions (NARC, 2004). Thus there is potential to increase yields significantly in
the coming years but this is subject to innovations in agriculture that include not
only the adoption of new cane varieties, but also the adaptability of these varieties
to local conditions, and improvements related to labour and logistics (Plate 13).

A study by the Nepal Agriculture Research Council (NARC) in 1998/1999
showed that 100,000 farming families were employed in sugar cane cultivation and
earned NPR2,760 million[5] from sugar cane farming (NARC, 2001). Nepal can
produce 171,000 tonnes of sugar per year from the available 1.8 million tonnes of

sugar cane, considering an average sugar extraction rate of 9.5 per cent. Previously, Nepal was reported as being self-sufficient in sugar production but the country has recently experienced the equivalent of a two-month (on an annual basis) sugar deficit.[6] As a result, sugar is being imported from India. Due to the open border for imports and lack of statistics, it is difficult to determine the actual amount of sugar imported. However, imports can be significant given the low prices of Indian sugar. In India, the government controls the price of sugar cane by fixing the minimum price for payment to farmers. There is no governmental regulation for sugar cane prices in Nepal. However, at times of sugar shortage, the government regulates custom duties for imported sugar, and arranges to sell sugar at lower prices from government-owned corporations, particularly during festive seasons.

Addressing efficiency in energy conversion

In a recent study, the entire life-cycle energy inputs inventory (both fossil and renewables) was considered, including sugar cane farming (human labour, irrigation and chemicals), transportation, sugar cane milling, fermentation, distillation and dehydration to produce anhydrous ethanol (EtOH, 99.5 per cent (v/v) ethanol) in Nepal (Khatiwada and Silveira, 2009). Bagasse, as a source of renewable energy input, is used to generate the heat and electricity required for sugar cane milling, distillation and dehydration processes. Molasses is converted into anhydrous ethanol fuel (EtOH) through the conventional process of fermentation, distillation and dehydration. Distillery wastewater effluent (spent wash or vinasse) is treated prior to disposal, since treatment is essential from an environmental point of view. An anaerobic effluent treatment plant (also called an anaerobic digestion process – ADP) generates biogas, which is used as fuel in boilers.

To produce 1 litre of anhydrous bioethanol (99.5 per cent EtOH), the life-cycle energy input analysis shows that renewable energy contribution amounts to 91.7 per cent (31.42 MJ per litre), since most of the operations use bagasse, biogas and non-motorized transportation. The only exceptions are the application of fertilizers/chemicals and irrigation. Fermentation/distillation consumes 12.63 MJ/L, which is the most energy-intensive part of the process, followed by sugar milling, which consumes 10.46 MJ/L (see Figure 15.5). The fossil fuel required to produce 1 litre of molasses-based bioethanol (EtOH) is 2.84 MJ, giving a good energy yield ratio (7.47). The net renewable energy value (NREV) is 18.36 MJ/L,[7] and the high value indicates the low amount of fossil fuels used in the production cycle of ethanol in Nepal. Thus, bioethanol production is energy efficient in terms of the amount of fossil fuel used to produce it. The total energy (fossil and renewable) requirement is 34.26 MJ/L, which is higher than the energy content of 1 litre of bioethanol (i.e. 21.2 MJ/L), giving a negative NEV (−13.05 MJ/L).[8] However, low-quality biomass feedstock (in terms of market and energy values) – i.e. molasses – is converted into the high-quality modern renewable transport fuel. In addition, there is plenty of room for significant improvements along the production chain. Improvements can be made by: (1) increasing cane yields through the adoption of modern agricultural

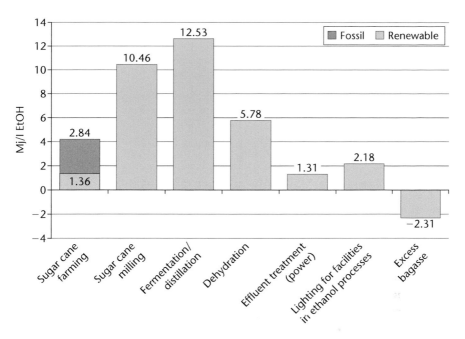

FIGURE 15.5 The contribution of fossil and renewable energy required to produce 1 litre of EtOH in Nepal, at each stage of the ethanol production chain (source: Khatiwada and Silveira, 2009)

practices; (2) efficient use of cane bagasse and trash to generate bioelectricity; and (3) upgrading and optimizing industrial operations.

There also exist many opportunities for energy savings in sugar milling and ethanol production: a 10 per cent reduction in energy consumption helps to increase NEV by 33.5 per cent. An improvement in energy efficiency of 30 per cent of the plant will result in a break-even situation for NEV; i.e. a point where the total energy input in the system would be equivalent to the output in terms of energy content.

Sugar industries in Nepal are currently self-sufficient in their energy requirements. Bagasse generates the heat and electricity required to run the whole plant. However, excess bagasse can be used to provide surplus electricity to replace diesel-powered electricity to local industries. With more efficient utilization of bagasse and cane agricultural residues, surplus bioelectricity can be generated not only for the industrial sector but also to promote electrification throughout the country, where 61 per cent of the population has no access to electricity.

Although it is rather complex to compare studies on energy balances made in different countries due to different system boundaries and varying methodologies, typical outcomes could be obtained in LDCs, as in the case of Nepal, depending on the extent to which relatively similar conditions prevail. The low fossil fuel input in the whole production chain results mainly from traditional agricultural practices and large manual labour inputs. While modern production is characterized by increased use of fossil fuels, the analysis in Nepal shows that significant improvements can be achieved in the total energy balance of bioethanol production if modern technologies are properly adopted and all by-products are efficiently exploited. An important aspect for developing countries is the possibility of capitalizing on proven technologies that are readily available and using accumulated long-term production experience, which facilitates investments and offers fewer risks. In this context, bioethanol production offers substantial opportunities for many LDCs.

Linking national gains with global environmental goals

The need to improve energy security is high on the agenda of most countries, especially those that are oil importers (see Chapter 14). For developing countries, finding alternatives based on local resources means not only the opportunity to increase energy security but also the opportunity to develop local economies and improve the trade balance. From regional and national solutions, there will also arise opportunities for expanding towards global solutions, as international markets for biomass products expand.

In spite of the international debate on biofuels, the bioethanol production in Nepal does not pose immediate threats to food security since it uses as feedstock the low-value agro-industrial by-product (molasses) obtained in the sugar cane milling process. Nepal can produce 18 million litres of bioethanol annually without compromising the production of food products, and savings of US$10 million could be possible through the implementation of the E20 blend in the Kathmandu Valley to replace conventional gasoline (Silveira and Khatiwada, 2010). Vehicles running on ethanol blends (E10 or E20) also release lower levels of air pollutants compared to pure gasoline.

Considerable bioethanol potential exists in many LDCs in Africa, just as in the case of Nepal. Mozambique could extract 68 million litres of bioethanol from sugar cane by 2010 (Batidzirai et al., 2006). Current ethanol production in Malawi is 30 million litres of ethanol derived from sugar molasses, which is used to complement the imported fuel (Amigun et al., 2011). Zimbabwe produced 22.7 million litres of ethanol in 2004, although the lack of blending mandates has led to this ethanol being directed in recent years to potable or industrial markets for export rather than domestic fuel markets (see Chapter 8). Amigun and colleagues (2011) have reviewed biofuels and sustainability in African countries, showing that many LDCs can produce ethanol from sugar cane molasses.

The development of this potential in African countries can similarly help save hard-cash foreign currency, reduce pollutants and address health problems. The substitution of lead in gasoline by using ethanol as an octane enhancer would help address a long-standing environmental issue that has already been solved in most of the world but still afflicts some African regions. Moreover, this substitution provides such health benefits at attractive cost savings due to favourable economic conditions for bioethanol production (Thomas and Kwong, 2001).

In addition to the national and regional gains, there are global environmental gains related to the bioethanol option. Bioethanol production and fuel substitution can help to avoid CO_2 emissions and mitigate climate change. Kathmandu Valley consumed 71,338 m^3 of gasoline, 46,003 m^3 of diesel, and 5,400 tonnes of LPG in the transport sector in 2006/2007. The total CO_2 emissions resulting from the consumption of these transport fuels were estimated to be 304,000 tonnes (petrol – 54.4 per cent; diesel – 40.3 per cent; LPG – 5.3 per cent). The introduction of E10 can avoid 11,283 tonnes (7 per cent of total gasoline emissions) of CO_2 emission, while E20 can contribute to avoiding 23,397 tonnes of CO_2 emissions, which is 14 per cent of the total gasoline emission in the year 2006/2007 (Khatiwada, 2010).

The total life-cycle emissions from 1 m^3 of bioethanol are 432.5 kg CO_2eq (i.e. 20.4 g CO_2eq MJ^{-1}), which is a 76.6 per cent reduction in GHG emissions compared to conventional gasoline from a life-cycle perspective. Table 15.2 shows the

TABLE 15.2 Life-cycle GHG (CO_2eq) balance of molasses-based ethanol (EtOH) fuel

Activities and constituents	Emissions (kg CO_2eq m^{-3} EtOH)
Fertilizers production	
Phosphorous (P_2O_5)	5.95
Potash (K_2O)	3.89
Nitrogen (N)	48.92
Chemicals production	
Insecticides/pesticides	38.66
Herbicides	4.16
Diesel (irrigation): production and combustion	95.57
Diesel (transport): production and combustion	12.96
Fertilizers application	
N_2O from fertilizer N-application	76.11
CO_2 from fertilizer N-application	22.37
Human labour (fossil energy inputs)	14.25
Cane trash burning	22.90
Returned residues N_2O (spent wash/stillage)	1.06
N_2O (filter cake/mud)	9.13
N_2O (unburnt trash)	7.48
Bagasse combustion in boilers (for heat and power)	35.57
Biogas combustion in boilers	8.56
Sub-total (emissions along the ethanol production chain)	407.53
Emissions from combustion of 1 m^3 ethanol (EtOH) in vehicles	25.00
Total life-cycle emissions (production and combustion of EtOH)	432.53
Emissions from production and combustion, 0.658 m^3 gasoline (= 1 m^3 of EtOH)	1,850.95
Net avoided emissions (kg CO_2eq m^{-3} EtOH)	1,418.42
% reduction in life-cycle GHG emissions (avoided emissions)	76.63%

Source: Khatiwada and Silveira, 2011.

estimation of GHG balances for different activities along the bioethanol chain. The production and consumption of bioethanol saves $1,418 \text{kg CO}_2\text{eq m}^{-3}$ of GHG emissions when an equivalent volumetric amount (i.e. 1m^3 ethanol $= 0.658 \text{m}^3$ gasoline) of gasoline is replaced by ethanol as the transport fuel in Nepal.

In relation to the estimation of the life-cycle GHG balance, scenario analysis in the choice of wastewater treatment plants between the anaerobic digestion process (ADP) and pond stabilization (PS) shows that ADP with biogas recovery significantly reduces GHG emissions, particularly if leakages are avoided. The pond stabilization (PS) treatment process increases GHG emissions quite significantly and should therefore be avoided. Excess bagasse can be used to generate and supply electricity to the grid. Part of the electricity produced by diesel power plants can be replaced by this renewable bioelectricity with a consequent reduction in GHG emissions (Khatiwada and Silveira, 2011). It should be noted that boilers and turbines with their current installed capacities could operate in such a way as to generate excess electricity, though improved efficiency would add both economic and environmental value to the process. Surplus bagasse can be used to generate excess electricity and substitute electricity credits for GHG emission reductions, paving the way for the replacement of diesel power plants. In this way, system expansion following the sale of surplus electricity obtained from the combustion of excess bagasse can help reduce GHG emissions still further (Khatiwada and Silveira, 2011).

Favourable governmental policies such as mandatory bioethanol blends and incentives/subsidies for sugar cane farmers and private investors could play an important role in the realization of the bioethanol potential in Nepal and other LDCs. Proper institutional mechanisms and coordination amongst concerned stakeholders, including the private and public sectors, are required for bioethanol production and its commercialization. Both the political and institutional concerns have become the most urgent issues to address at this stage when mature conversion technologies are already available and accessible.

One of the aims of the MDG (Millennium Development Goals) is to develop a global partnership for development to address the pressing needs of LDCs. The role of donors is key in this context. Nepal and other LDCs are largely dependent on development aid, so there is a great need to sensitize donors about the real potential that these countries have. In addition, there is a possibility to frame bioethanol development as described in the Nepalese case within the framework of a CDM scheme (Silveira, 2005b). By improving sugar cane yields and the total energy balance of ethanol production in Nepal, and by using ethanol-blended gasoline in the transport sector, Nepal could earn certified emission reductions (CERs). This could also improve the economic performance of bioethanol plants and help promote more efficient technologies (Gnansounou et al., 2005). The CDM scheme could then serve to attract part of the investment necessary to expand ethanol production capacity in the country and implement the fuel substitution policies. This would provide an important contribution to the development of bioenergy in Nepal and improved access to modern sustainable energy services using indigenous energy sources. Bundling projects in poor countries and regions is likely to be important as a way to reduce transaction costs and facilitate such CDM projects. Unfortunately, although the global agenda provides the CDM framework and encourages a move towards renewables and low-carbon emissions, it has so far failed to promote significant amounts of CDM activities in LDCs. Bioethanol projects provide a concrete opportunity for CDM in these countries.

Discussion and conclusions

The story of Nepal is similar to the story of landlocked countries in sub-Saharan Africa that are fully dependent on oil imports, creating a significant drain in their economies. At the same time, their agricultural systems need modernization, their industries wait for a dynamic push of markets, and their populations need jobs, income and electricity. Despite the opportunities right at hand, many LDCs depend largely on the support of the international community to realize their bioenergy potential. Many of these LDCs are strongly dependent on development assistance to reform institutions and make the initial infrastructure investments. However, donors have not prioritized bioethanol production. At the same time, developing countries need to put proper government policies in place to actually trigger bioethanol production. This includes cutting subsidies on oil, improving the regulatory framework to favour biofuels, and creating fiscal incentives for sugar cane farmers and private investors to explore the bioethanol potential.

Thus proper institutional mechanisms and coordination amongst concerned stakeholders, including both private and public sectors, are required for the production and commercialization of bioethanol in Nepal and in other developing countries. Mature conversion technologies exist and are often available and accessible in the regions concerned. The insight provided using the example of Nepal could serve as a motivation for assessing the potential for production of bioethanol in other LDCs in Africa.

Sovacool and Brown (2009) discuss the need to differentiate the scales of action when it comes to climate change. Often the global dimension of the problem is emphasized and policy makers deal with it as if the global issues were all-encompassing. However, there are important connections to local, national and regional impacts, especially in LDCs: thus climate mitigation strategies need to address multiple scales. In other words, both top-down and bottom-up approaches are needed to address climate change. The implication of understanding the multiple scales of action is that costs and benefits are distributed differently. Bottom-up actions can favour diversity and innovation that is suited to local or national needs and conditions.

As shown in this chapter, bioethanol production and use could play an important role in the development of Nepal. The bioethanol industry could immediately help improve the economy of the country through the reduction of oil imports and promotion of electricity access with the use of residues from the sugar/ethanol production. Environmental benefits can be achieved through fuel substitution and reduction of greenhouse gas emissions. The development of the ethanol industry could serve as an incentive to modernize agriculture in the country and promote rural development. All these benefits are highly correlated to the development agenda of Nepal.

Many African countries are in a similar situation. They have good conditions for starting to produce their own fuel and improving their national economies. It is very important for the international community to acknowledge this and provide the support necessary, in the form of technical, institutional and economic assistance, to help poor countries realize their bioethanol potential. Bioethanol production is attractive both in terms of national development and climate mitigation measures, and should therefore be supported as a development opportunity for LDCs.

Notes

1 The Gini coefficient is a commonly used measurement of inequality, with zero reflecting complete equality and 100 per cent indicating complete inequality. More information on the development of Nepal can be found at www.adb.org/media/Articles/2007/12324-nepalese-poverties-reductions/ (last consulted 10 October 2010).

2 Note that, as per the definition used by the Central Bureau of Statistics in Nepal, only residents of five municipalities located inside the Kathmandu Valley are considered urban. However, if we also consider the Village Development Committees (VDCs), approximately 47 per cent of the total urban population in Nepal lives in the Kathmandu Valley. Some VDCs have similar infrastructure, organization and mobility as municipalities, which is the reason why the government is considering merging them with existing municipalities or creating new ones.

3 For comparison, per capita primary energy consumption in 2007 was as follows: US – 331.38 GJ; Sweden – 242.76 GJ; India – 20.58 GJ; and China – 55.44 GJ in 2007 (IEA, 2007).

4 The consumption of petrol and diesel in Kathmandu Valley have been estimated based on sales records available at Nepal Oil Corporation Ltd (NOC) and district sales distribution. In addition, the use of gasoline and diesel is verified with estimations on the demand side considering the number of vehicles, annual distance coverage (km) and fuel economy (km/litre).

5 In January 1999, the amount of 2,760 million Nepalese rupees (NPR) was equivalent to about US$40.6 million at an exchange rate of 0.0147 NPR/USD.

6 Personal information gathered during field study at Sri Ram Sugar Mills Pvt. Ltd (SRSM).

7 Net renewable energy value/balance (NREV) is calculated as follows:

$$NREV = E_F - NE_I$$

where E_F is the energy content (lower heating value) of ethanol and NE_I is the non-renewable energy or fossil fuel input.

8 The net energy value or balance (NEV or NEB) of bioethanol (EtOH) is the difference between the energy content of the bioethanol produced and the total primary energy inputs (fossil plus renewable) in the entire fuel production cycle.

$$NEV = E_F - E_I$$

where E_F is the energy content (lower heating value) of ethanol and E_I is the total amount of primary energy inputs.

References

ADB (2006) "Environment Assessment of Nepal – Emerging Issues and Challenges", Chapter 7 – "Air pollution and climate change", Asian Development Bank (ADB), available at www.adb.org/documents/books/emerging-issues-challenge/chap7.pdf (last consulted 12 October 2010).

AEPC (2008) "Report on Assessing the Economic, Technical and Environmental Aspects of Using Petrol-Ethanol Blend in the Automobiles in Nepal", Alternative Energy Promotion Centre – MOEST, Government of Nepal and CRE (Centre for Renewable Energy), Kathmandu.

AICC (Agriculture Information and Communication Centre) (2006) Ministry of Agriculture and Cooperative (MOAC), Government of Nepal, www.aicc.gov.np/database/agriculture_ststistics/national_level_data/1executive-summary per cent20of per cent20agriculture per cent20statistics.pdf, last consulted 10 October 2010.

AICC (2007) "Area Production and Yield of Cash Crops in Nepal", Ministry of Agriculture and Cooperative (MOAC), Government of Nepal.

Amigun, B., Musango, J.K. and Stafford, W. (2011) "Biofuels and sustainability in Africa", *Renewable and Sustainable Energy Reviews*, 15, pp. 1360–1371.

Batidzirai, B., Faaij, A.P.C. and Smeets, E. (2006) "Biomass and bioenergy supply from Mozambique", *Energy for Sustainable Development*, vol. 10, no. 1, pp. 54–80.

CAI-Asia Clean Air Initiatives for Asian Cities (2005) "Slight improvement in Kathmandu Valley's air quality", www.cleanairnet.org/caiasia/1412/article-59889.html, last consulted 11 October 2010.

CBS (Central Bureau of Statistics) (2008) Government of Nepal.

FAO (The Food and Agriculture Organization of the United Nations) (1999) Fiji/FAO Asia Pacific Sugar Conference, ftp://ftp.fao.org/docrep/fao/005/x0513e/x0513e03.pdf, last consulted 10 October 2010.

Filho, J.M., Burnquist, H.L. and Vian, C.E.F. (2006) "Bioenergy and the rise of sugarcane-based ethanol in Brazil", *Choices – The Magazine of Food, Farm, and Resource Issues* (American Agricultural Economics Association), vol. 21, no. 2 (2nd quarter), www.wilsoncenter.org/news/docs/bioenergy per cent20and per cent20the per cent20rise per cent20of per cent20ethanol per cent20in per cent20brazil.pdf, last consulted 10 October 2010.

Gnansounou, E., Bedniaguine, D. and Dauriat, A. (2005) "Promoting Bioethanol Production through Clean Development Mechanism: Findings and Lessons Learnt from ASIATIC Project", Proceedings of 7th IAEE European Energy Conference, Bergen, Norway, August.

IEA (International Energy Agency) (2007) "Key World Energy Statistics 2007".

Khatiwada, D. (2007) "A Comparative Environmental Life Cycle Assessment of Ethanol Blended (E10) and Conventional Petrol Fuel Car – A Case Study in Nepal", Poster presented at European 14th LCA Case Studies Symposium of "LCA of Energy – Energy in LCA", Society of Environmental Toxicology and Chemistry (SETAC), Gothenburg, Sweden.

Khatiwada, D. (2010) "Assessing the Sustainability of Bioethanol Production in Nepal, Licentiate Thesis", Division of Energy and Climate Studies, KTH.

Khatiwada, D. and Silveira, S. (2009) "Net energy balance of molasses based ethanol: the case of Nepal", *Renewable and Sustainable Energy Reviews*, vol. 13, no. 9, pp. 2515–2524.

Khatiwada, D. and Silveira, S. (2011) "Greenhouse gas balances of molasses based ethanol in Nepal", *Journal of Cleaner Production*, vol. 19, no. 13, pp. 1471–1485.

KVEO (2007) Kathmandu Valley Environment Outlook, ICIMOD, MOEST/GoN, UNEP, Kathmandu, Nepal, available at www.roap.unep.org/pub/KVEO_full_report.pdf, accessed on 12 October 2010.

MOEST (2005) "Report on Ambient Air Quality of Kathmandu Valley 2003–2004", Ministry of Environment, Science and Technology (MOEST), Government of Nepal.

NARC (Nepal Agriculture Research Council) (2001) "Annual Technical Report – 2001", Kathmandu, Nepal.

NARC (2004) *A Quarterly Newsletter of Nepal Agricultural Research Council (NARC)*, vol. 11, no. 2, Kathmandu, Nepal. Available at www.narc.org.np/publicaton/pdf/newsletter/Vol per cent2011 per cent20No per cent202.pdf, last consulted 12 October 2010.

NEA (Nepal Electricity Authority) (2007) "A Year in Review, Fiscal Year 2006/07", Kathmandu.

NOC (2008) "Import and Sales Record of Petroleum Products in Nepal", Nepal Oil Corporation Ltd, Kathmandu.

Purohit, P. and Michaelowa, A. (2007) "CDM potential of bagasse cogeneration in India", *Energy Policy*, vol. 35, no. 10, pp. 4779–4798.

Silveira, S. (ed.) (2005a) *Bioenergy – Realizing the Potential* (ed.), Elsevier Publications, Amsterdam.

Silveira, S. (2005b) "Promoting bioenergy through the Clean Development Mechanism", *Biomass and Bioenergy*, vol. 28, no. 2, pp. 107–117.

Silveira, S. and Khatiwada, D. (2010) "Ethanol production and fuel substitution in Nepal – Opportunity to promote sustainable development and climate change mitigation", *Renewable and Sustainable Energy Reviews*, vol. 14, no. 6, pp. 1644–1652.

Sovacool, B.K. and Brown, M. (2009) "Scaling the policy response to climate change", *Policy Society*, 27, pp. 317–328.

Thomas, V. and Kwong, A. (2001) "Ethanol as a lead replacement: phasing out leaded gasoline in Africa", *Energy Policy*, vol. 29, no. 13, pp. 1133–1143.

UNCTAD (2008) "The Least Developed Countries Report 2008", United Nations Conference on Trade and Development.

UNCTAD (2009) "The Least Developed Countries Report 2009 – The State and Development Governance", United Nations Conference on Trade and Development.

UNDP (2006) "Human Development Report – 2006", United Nations Development Program.

WECS (2006) "Energy Synopsis Report 2006", Report no. 7, seq. no. 489, Water and Energy Commission Secretariat, Government of Nepal, Kathmandu.

16

BIOFUELS TECHNOLOGY CHANGE MANAGEMENT AND IMPLEMENTATION STRATEGIES

Lessons from Kenya and Benin

Sergio C. Trindade, Maurizio Cocchi, Alain Onibon and Giuliano Grassi

Introduction

Biofuels have been part of the daily life of humankind since men learned to make fire. Over thousands of years the technologies changed little. More recently, in the past hundred years or so, more advanced processes have come into use to transform solid biomass into gases and liquids through pyrolysis and gasification. Starch and sugar-containing agricultural materials, which had long been fermented for potable uses, acquired a new market in the form of fuel ethanol. Oil-bearing biomass began to be used as diesel fuel substitute, directly or transformed into biodiesel. Technology change in the biofuels space continues with the promise of second-generation processes that are based mainly on non-food feedstocks.

Technology is but one of many factors influencing the market penetration of biofuels at national and international levels. Besides its scientific and technical components, technology also embodies a host of choices made, relative to social, economic and environmental policies. Such choices include the selection of feedstocks for biofuels and scales of production, urban versus rural development, public versus personal transportation; labour intensiveness, capital intensiveness; and openness to international trade and foreign investment.

Technology change and technology transfer results from many day-to-day decisions engaging the relevant stakeholders who participate in the decision-making process regarding strategy, investment, international trade and market opportunities. Cultural preferences, consumer awareness, social values, lifestyles, corruption, competition and so on are also reflected in technology transfer (Trindade, 2000). Crucial to managing the flow of technology change and transfer is the existence of national systems of innovation – that is, the institutional and organizational structures – that support technology development and innovation. The innovation system includes, inter alia, education, technology and investment financing, standards and norms and intellectual property rights protection. Governments can build or strengthen scientific and technical educational institutions and modify the form or operation of technology networks; that is, the interrelated organizations generating, diffusing and utilizing technologies (Mckenzie-Hedger *et al.*, 2000).

Technological change

Technological change does not take place in a vacuum. Therefore, it is important to understand the context in which it evolves, in particular within the frameworks of national and international initiatives. The process of transferring technology and know-how can be better understood as a process of managing technological change. This encompasses the broad set of processes covering the flows of knowledge, experience and equipment amongst different biofuels stakeholders such as governments, private sector entities, financial institutions, NGOs, research and educational institutions and labour unions. This inclusive concept covers diffusion of technologies and technology cooperation across and within countries. It also includes the process of learning to understand, utilize and replicate biofuels technologies, including the capacity to choose and adapt them to local conditions and even to sell back to the original source an improved technology (Trindade, 2007).

The global context for biofuels

Biofuels technology change has become a global process. Biofuels technology transfer does not just occur from North to South. It happens mostly within and between industrialized countries, but also increasingly within the developing world and even from South to North. This is illustrated, for instance, in the joint statement by President Luiz Inácio Lula da Silva and Prime Minister Tony Blair in London on 9 March 2006, where they said that:

> We note the importance of the current joint study of the global potential of bioethanol production from sugar cane. Brazil is a world leader in bioethanol production and use. We agree that spreading these technologies to developing countries, in particular in Africa and the Caribbean, could significantly contribute to poverty alleviation; assist in slowing climate change; and help to develop a global market for bioethanol.
>
> *(UK in Brazil, 2006)*

The technology embedded in Brazilian ethanol distilleries, for example, has been transferred to a number of countries, including Costa Rica, Kenya and Paraguay. Similarly, technology from Indian distilleries has been brought to Colombia. On the occasion of the visit to Brazil by India's Prime Minister, Manmohan Singh, a joint statement was issued on 15 April 2010, which emphasized enhancement of cooperation in biotechnology and strengthening bilateral cooperation in the energy sector, including in hydrocarbons and new and renewable energies (Government of Brazil, 2010). The statement reaffirmed the leaders' intention to encourage better coordination of their positions on the issue of biofuels at multilateral fora.

Biofuels energy systems and technologies

Biofuels resources (primary biofuels sources), conversion, carriers, storage, transportation, distribution and end use should not be analysed separately. They belong to a chain of interacting elements that constitute the biofuels matrix, in this case adapted for sugar cane, as shown in Figure 16.1 (Trindade and Yang, 1982; UN, 1991).

The biofuels matrix also gives rise to the notion of the biorefinery, assembling the land base component that supports agriculture and forestry, and the transformation components, which involve processing of the feedstocks into biofuels for delivery and consumption in the market place. Each of these components employs

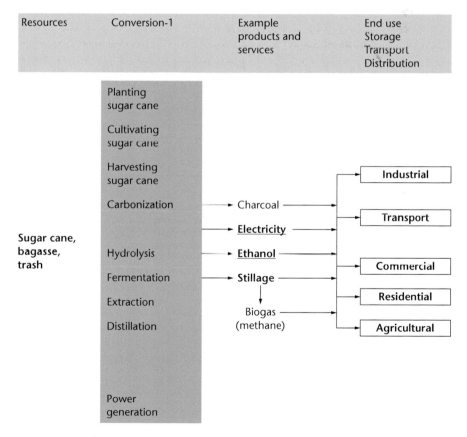

FIGURE 16.1 The sugar cane energy technology system (sources: adapted from Trindade and Yang, 1982 and UN, 1991)

specific technologies that make the system work as a whole. The concept of a biorefinery or sugar cane agro-industrial complex is discussed briefly in Chapter 7, which elaborates on the many different co-product options that are potentially available.

The biofuels matrix given in Figure 16.1 is not exhaustive; a more comprehensive version would include the entire *ensemble* of potential products, processes and markets in the biofuels (sugar cane) universe and identify the key technologies required to transform resources from raw materials to end uses in the market place.[1]

In addition to price, energy resource availability also depends on the state of technology. In the case of biofuels, the technologies refer to agriculture (terrestrial and marine) and forestry, and to the recovery of organic residues from agricultural, urban and industrial sources, and, not least, to political, social and economic constraints impinging on technological innovation.

The issue of availability acquires a different meaning in the case of biofuels conversion process technologies. Here the concept implies the existence of a body of scientific and technological knowledge that can be practically applied for the design and operation of plants to convert energy resources to socially and environmentally acceptable forms. Availability in this context is seemingly easier to track since it is a direct result of human efforts and investment in science and technology. Nevertheless, resource conversion process availability depends on capital investment in research and development and to some extent on specific stakeholders' perceptions.

The energy industry traditionally has taken a relatively conservative view that a given conversion technology is commercially available only when proven on a large enough scale. There has been ample evidence of biofuels technology learning curves, driven by expanding markets and resulting from process design and development, aided by information and communication technologies, resulting in sharply decreased operational costs and investment requirements (Trindade, 1994). The decreasing ethanol production costs in Brazil and the USA, with increasing cumulative production over a period of 30 years, makes this point very clear. Given enough time and increasing global output, the future costs of biodiesel are likely to decrease as well.[2]

The technology change and transfer process

Technology transfer is embodied in the actions taken by individuals and organizations, such as investment and trade decisions made by firms; acquisition of knowledge and skills by individuals through formal education and on the job; purchase of patent rights and licences; assimilating publicly available knowledge; migration of skilled personnel with knowledge of particular technologies; and stealing. Technology flows can also be influenced by government policies and financial aid and development programmes (Trindade, 2000). The rate of such technology flows is affected both by the motivations of the relevant stakeholders and the barriers that impede them, both of which are influenced by government policies, including environmental and global warming policies.

Managing technological change and transfer requires an interdisciplinary approach. This results from the nature of the innovation system, operating as it does through diverse flows of knowledge, investment and trade among different stakeholders: governments, private sector entities, financial institutions, NGOs, research and teaching institutions and labour unions.

Other, less interactive pathways that do not necessarily include the provider are piracy, industrial espionage, third country diversions and reverse engineering. Not every pathway leads to the same level of learning and ultimately the degree of technology-as-knowledge transfer needs to go far beyond simple hardware transfer if it is to be successful.

African case studies

During the past three to four decades, a number of countries in Africa considered and/or engaged in the production and use of biofuels, ethanol in particular. These countries include Benin, Kenya, Malawi, Mozambique, Nigeria, South Africa, Zambia and Zimbabwe.

Two case studies are presented and discussed here, based on the authors' long experience with national biofuels programmes worldwide, especially ethanol fuels. Synopses of an early experience in Kenya, with updates to the present time, and a recent initiative in Benin are presented and discussed to draw conclusions that might be useful for other African countries, with respect to managing technology change.

As suggested in Figure 16.2, the case narratives are organized along the following basic process steps: assessment, agreement, implementation, evaluation, adjustment

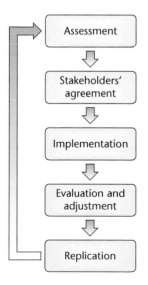

FIGURE 16.2 Specific technology transfer – five basic stages

and replication. Three main areas that impact on the competitiveness of biofuels relative to conventional fuels are: primary feedstock productivity; crop modification; and the nature and scale of the conversion process.

Fuel ethanol in Kenya[3]

Assessment and needs identification

Fuel ethanol, then called power alcohol, was not the priority issue in Kenya's energy agenda in the early 1980s. The energy problem in Kenya was first and foremost one of supplying adequate firewood to the increasing residential demand derived from the country's very high population growth rate.

Second, Kenya's energy problem related to displacing fossil fuel consumption in electricity generation, as well as supplying (or finding alternatives to) the fast-increasing incremental demand for automotive diesel oil. Alcohol substitution for gasoline therefore represented mainly a short-term opportunity for petroleum substitution and self-reliance.

Sugar cane has been cultivated in Kenya's highlands and could, in principle, provide the molasses required for conversion into fuel ethanol. This is why the three existing projects in 1982, in Muhoroni (Agro-Chemical and Food Company – ACFC), Kisumu (Kenya Chemical and Food Corporation – KCFC) and the Riana project were all located in the Highlands area. The availability and low opportunity cost of the raw materials, in the form of molasses or sugar cane juice, made inland locations with limited logistical infrastructure especially attractive (Trindade, 1982). Table 16.1 summarizes the basic facts about the Kenyan ethanol projects in 1982; in the case of ACFC and KCFC, the expected markets for other co-products contributed to overall economic feasibility, which is in general a key feature of sugar cane, due to the many co-product options available (see Chapter 7).

Sugar cane alone was not an adequate sole source of fuel ethanol in Kenya. Any consideration of a truly national fuel ethanol programme would have required, depending on the scenario considered, between two and five additional distilleries by 1995, which did not necessarily need to be based entirely on sugar cane (Trindade, 1982). Cassava and sweet sorghum cultivated on marginal lands would have been likely alternatives to water-demanding sugar cane agriculture. Assessment and needs identification were, to say the least, inadequate to support a national fuel ethanol programme in Kenya.

Agreement

The relevant stakeholders in Kenya would have been the potential fuel ethanol producers – ACFC, KCFC and the Riana project; the feedstock-growing farmers; the consumers; the motor trade and allied industries; the oil marketers; the Kenya Pipeline Company; the East African Oil Refineries – EAOR Mombasa refinery;

TABLE 16.1 Summary of Kenyan ethanol projects, 1982

Item	ACFC	KCFC	Riana
Location	EASI, Muhoroni	Kisumu	South Nianza
Product mix	Ethanol	Ethanol	Ethanol
	Baker's yeast	Yeast	
	Fodder's yeast	Citric acid	
	Concentrated stillage	Methane	
	Fusel oil	Gypsum	
		Ammonium sulphate	
Ethanol capacity, m³/yr	18,300	20,000	45,570
Investment per daily capacity, $/litre/day	211	383	437
Equipment provider	Vogelbusch	Conger/PEC/Sorigona	FCB/Sodecia
Capital structure	ADC, ICDC (28% each)	GoK (51< %)	T.J. Cottington
	IIC (34%), VEW (10%)	Chemfocd Inv. and Advait (49%)	OTH International SUDE

Sources: (Trindade, 1982; Baraka, 1989)

Notes

ACFC – Agro-Chemical and Food Company.
ADC – Agricultural Development Corporation, a Kenyan parastatal.
Chemfood Investment Corporation, SA and Advait International, SA, Kenyan corporations.
Conger – a Brazilian distillery manufacturer.
EASI – East African Sugar Industries Ltd, sugar factory adjacent to ACFC.
FCB – French sugar processing equipment manufacturer.
GoK – Government of Kenya.
ICDC – Industrial and Commercial Development Corporation, a Kenyan parastatal.
IIC – International Investment Corporation, a foreign financing company.
KCFC – Kenya Chemical and Food Corporation.
OTH – French consulting and construction firm.
PEC – Process Engineering Company, Zurich.
SODECIA – French distillery equipment manufacturer and engineering company.
SORIGONA – Swedish equipment manufacturer.
SUDE – Sugar and Development, French implementation company.
T.J. Cottington & Partners Ltd – Kenyan engineering and industrial manufacturing company.
VEW – Vereignigte Edelstahlwerke AG, Vogelbusch's parent company.
Vogelbusch – Austrian distillery manufacturer.

the Kenyan scientific and technical community. Unfortunately, the framework for a relevant dialogue process among such stakeholders was not in place.

Those stakeholders ultimately involved in decision making were relatively few. There was no national fuel ethanol programme in practice. No clearly formulated policy, legislation and regulations existed that would define such a programme and provide the basis for appropriate institutional machinery to implement it. There was no clear system to appraise, approve, monitor and control the execution of projects.

Project economics alone would not suffice to justify fuel ethanol development. However, other dimensions were never factored in, such as balance of payments shortfalls, strategic need, energy security, domestic jobs creation, the social value of foreign exchange over and above official exchange rates, etcetera.

The lack of adequate domestic funding encouraged foreign sources of financing. Due to the government guarantees offered, this not only impinged on sovereign debt but brought in partners with limited commitment to project success. Furthermore, scant government supervision capacity provided opportunities for private gain with only limited social benefits.

Implementation

In the absence of a formal Kenyan national fuel ethanol programme, one can only discuss the implementation of specific projects. The fuel ethanol projects in 1982 had resulted from isolated, disconnected efforts by separate groups of promoters, foreign and national, with government support. At the time, the existing projects, in decreasing order of completion, were the ACFC, the KCFC and the so-called Riana Project. In the first two cases, ethanol was not the sole product to be manufactured.

Apparently, the original interest in fuel ethanol manufacture in Kenya derived from the availability and low opportunity cost of molasses produced by the sugar mills in the sugar belt in the west of the country. The lack of a steady market for molasses raised environmental issues associated with its disposal. The increased availability of molasses resulted from the increased production of sugar at the time, which reversed Kenya's sugar net import position. The low value of molasses in the highlands reflected the cyclic and low netback potential payments to sugar mills, due to the high logistical costs required to bring molasses to international markets.

The particularly long harvesting campaign in Kenya (Trindade, 1982, p. 36), in the order of 300 days per year, constituted another strong inducement to produce ethanol in the country: for the same annual output, Kenyan projects could employ smaller distilleries than countries such as Brazil where distilleries sit idle for many months due to the shorter harvest season. While it is true that unit capital costs are higher for smaller distilleries, total investment cost is lower, an important factor to consider when foreign exchange is borrowed to purchase equipment, in the context of scarce upfront investment funds in general.

The key constraints on technology transfer, development and technology change derived from:

1. financing via suppliers' credits tied to turnkey packages of imported equipment being the only source available to Kenya at the time; and
2. lack of trained personnel able to absorb technologies embodied in the turnkey packages.

On the other hand, Kenya might have profited from opportunities to improve the situation described above by virtue of the:

1. ongoing ethanol production and utilization in the country, as the programme was implemented;
2. need for technical extension support and training activities;
3. longer-term need for research, development and demonstration activities ranging from production of agricultural feedstocks through end use of fuel ethanol.

Nevertheless, the 1982 situation was dominated by the impact of suppliers' credits tied to turnkey packages of imported equipment, which did not allow any say by Kenyan society on the choice of fuel ethanol technologies. The use of local labour and inputs rather than a turnkey package was in fact shown to be a key factor in the economic feasibility achieved in the ethanol programme that had been initiated in Zimbabwe just a few years previous, in 1979 (Scurlock et al., 1991). The turnkey technologies brought into Kenya did not fit well with the country's human and material resource base or with its market structures, as further elaborated in the specific case discussions below.

Agrochemical and Food Company, Ltd (ACFC)

In 1982, the ACFC distillery (18 million litres/yr) and yeast plant (1,800 tonnes/yr) was situated across the fence from the sugar mill of East African Sugar Industries Ltd (EASI). Trial runs took place during 9 June–14 July 1982, but the target of 72 hours of continuous production was not met due to power supply interruptions. ACFC managers were the Mehta Group International Ltd, which also managed EASI and South Nyanza Sugar Company Ltd (SONY), through Uketa Development Corporation Ltd.

From the technological point of view, ACFC adopted advanced process technologies, which included continuous (rather than batch) fermentation. As a point of reference, Brazil's 300 or so distilleries operated batchwise at this time. ACFC also concentrated the waste stream of stillage or vinasse, a relatively advanced approach. Again, Brazil had only two such plants at that time, neither of which was entirely successful in feed and fertilizer markets. But the bottom line here is that the benefits of ACFC's choice of technology were not evident from the data available, when direct materials and energy costs were compared between ACFC and

Brazilian cases, or even compared to the ETHCO distillery in Malawi, where more conventional, simpler and cheaper ethanol-making approaches were used.

Furthermore, the introduction of the latest technology into a market without a background in fuel ethanol production complicates the implementation and operations and brings with it the associated costs of expatriate technical support. Nevertheless, the ACFC was completed and ready to operate.

The KCFC Project

The Kenyan Chemical and Food Corporation Ltd was conceived as a molasses-based biorefinery complex, whose construction had been halted in May 1982. It was designed to produce industrial and fuel ethanol, citric acid, yeast, vinegar and by-products such as biogas from stillage digestion, gypsum, ammonium sulphate and carbon dioxide. It is situated in Kisumu on the banks of Lake Victoria. However, the nearest molasses available was situated far away in Mumias.

The project manager was Eximcorp with the assistance of the Madhvani Group, which is the major investor in Advait International.

The original investment estimate in July 1977, when the joint venture agreement was signed, amounted to US$61.4 million. As of May 1982, when construction came to a halt, KCFC had already spent US$125 million, and the complex remained unfinished. In August 1982, the ethanol distillery was 95 per cent complete, the yeast plant was 85 per cent finished, the citric acid unit was 25 per cent complete, and the water and effluent treatment plant were 95 per cent concluded.

Funding for the project had come from foreign loans and overseas credits. The Kenyan government was directly or indirectly, through the Kenya Commercial Bank, the guarantor of most foreign loans. There were serious questions about the financial viability of the KCFC project at the time.

The Riana Project

The so-called Riana project considered implementing a $150\,m^3/day$ sugar cane juice ethanol distillery in an area in South Nyanza that in 1982 lacked adequate infrastructure. Total investment was estimated at US$65 million, with a unit capital cost for the 150,000 litre/day distillery alone of US$437/litre-day.

The fuel ethanol demand scenario in 1982 called for a third sugar cane ethanol distillery only by 1990. Consequently, given the high estimated capital costs of the Riana project, the lack of public infrastructure in the proposed area for it and the need for a learning period in Kenya, the implementation of the Riana project would be feasible only much later than originally proposed.

Evaluation and adjustment

The ACFC distillery finally began to ship anhydrous fuel ethanol to oil depots for blending with gasoline at 10 per cent (E10). This lasted for about ten years. Blending was discontinued in 1993 due to persistent losses as a result of uncompetitive pricing, another effect of the lack of a national fuel ethanol policy. Other complications resulted from poor management, oil companies' reluctance to cooperate and the loan service burden (Batidzirai, 2007). Export opportunities for ethanol developed, and Kenya missed the opportunity of saving on oil imports.

In 1982, the KCFC complex was already stalled due to massive cost over-runs. The fundamental error was to site a molasses-based manufacturing system where there was no readily available molasses. Furthermore, there was no sugar cane bagasse to fire the boilers to produce process steam and electricity for the plant and perhaps for the commercial electrical grid. There appeared to be political considerations, since the location is the region of the Luos, who are the second most important ethnic group in Kenya.

Spectre International, a South African company, took control of the assets in place in Kisumu in 2003. Energem Resources, a Canadian company formerly known as Diamondworks Ltd, with many activities in Africa including diamond mining, acquired for US$2 million a 55 per cent controlling interest in the Kenyan Spectre, to expand business opportunities in the Kenyan market. The distillery was revamped at some cost. By mid-2004 the distillery was ready to go and by 2008 was producing some 75,000 litres/day (Crilly, 2008), mainly for the pharmaceutical and potable ethanol markets.

It seems that decades after its inception and following ten years of experience (1983–1993), Kenya is likely to bring fuel ethanol back into use. The plans are for the Kenya Pipeline Company's depots in Eldoret, Kisumu and Nakuru to begin blending 10 per cent ethanol with gasoline. By now, to meet the national requirement of 10 per cent ethanol blend, Kenya's national production will need to triple. Currently, the two largest producers of ethanol in Kenya are Spectre International in Kisumu, with 27 million litres/year, and ACFC in Muhoroni, with 22 million litres/year.

Replication

The expanding market for fuel ethanol in Kenya is prompting other players to enter it; previous experiences have highlighted the need for consistent national policies so as to allow for steady rules of the game. A process of privatizing the five state-owned or controlled sugar mills is anticipated by 2012, which should affect fuel ethanol prospects. Mumias Sugar Company has plans for a 100,000 litre/day, US$45 million molasses-based fuel ethanol distillery due to start up in 2012 (Odhiambo, 2009). In August 2009, Mumias began generating electricity from bagasse-fuelled boilers. Mumias has recently been engaging in other acquisitions, by associating with neighbouring sugar mills and considering purchases in Tanzania and Uganda.

Production is expected to commence by July 2011, and to include eventually a variety of products such as neutral alcohol, anhydrous ethanol, food-grade and fuel ethanol for blending with premium fuel (Odhiambo, 2009). Half of the distillery project funds would be drawn from the miller's retained earnings while the rest would be sourced from financiers. The distillery is scheduled to run on the 100,000 tonnes of molasses Mumias produces every year, and has a target production of 25 million litres of ethanol per annum.

Other projects announced in 2009 included Chemelil, Nzoia and SONY sugar factories jointly to produce up to 140 million litres/year of ethanol (Anonymous, 2010).

Feedback

Kenya is no exception to the requirement for learning time to develop and accumulate experience in implementing a new energy source such as fuel ethanol. The difficulties encountered – institutional, technical and otherwise – in implementing a fuel ethanol programme are an integral part of this learning process. An awareness of such a "learning curve" helps expedite progress towards the market penetration of fuel ethanol.

It seems that despite all the learning that has taken place over the past 30 years or so, there is still much to do to achieve a consistent national fuel ethanol programme in Kenya. Such a programme must be discussed among the relevant stakeholders and translated into legislation and regulations.

A crucial development impacting on the prospects of ethanol production in Kenya is the ending in 2011 of the preferential trade terms on sugar with other producers within the Common Market for Eastern and Southern Africa (COMESA). This will leave local millers open to all-out competition (Odhiambo, 2009). Under the existing COMESA regime, Kenyan sugar companies enjoy protective tariffs that enable them to charge about twice the price of imported sugar. However, if and when protection disappears, then the economics for ethanol production will improve (see Chapter 8 for a detailed discussion on the market economics of sugar vs ethanol). The income per hectare for the co-production of sugar and ethanol (from molasses) was in fact estimated as 3 per cent less than that of producing only ethanol, using sugar cane juice (Endelevu/ESD, 2008).

Fuel ethanol in Benin

The case of Benin is a study in contrasts with the case of Kenya. The learning curve in Kenya has run for at least 30 years, whereas in Benin the curve is just beginning. The approach towards a national biofuels programme in Benin has been slow, but systematic. It has paid attention to the views of the relevant stakeholders and taken a holistic view of the country's biofuels prospects. Benin does not yet have a national biofuels programme, although it has made advances, including on the legislative front, with the creation in 2008 of a National Commission for the Promotion of Biofuels and a set of rules for companies interested in producing biofuels in the country. The development of such a programme in Benin has been supported by the World Bank system. Under this umbrella, the government decided to run a competitive bidding process for selection of a consulting group to conduct basic studies to inform its decision-making process. ETA Florence of Italy was selected to carry out such studies, from 2007, with the participation of the four authors of this chapter and others. ETA Florence has discussed its reports (ETA, 2009a, 2009b) with the government and has participated in Benin in a round of stakeholders' dialogues on its main conclusions. Still, in early 2011 there were still no liquid biofuels produced or used in Benin as the country travels through the early steps of the biofuels learning curve.

Assessment and needs identification

The Republic of Benin has a surface area of 114,763 km^2 and a population of 7,833,744. As with many other developing countries, agriculture plays a vital role in the national economy, employing nearly 75 per cent of active populations and contributing 38 per cent to GDP.

Given the country's growing population and developing economy, energy demand is rising. Transport in particular accounts for 23 per cent of the total demand, with an average growth rate of nearly 11 per cent per year for gasoline and 7 per cent per year for diesel (average 1999–2005).

Even though some recent oil finds have shown the potential of offshore oil reserves within the territorial sea of Benin in the Guinea Gulf, the current demand for fossil fuel is completely satisfied by imports. In the current scenario of rising oil prices, the country's expenditure for energy provision poses a serious constraint to its economic development

In this context, the introduction of mandatory biofuel blends in conventional fuels sold at service stations could help to reduce fossil fuel imports, as is already being developed in some African countries (Plate 18).

The development of a national biofuel programme in Benin has occurred in the context of a major energy crisis for the country, characterized by continuous power shortages and constant increases in fossil fuel prices, in a country heavily dependent on oil imports.

Agreement

In 2007 a feasibility study was commissioned and co-financed by the government of Benin via the Ministry of Mines, Energy and Water of Benin and by the IDA, the International Development Agency of the World Bank. The project was carried out by the authors of this chapter with the technical support of local and international experts, and was divided into two distinct phases: preliminary feasibility study and promotion strategy, and action plan for the implementation of the national biofuels programme of Benin (ETA, 2009a, 2009b).

Implementation

The first phase was the implementation of a preliminary feasibility study, which investigated six main issues, as follows.

- Ongoing experiences and current biofuel market trends in the world.
- Assessment of market potential for biofuel (ethanol and biodiesel) in the scenarios of national usage and export.
- Techno-economic analysis of biofuel production from different feedstocks based on estimates of production costs and financing.
- Identification and evaluation of possible equipment for the end use of biofuels in households (lighting and cooking) in order to reduce the use of traditional biomass and fossil fuels.
- Analysis of environmental issues related to the production of biofuels.
- Assessment of socio-economic issues and identification of potential financing schemes.

Preliminary estimates showed that, under current growth rates for gasoline and diesel consumption, the introduction of E10 blends would have required production of nearly 90 million litres of ethanol in 2010 and more than 400 million litres in 2025.

The introduction of biofuel blends would not only contribute to a reduction in fuel imports, but would also bring environmental benefits and provide a potential market for the development of a national agro-industrial sector, thus stimulating rural development and diversification of agricultural activities.

In terms of export potential, Benin could take advantage of the Cotonou Agreement signed in 2000 amongst USA/EU and ACP countries (Africa, Caribbean, Pacific), which would allow a total exemption of import tariffs for ethanol and biodiesel into the EU and the USA.

Despite some limiting factors, the agricultural sector generally covers the country's food demand, the main food crops being cassava, maize, grain sorghum and groundnuts, whereas the two main cash crops for export are cotton and cashew nuts.

The introduction of E10 blends would require the cultivation of 20,000–25,000 hectares of energy crops in 2010, increasing to 80,000–90,000 hectares in 2025.

Benin's climate conditions vary among different regions, ranging from the sub-equatorial climate of the south (1,300–1,500 mm of annual rainfall with two rainy seasons) to the semi-arid conditions of the north (700 mm and one rainy season).

As far as ethanol is concerned, three main feedstock options were identified: sugar cane (*Saccharum officinarum*), sweet sorghum (*Sorghum bicolor*) and cassava (*Manihot esculenta*).

Sugar cane is certainly a suitable option for ethanol, well known worldwide, which could yield up to 6,000 litres per hectare if cultivated under the right conditions (preferably in the southern agro-ecological zones with a sub-equatorial climate and higher annual rainfall, as long as it does not displace food crops). Moreover, ethanol from sugar cane has a very positive energy balance as its lignocellulosic biomass (bagasse) can fuel boilers for heat and power generation and even export electricity to the grid. At present, 4,000–5,000 hectares of sugar cane are cultivated in Benin in a single irrigated complex for the production of sugar.

Although sweet sorghum is not yet cultivated in Benin, traditional grain sorghum (which has similar cultivation techniques and requirements) is well introduced into the farming system, being an important food and fodder crop, especially in the northern regions with semi-arid conditions. Sweet sorghum could represent a very promising feedstock for ethanol production. Compared to sugar cane, sweet sorghum has a shorter growth cycle (4–5 months instead of 12–18); it is an annual crop, although under proper conditions it could be cultivated in two or even three cycles per year. In optimal conditions the potential ethanol yield of sweet sorghum is comparable to that of sugar cane (more than 5,000 litres per hectare).

Cassava is widely cultivated throughout Benin and is an important food crop for the country, especially in rural areas; it is estimated that at least 68 per cent of arable lands are suitable for this crop. Recent developments promise a yield of 20 tonnes/ hectare of fresh roots (ETA, 2009a). But, as a staple crop, it may not be the optimal choice for conversion into fuel ethanol.

Thus, the implementation of a programme for the development and production of biofuels in Benin will require the cultivation of thousands of hectares of energy crops; for this reason the identification of an optimal agricultural development model will be necessary, in order to ensure a reliable and sustainable production of feedstock.

With regard to this issue several options could be adopted, each one having its advantages and disadvantages, as summarized in Table 16.2.

An integrated approach to these three models (contract farming, village-scale community-managed plantations and agro-industrial complexes) would help to establish a balance between the need to achieve a high production capacity of cost-competitive, high-quality standardized biofuels (mainly through the adoption of the complex model, at least at the beginning), and the need to promote rural development, farmers' entrepreneurship and organizational and productive capacity (mainly through contract farming and village-scale plantations).

Evaluation and adjustment

Based on the results of the feasibility study, a proposal for a Biofuel Promotion Strategy was discussed by a large group of local stakeholders (ministries and institutions, research centres, farmers' associations, chambers of commerce and agriculture, NGOs). Their various recommendations, concerns and suggestions were collected and integrated in the draft strategy to develop a consensus document.

Forming the base for a future framework policy document that will be adopted by the state, the proposal is focused on a general vision that is "to develop biofuel production and supply chains, as a driver for economic growth and poverty reduction, with positive effects on food crops and environment" and on an objective, that is to "contribute to the growth of the agricultural GDP, the improvement of trade balance, the increase of farmers' incomes, and to the reduction of the human pressure on forestry resources".

TABLE 16.2 Feedstock agricultural development models for biofuels in Benin

Item	Contract farming	Village-scale community mgt	Agro-industrial complex
Crop management	Benin cotton experience	Single-minded focus, mgt ctrl	Complex multi-crop management
	Farmers likely to get higher share of value added	Farmers likely to get lesser value added	Farmers' share of value added varies with crop
Technology change	Via experience and training	Higher rate of tech change likely	Rate of tech change varies with crop
Risk to steady supply	Higher	Medium	Lower
Risk to food supply	Higher	Medium	Lower
Overall farmers' development	Higher	Lower	Medium

The strategy proposed would establish a mandatory biofuel blend from B5 and E5 up to E10 and B10 by 2020 to implement a national biofuel market. At the same time an effort should be made to develop reliable partnerships with foreign countries and investors to promote Benin's export capacity.

The proposed strategy will focus on the following initiatives, each one introducing specific measures addressing the different stakeholders and different segments of the supply chain in order to develop the biofuel industry:

- actions to promote development of a favourable institutional environment;
- promoting the country's capacity of biofuels use, storage and distribution;
- promoting the production of agricultural feedstocks;
- promoting the establishment of a network of processing plants.

The strategy aims at:

- reduction of energy dependency on oil imports;
- increasing agricultural contributions to the improvement of trade balance through reduction of fossil fuel imports and promotion of biofuels exports;
- improvement of small farmers' incomes;
- job creation;
- diversification of agricultural production;
- reduction of GHG emission.

In June 2008 major steps towards the development of the National Regulatory Framework for biofuels were achieved, through the following legislation:

1. Decree 360/2008: Nomination of a National Commission for the Promotion of Biofuels;
2. Decree 361/2008: General conditions for the installation of biofuel companies in Benin.

Decree 361 defines the general conditions for biofuel production in Benin and introduces the following important statements.

1. To ensure value added in Benin, the export of raw agricultural feedstock to foreign countries for biofuel production is forbidden.
2. The State decides which food crops can be used as energy crops for biofuel production.
3. To promote family agriculture, priority should be given to small rural farms for the production of energy crops, preferably in a contract farming model.
4. Research, production, processing and trade of seeds, feedstock and biofuels in Benin are all subject to authorization by the state, upon proposal of the National Commission for the Promotion of Biofuels.

This Commission is introduced by Decree 360 and is composed of members of different ministries and institutions. This group of experts is subdivided into three sub-committees (agricultural production, biofuel production and trade, support measures) and supervised by a coordination group.

This Commission is charged with the implementation of the future National Biofuel Plan and the technical evaluation of requests submitted by private companies for the production of energy crops and biofuels in Benin.

Replication and feedback

This proposal will support the actual National Biofuel Plan of Benin. The adoption of a detailed "Action Plan" originating from this strategy will be an important step forward to support a coordinated approach for the development of the biofuel market and industry in Benin. Hopefully it could emulate similar initiatives in other countries, especially in Africa.

Thus, the proposed initiatives to promote biofuels in Benin include promoting:

1. a favourable institutional environment for action;
2. the country's capacity of use, storage and distribution of biofuels;
3. the production of energy crops;
4. the establishment of a network of processing industries.

Overall conclusions

Lessons learned from the efforts in Kenya and Benin to promote biofuels on a national scale could be useful to other African countries. Kenya's approach has relied mostly on private initiative backed by state support in various forms, but generally lacking in systematic government policy, legislation and regulation. No particular attention has been given to the management of technology change. Benin has opted to launch its biofuels with a strong role for the state as legislator and regulator, attempting to implement policies that had in principle been vetted by relevant stakeholders in the country. However, like Kenya, Benin appears not explicitly concerned with technology change and technology absorption, except for its emphasis on small-scale family agriculture as a source of biofuels feedstock.

With respect to establishing national biofuels economies in any country, but especially African countries, the Kenya and Benin experiences suggest the following recommendations:

- *Improve government and institutional support to provide an enabling environment, but not necessarily control.*
  Benin is trying to set up the enabling environment, but should consider the extent of government involvement and its effect on the ability to attract private investment. Kenya, on the other hand, has been able to launch projects but has failed to create a national programme for lack of adequate institutional support and coordination. Other countries in Africa should consider the appropriate

balance of private and government engagement to ensure the sustainable implementation of their national biofuels programmes.

- *Develop clear, consistent, sustainable policies.*

As a late entrant to the biofuels economy, Benin is pursuing initiatives that might lead to consistent and sustainable policies, although it may wish to review the need for appropriate balance between government and private initiatives. Kenya, an early entrant into the biofuels economy, has run through its learning curve, but still has work to do to achieve sustainable policies.

- *Provide capital and pricing incentives.*

In the history of national biofuels programmes, there is no example of success without an initial government push that reduces the risk of private capital investment in the biofuels market. But, although such incentives are crucial for the launching stage, they should be lifted gradually over time, lest the will to innovate and improve competitiveness is dampened.

- *Promote stakeholders' dialogues and transparent public–private partnership.*

As Benin wisely understood, the engagement of the relevant stakeholders in the design and implementation of national biofuels programmes is a necessary condition for their sustainability. Such dialogues cover all kinds of concerns, including those about technology change and absorption, allowing them to be expressed, considered and acted upon.

- *Build up local construction and technical and managerial capacity.*

Although at first foreign designs, project management and know-how may be required to launch a national programme, the quicker local construction, technical and managerial competences are developed, the better the chances of achieving sustainability. This is why it is important to factor in the technology transfer and absorption components of a national biofuels programme early on.

- *Encourage simpler designs and staged expansion.*

To facilitate the implementation of projects and the first steps in the technological learning curve, consideration should be given to the simplest economic process designs available for biofuels production. This approach also facilitates the engagement of local equipment manufacture, the training of operators and the management of technology change and absorption. As the programme evolves, the successive capacity expansions can consider adopting more advanced economic technologies.

- *Make sure sustainable feedstock cost and supply are maintained.*

The single most important cost item in the final cost of biofuels is the cost of feedstock, which can range between 50 per cent and 80 per cent of total cost. Therefore, the economic feasibility of biofuels depends heavily on the reliable and economic supply of feedstock. The capital costs of equipment idle due to lack of feedstock supply can easily kill even the best project.

- *Develop strategies to meet feedstock shortages.*

The reliable supply of economically priced feedstock is so crucial that one useful strategy to consider, as exemplified by the case of Benin, is the utilization of multiple feedstocks, such as sugar cane and sweet sorghum or cassava.

Notes

1 The biofuels matrix can also be generalized to show all energy resources available and their respective process technologies. Alternatively, as in this case, it can be made specific to one crop or feedstock, i.e. sugar cane.
2 Since biodiesel is a new entrant to the biofuels arena, the shape of its learning curve will be defined in the years to come. It may be different from the ethanol curve due to the different sources of feedstocks and the different organization of vegetable oil production and conversion to biodiesel. The relatively mature rapeseed-based German biodiesel industry vs the emerging Malaysian palm oil biodiesel programme illustrates such differences.
3 In 1982, one of the authors was retained by UN/TCD to review the institutional framework in Kenya: the raw materials for alcohol production; the current power alcohol projects; the pricing of alcohol, gasoline and alcohol–gasoline blends; and standards and norms applicable to fuel ethanol. In addition to visiting Kenya in August 1982, there was also a study visit to Malawi, which had recently started using fuel ethanol.

References

Anonymous (2010) "Kenya targets E10 blend", 26 July Biofuels Platform News, http://www.biofuels-platform.ch/en/news/2012.
Baraka, M. (1989) "Alternative Energy for Transport Industry – The Kenyan Experience with Ethanol", internal document, Sales and Technical Services, Caltex Oil (Kenya) Limited, 5 September.
Batidzirai, B. (2007) "Bioethanol Technologies in Africa", UNIDO/AU/Brazil First High-Level Biofuels Seminar in Africa, Addis Ababa, 30 July–1 August 2007.
Crilly, R. (2008) "Kenya taps into Brazil's ethanol expertise", 14 November, www.csmonitor.com/World/Africa/2008/1114/p11s01-woaf.html?sms_ss=buzz.
Endevelu/ESD (Endelevu Energy and Energy for Sustainable Development Africa) (2008) "A Roadmap for Biofuels in Kenya – Opportunities and Obstacles", a report to GTZ/Kenya Ministry of Agriculture, 17 May.
ETA (2009a) "Étude de Faisabilité de la Production de Biocombustibles Modernes (Bioethanol et Biodiesel) au Bénin et Élaboration de la Stratégie de leur Promotion – Rapport Final de la Phase 1 – Analyse de la Faisabilité Technique, Financière, Économique, Institutionelle et Environnmentale, Cotonou, février", ETA, Florence.
ETA (2009b) "Stratégie pour la Promotion des Biocarburants au Bénin, Cotonou", ETA, Florence.
Government of Brazil (2010) "Declaração Conjunta Brasil–Índia – Brasília, 15 de abril de 2010" (Joint declaration of Brazil–India, Brasilia, 15 April 2010), www.itamaraty.gov.br/sala-de-imprensa/notas-a-imprensa/2010/04/15/declaracao-conjunta-brasil-india-brasilia-15-de/?searchterm=Brasil-India Memorandum de Entendimento.
Mckenzie-Hedger, M. (2000) "Enabling environments for technology transfer", in B. Metz, O. Davidson, J.-W. Martens, S. Van Rooijen and L. Van Wie Mcgrory (eds), *Methodological and Technological Issues in Technology Transfer*, Intergovernmental Panel on Climate Change, Cambridge University Press, pp. 105–141.
Odhiambo, A. (2009) "Mumias eyes new revenue stream from ethanol plant", *Business Daily*, 16 December, www.businessdailyafrica.com/-/539552/823086/-/69e6tt/-/.
Scurlock, J., Rosenschein, A. and Hall, O.D. (1991) "Fuelling the future: power alcohol in Zimbabwe", African Centre for Technology Studies (ACTS), Nairobi.
Trindade, S.C. (1982) "Review of Power Alcohol Issues in Kenya. Report on Integration of Power Alcohol Operations with the UN/TCD Project KEN/76/005", Energy Planning, Policy and Programming in Kenya, September.

Trindade, S.C. (1994) "Transfer of clean(er) technologies to developing countries", in R.U. Ayres and U.E. Simonis (eds), *Industrial Metabolism: Restructuring for Sustainable Development*, United Nations University Press, New York, pp. 319–336.

Trindade, S.C. (2000) "Managing technological change in support of the climate change convention: a framework for decision-making", in B. Metz, O. Davidson, J.-W. Martens, S. Van Rooijen and L. Van Wie Mcgrory (eds), *Transfer*, Intergovernmental Panel on Climate Change, Cambridge University Press, pp. 47–66.

Trindade, S.C. (2007) "Transfer of technology and expertise", in Worldwatch Instititute (ed.), *Biofuels for Transport: Global Potential and Implications for Sustainable Energy and Agriculture*, Earthscan, London, pp. 263–275.

Trindade, S.C. and Yang, V. (1982) "Resources and resource conversion", Proceedings of the Fifth International Alcohol Fuel Technology Symposium, 4, Auckland, New Zealand, 13–18 May, pp. 173–192.

UK in Brazil (2006) Joint Statement by President Luiz Inácio Lula da Silva and Prime Minister Tony Blair, 9 March 2006, http://ukinbrazil.fco.gov.uk/en/news/?view=News&id=2056244.

UN (1991) "Energy systems, environment and development – a reader", *Advanced Technology Assessment System*, 6, New York, Autumn, p. xvi.

17

FINANCING AND INVESTMENT FOR SUGAR CANE AND BIOENERGY IN AFRICA

David Bauner, Melinda Fones-Sundell,
Karoli Nicholas Njau, Tom Walsh and Pontus Cerin

Introduction

Investment is an indispensable element in the realization of any venture, ranging from large commercial enterprises to undertakings in the informal sector, such as the building of a rural house or hut; all such activities require the use of human resources, land, tools and raw materials. Smaller projects may require only the use of locally available resources made accessible to the project owner by legal rights or barter. Beyond the small local investments, capital becomes part of the project realization; this capital can come from national or foreign grants, loans or equity, or more typically a combination of these. This chapter focuses especially on investment and financing for sugar cane-based bioenergy, which in some respects is a sub-sector of biomass-based energy, but is also related to agricultural investment. Although sugar cane has a long history in some parts of Africa, investments on the energy side of sugar cane remain quite limited, but have received greater attention in recent years due to concerns over climate change, energy security and rural development.

Sustainable investments are those that societal and environmental impacts into the financial calculus of the investor. This investment segment has grown to a non-negligible size during the last decade and has found many advocates within as well as outside the investment community, including academics, analysts, commercial investors, NGOs and fund trustees (Cerin and Dobers, 2008). Currently, the UN initiative for responsible investments, the Principles for Responsible Investments (PRI), established in 2006, has already gathered signatories with a total of US$19 trillion under asset management, involving some 210 asset owners and 437 investment managers. For investments in bioenergy to be sustainable, the allocated capital in the developed projects will need to respect socio-economic and environmental issues (FAO, 2008). The need for "informed" investments implies that investments should be preceded by a systematic collection and analysis of the information required for a sustainable project design.

Section 1: investment linkages to agriculture, land resources, industry, sustainable livelihoods and environmental management

A larger bioenergy project, such as a sugar cane plantation of 10,000 or 20,000 hectares, with combined ethanol production and heat and power production, is a major industrial undertaking and a major investment. It will invariably affect the region it is located in with respect to land ownership, employment, water supply, economic development and the environment. Depending on how the project is tailored and carried out, this may constitute an opportunity or a risk in financial terms.

Investment in biofuels has the potential to provide a new source of agricultural income and economic growth in rural areas. Biofuels production projects can also act as stimuli to improve agricultural productivity, which, in sub-Saharan Africa, has improved much less than in other regions over the last 40 years. This stagnating productivity trend is a significant brake on the development of local infrastructure as well as broader economic development. Biofuels offer other economic benefits, such as access or increased security of energy supply and access to modern energy services in rural areas (see Chapter 14). Access to energy services is critical for achieving the Millennium Development Goals (Modi *et al.*, 2005). Bioenergy projects and agricultural technology transfer in Africa could become a significant source of economic growth. On a national level, increased availability of foreign reserves and export earnings may follow. With a focus on outgrowers for biomass supply, a biofuel project need not alter the traditional smallholder structure of a region, but could instead provide a cash crop to subsistence farmers.

An UNCTAD report discussed the increasing agricultural investments by China in Africa, resulting in quality improvement, technology transfer and increased agricultural exports from some African countries (UNCTAD, 2004). Traditionally agricultural exports from Africa have been mainly raw materials, exposing African countries to worsening terms of trade, but a UNDP report notes that the Chinese government encourages well-established Chinese enterprises to invest in African agriculture. Although there is a considerable amount of foreign direct investment (FDI) from Chinese enterprises, most investments are joint ventures with African counterparts (UNDP, 2007). India and Brazil have also taken an interest in agricultural and bioenergy investment in Africa.

There is some concern that South-South investments introduce a different set of ethical standards compared to Western investments (cf. Swanström and Cerin, 2006; UNDP, 2007). A recent publication on biofuels investment in Tanzania (Habib-Mintz, 2010) has concluded that without strong regulatory frameworks for land, investment management and rural development, biofuel industrialization could further exacerbate poverty and food insecurity. Furthermore, production of biofuel that is not seen as strengthening the local community may alienate intended customers in importing countries in the EU and elsewhere. These customers are also voters with the power to affect the decisions made by their politicians and policy makers, which may affect support for FDI.

Although many biofuel investments involve large plantations, biofuel production can also be carried out by smallholder farmers as well as through outgrowers or locally contracted farmer arrangements. Smallholders typically combine cultivation of subsistence crops with cash crops, but with only 1–2 hectares of land production capacity is very limited from an industry point of view. While it is not surprising that industry is sometimes sceptical about receiving a significant amount of feedstock from outgrower schemes, the outgrower contribution to agro-industries like sugar cane is important in Africa. Such outgrowers have much lower administration costs compared to operating plantations, and are valuable during the pre-investment phase, when environmental permits and land leases are being sought.

Section 2: international investment climate for sugar cane and ethanol industries: lessons from Brazil

Typically, sugar cane industries initially focus on meeting domestic sugar demand. Supplying the EU and USA preferential sugar markets is attractive when this is possible (see Chapter 8). Only when these possibilities are exhausted or when the economics change does the sugar cane industry become more motivated to shift into ethanol production for domestic use or export, provided that such production is either economically viable or is supported by subsidies or fiscal incentives to improve its competitiveness vis-à-vis other alternatives. The only large-scale experience with ethanol from sugar cane is in Brazil, and some lessons from the Brazilian investment are discussed briefly below.

In Brazil, over 400 sugar and ethanol factories process sugar cane grown on seven million hectares of land, which is less than 10 per cent of total agricultural land. Roughly half of the cane is used to produce sugar, and the other half is processed to an annual yield of around 18 million m^3 of ethanol (Goldemberg, 2008). Brazil has become the world's most economically competitive producer of sugar and ethanol; production cost estimates are between 25 US cents per litre and 30 US cents per litre, corresponding to an oil price of between US$36/barrel and US$43/barrel (BNDES, 2008).

Figure 17.1 shows that in recent years, sugar cane bioethanol has made a good return for the producer, without including taxes or subsidies. A comparison must also take into account the significant depreciation of the US dollar, starting in 2005; the dollar lost close to 30 per cent of its value over two years, leading to some overestimation of the economic value of Brazilian bioethanol.

The rapid development of ethanol in Brazil began in 1975 with the Brazilian alcohol programme, known as PROALCOOL. Anhydrous ethanol was blended with gasoline, but after the second oil shock in 1979, the government decided to expand production to include hydrated ethanol to be used as neat fuel in modified engines. Ethanol production in Brazil was initially highly subsidized: the price paid

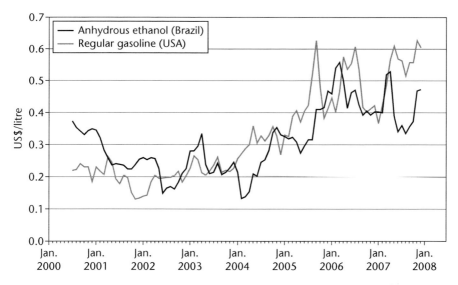

FIGURE 17.1   Evolution of prices paid to producer, excluding taxes (source: BNDES (2008))

to producers in 1980 was US$700 for 1,000 litres. In the intervening years, with technology improvements and economies of scale, the production cost has been continuously reduced and there are no longer any production subsidies.

Development in Brazil has gone from being a family business to a more diversified and industrialized capital investment base through a range of companies as well as strategic national and foreign investors. Foreign financial investors such as Goldman Sachs, Adeco and Merrill Lynch have increased their presence, either individually or in consortium with sugar cane operators. The latter include investment groups formed specifically to implement platforms for the production and sale of sugar cane bioethanol. The business model based on foreign capital typically includes Brazilian partners (BNDES, 2008). Where rainfall and other conditions in Africa allow for sugar cane expansion, the Brazilian investment model might be applied: public lines of credit and provisions to make private capital available could spur cane ethanol and bioenergy development.

One risk lies in disturbing current rural production conditions and impacting food production and employment for local smallholders. Other risk factors include water shortages, transport logistics, provision of hardware and chemicals, animal migration and several other factors (see Section 3 below). Mechanization of sugar cane production requires more investment, but can improve productivity and thereby facilitate expansion. Mechanization is often desirable from an environmental point of view as manual harvesting is often accompanied by burning of cane (see Chapter 4 and Chapter 11). From an African viewpoint, mechanization represents a challenge for outgrower-based investment, given the economic advantage of low-cost manual labour. However, if irrigation infrastructure and mechanized harvesting equipment are provided by the collaborating industry, outgrowers or "block farms" may be feasible, as is common in Tanzania.

Partly due to the financial crisis of 2008–2009, bioenergy projects that were under way have struggled with funding, and those that have been successful have relied on a combination of grants, loans and a smaller fraction of debt. The more difficult investment climate of the post-crisis period is likely to continue in 2011–2012 and beyond.

Section 3: investments in bagasse cogeneration and bioethanol plants

Harvested sugar cane is processed at the factory and yields sugar, molasses, bagasse, ash, filter mud and various other co-products for which markets might also be developed (see Chapters 5, 6 and 7). Investment in sugar cane may be for the production of sugar, bioethanol or both, which may also be accompanied by electricity export from a bagasse cogeneration plant. Historically, production capacity to meet the national or regional demand for sugar is developed first, and ethanol production is based on final molasses. Brazil is the only country where cane juice rather than molasses has been used, based on the technical parameters and the economic logic for diverting some cane production into ethanol (discussed in Chapter 6).

Bagasse can be used as fuel for cogeneration of heat and electricity, and typically suffices for plant operations during the harvest season (BNDES, 2008). The size of a bagasse cogeneration plant depends on whether electricity is to be produced during harvest season only, or all year round. In the first case, the plant would require at a minimum the amount of bagasse necessary to produce heat and power for the sugar and ethanol production process itself. Where there is external but seasonal demand, a larger and/or more efficient plant might be built to allow for sale of surplus power. In the second case, a larger or additional power plant could burn surplus bagasse and other feedstock throughout the year for sale of electricity and possibly heat to industrial customers and/or to the electric grid (see Chapter 5). If a region has some deficit in electric power or otherwise requires marginal capacity, there may be power sector incentives for investment in bagasse cogeneration at the sugar factory.

While entailing higher capital cost, energy is ideally generated using high-efficiency turbines rather than mechanically driven equipment. Processing a tonne of sugar cane into sugar with 100 per cent electric-powered equipment takes around 30–35 kWh of electricity besides a steam consumption of around 400–600 kg, while these requirements are lower for cane conversion into ethanol. Therefore, it is feasible for high-efficiency sugar mills to generate up to 130 kWh of surplus electricity per tonne cane crushed, depending on the technical configuration of the boiler (see Chapter 5 for a more detailed analysis). Gasification technology, if successfully adapted to cane fibre resources, is expected to at least double the electricity generation potential compared to conventional systems. The "Cogen for Africa" initiative, supported by the GEF via the United Nations Environment Programme

(UNEP) and the African Development Bank (AfDB), estimates that the potential for cogeneration with high-pressure systems amounts to 17.5 per cent of the installed electric power capacity in the selected countries – Ethiopia, Kenya, Malawi, Sudan, Swaziland, Tanzania and Uganda (UNEP, 2010).

According to Capoor and Ambrosi (2008), the demand for investment capital is huge in terms of power generation alone: US$4 billion will be needed in generation, transmission and distribution networks and in off-grid infrastructure to increase electricity access in sub-Saharan Africa to 35 per cent by 2015 and to 47 per cent by 2030. This is in addition to the financing needs for expansion of the electricity networks to maintain economic growth.

External opportunities and risks

There are a number of <u>external</u> or exogenous factors that could affect an investment. The decision to go ahead with a given project must take these aspects into consideration. A brief overview of examples of such factors is presented below, focusing on those cases most relevant for investment in sugar cane energy and/or co-products.

Global economic climate

Even if there are sound economics in favour of a project, investors may bail out if there is a looming threat of a global or regional economic downturn, or a better alternative for investment. The global economic climate may affect the prospects for expected market demand. An example is the recent downturn in the development of the relatively efficient ethanol production in Brazil. The success years 2007 and 2008 led to investment plans of approximately US$50 billion, much of which stalled in 2009–2010 as prices dropped and the financial crisis reduced the possibilities of funding. Since the price for sugar has gone down somewhat, the economics for sugar cane have been impacted further, and consolidation within the industry is expected.

The former Agricultural Minister of Brazil, Roberto Rodrigues, estimates that today's 200 companies in the sugar cane industry will be reduced to 50 in five years' time. Only 25 of the planned 40 plants to come on stream in 2009 appeared to be in a position to go ahead. Also somewhat lower oil prices and the weaker US dollar have contributed to the stagnation, as many loans were taken in US dollars just as the value of the Brazilian real was increasing. Since the production cost of cane ethanol in Brazil is competitive with oil, at US$40–50 per barrel, most new projects are mothballed rather than abandoned (F.O. Licht, 2010a).

African countries were less affected by the financial crisis in comparison to other world regions, in part because there is less foreign investment in the region and therefore fewer stalled projects, which could be an advantage as the economy picks up. Several initiatives, such as Kofi Annan's Alliance for a Green Revolution in Africa, aim to increase agricultural investment.

Land use and food/fuel production

Land use conflicts tend to grab public attention worldwide, and where there is a potential impact on food production such conflicts can seriously challenge investment plans. Not surprisingly, both government officials and investors are wary of situations where the competition over agricultural land can be presented to the public as a struggle between smallholder food producers and large corporate interests promoting feedstock plantations for biofuels aimed at urban consumers in the North. To mitigate such real or perceived impacts, investors have looked towards promoting the intercropping of sugar cane with food crops, flexibility at the processing point or looking for crops that can be grown on land that is considered marginal for food crop production.

Climate and land use change

Biofuels projects are favoured for their carbon neutrality and ability to promote rural development; sugar cane-based bioethanol has higher GHG reductions due to the high yield and efficiency of sugar cane. Carbon reduction-driven measures such as the Clean Development Mechanism (CDM) and Joint Implementation (JI) projects of the Kyoto Protocol and its successors may induce further development of biofuel projects.

Land use change associated with deforestation and agricultural burning is a significant contributor to GHG emissions, and therefore has emerged in the biofuels debate as an important risk factor for investments. Direct land use change (DLUC) occurs when land is converted in order to grow biofuel feedstocks. Where the original land has been degraded in some way or consists of abandoned agricultural land, the resulting land use change will generally have little or no net GHG emissions. On the other hand, if the land for conversion to biofuel feedstocks is tropical rainforest, peatland or other areas with high carbon stocks, such conversion will result in significant GHG releases. These land conversion impacts are of course not limited to biofuel feedstocks; the effects are similar when such land is converted for production of food, fibre or feed (fodder). Indirect land use change (ILUC) occurs when agricultural land is switched to biofuels production and thus results in the need for additional land elsewhere in the world to compensate for the loss in food or other biomass production. The calculation of ILUC requires various complicated assumptions and/or some type of modelling framework, and the EC is considering how to incorporate ILUC into the certification of biofuels as "sustainable" under the terms of the Renewable Energy Directive (EC, 2009).

Energy/fuel demand

In the case of biofuel projects in Africa, the security of market demand for the product presents a significant project risk. Few African countries have blending mandates, which is the most common way to promote biofuels demand in the transport sector. It is possible for African producers to sell into EU markets, given that African states have favourable market access treatment, and that the EU has established renewable energy targets for the transport sector (EC, 2009).

Infrastructure deficit

Insufficient infrastructure is a common inhibitor of production and project development in Africa. In the case of rural projects, there may be an advantage in that local demand is enhanced if the region is landlocked and long truck transport is required for any goods. The higher costs of transport in landlocked African countries in combination with the general lack of infrastructure results in much greater economic incentives for local and national biofuel markets rather than export markets (see also Chapters 10 and 15).

Weak enabling environments

Few African countries or regions have the necessary enabling environment in place to promote bioenergy markets and/or exports, with notable exceptions like Ghana and ECOWAS. Legal or regulatory mechanisms to protect local rights and take account of local interests, livelihoods and welfare are often lacking. Improved dissemination about costs and benefits and strengthening of institutional capacities can reduce risks and transactions costs, and thereby encourage renewable energy investment. The various institutional models from other countries and regions, including the EU renewable energy directive, can be used as a reference case for policy and legal structures. National policies and incentives are a determining factor in shaping biofuel investment (see Section 2 above, and also Chapter 9).

The key private sector partners can only be effective in scaling up renewable energy investments if an enabling environment exists. The public sector thus plays an important role in setting the policy and regulatory framework for private sector interventions and contributing to investments in the early stages of a transformative programme.

Large capital investments

There remains a funding gap for renewable energy, as commercial lenders perceive such investments to be risky. Part of the risk lies in the limited financial resources of the buyers of renewable energy. The high capital costs of renewable energy investments further exacerbate the problem. When there are capital constraints, the tendency is to favour projects that may have lower upfront capital intensity, which can be the case for biofuel and bioenergy projects, since feedstock supply costs are the greatest cost component. Agriculture is difficult to predict and control, making it a "last resort" for industrial investors and financial institutions. Sugar cane, in particular, is viewed also as a food crop, which provides a (sometimes false) sense of security to investors.

Local/national opportunities and risks

How, then, are investments affected by the local or national enabling environment and the relevant policies and institutions? A number of factors come into play, as discussed further below.

National policy

Government policies may give priority to various issues or sectors (e.g. food, infrastructure, energy, tourism), which in turn impact investment and financing in biofuel production. One of the key risks to project finance is a lack of policy and regulatory clarity, and it can thus be favourable to instigate capacity-building actions between EU partners and African governments in areas such as sustainability criteria, policy instruments (e.g. feed-in tariffs for renewables), co-financing for infrastructure development and supportive taxation policy.

Complex land tenure systems

Private investment in agricultural land – for bioenergy or any crop produced on a large scale – is often complex in Africa. Most countries have a dual system for land tenure, including both the "modern" or market-oriented type and a "traditional" land tenure system in which local chiefs or elders are often involved in land allocation; the two parallel systems can sometimes be in conflict, especially when it comes to international investment, or they may operate in a sort of uneasy truce.

In many countries there is also a system of public lands, which at a minimum encompasses protected areas and national parks (e.g. Kenya), or in the most extreme case encompasses, theoretically at least, all land in the country (e.g. Ethiopia). The ability of the state to enforce existing regulations, and mediate when there is a conflict between systems, can also be limited.

The traditional land tenure systems that are still respected in many parts of Africa are often understood by the local population and governed through mechanisms such as traditional courts or local government entities. This means that the investor can gain a permit to access land which is legal and valid at the national level and then, when the local population begins to understand the potential impact on their own access, can be successfully contested through a time-consuming and potentially aggressive process. Ideally, investors will need to understand the relationship of local populations to the land they intend to cultivate.

Water rights and usage schemes

In addition to facing a complex land tenure system, investors intending on using crops requiring irrigation, such as sugar cane, will also have to explore water users' rights and allocation schemes. These are often complex at the local level, due primarily to seasonal variation and the dispute over the size of flows adequate for preserving specific environmental measurements. In addition, there are many informal or illegal users of water and an investor may be given formal rights to use water that does not really exist in the quality or quantity needed. Sugar cane cultivation requires considerable amounts of water, as mentioned in the brief case study on Tanzania (Section 6).

Public opinion

Local public opinion may enhance or prohibit the development of a bioenergy project. An example of this is the moratorium on biofuel development in Tanzania, established partly as a result of a report (Mwamila *et al.*, 2008) pointing to a lack of knowledge on the effects of larger-scale biofuel projects.

Involving smallholders may be an effective means to engage the local community. In a tentatively more fuel-scarce future, developing and operating fuel production projects that meet local energy needs and improve energy access may be instrumental for local economic development. Conversely, evicting smallholders to set aside land for larger renewable energy projects without allowing those evicted access to the new amenities may cause unrest.

Natural environment

Climatic factors can complicate bioenergy projects just as they complicate agriculture and land use more generally: African agriculture is plagued with periods of drought and flooding, with both extremes affecting production. Only 5 per cent of the land area is under irrigation, and in many places this makes sugar cane investments difficult; irrigation for cash crops is inappropriate where there is water scarcity, in contrast to Brazil where nearly all cane cultivation is rain-fed. In the case of rain-fed sugar cane cultivation and harvesting, rain periods determine the optimum harvest cycles, but field conditions may inhibit harvesting when roads and fields are too wet for trucks and harvesters.

There are various environmental risks associated with large-scale biofuel production, such as new animal (or human) migration patterns, drought, water use, wildlife, pollution, eutrophication and acidification. Environmental impact analysis is important for decisions regarding biofuel projects, but the complicated nature of the impacts can lead to negative media exposure, describing biofuels as a threat rather than an opportunity for domestic production and local security of supply.

Investment centre/capacity

Most developing countries have an institution that coordinates FDI nationally and provides investors with information, licences, ideas for possible joint ventures and land use advice. Such an institution would also advocate to government if new policies or regulations were required for the investment to advance. These institutions enhance the degree of control of capital entering the country, both to enable taxation and other national benefits, but also to avoid scams that would deter future investment and investors.

International trade vs domestic use

International trade (or lack thereof) has some special implications in the case of bioethanol investment. First, the opportunity to export a value-added product such as ethanol rather than raw biomass is important for developing countries. Second, there are many significant potential producers of bioethanol; in the absence of protectionist measures, potential market participation is in principle quite broad. Third, the most economical biomass source or feedstock, sugar cane, is found almost exclusively in the developing world, for climatic reasons. Fourth, unlike biomass or wood products, ethanol markets are impacted significantly by trade barriers and tariffs. Both sugar and ethanol are currently protected products in most markets. Preferential sugar prices have also been a disincentive for developing countries to switch to ethanol, since these countries can obtain more money from subsidized sugar exports or by supplying the national market (see Chapter 8).

Between 2010 and 2015, ethanol trade is expected to more than double (F.O. Licht, 2010b). More significantly, the number of exporting countries/regions will increase significantly, with countries other than Brazil and the USA making up about 30 per cent of the total, compared to less than 5 per cent in 2005. Exports are increasing as a growing number of countries are developing ethanol fuel policies and programmes.

Fulton and colleagues (2004) have noted the potential for large-scale ethanol production from sugar cane up to 2050, estimated at 633 billion litres/yr (14.5 EJ/ yr or about 20 per cent of the estimated projected world gasoline demand in 2050. This scenario considers only a maximum of 10 per cent of the cropland area to be used for sugar cane (excluding Brazil). Brazil accounts for nearly half of the total ethanol production in this scenario. It is estimated that 3,460 new industrial plants would have to be built up to 2050 (of which 1,720 will be in Brazil); the cumulative associated investment is estimated at US$215 billion. This appears to be an optimistic scenario in terms of total market share; on the other hand, the estimated amount of cropland required may in fact be less, given the historical improvement in yields and the possibility to focus production on the high-yielding regions and varieties best suited to those regions.

The EU Directive on biofuels came into force in May 2003, under which member states should ensure a minimum 2 per cent share for biofuels by 31 December 2005 and 5.75 per cent by December 2010. With the exception of a few cases, member states in general have made slow progress on these targets. The Renewable Energy Directive of 2009 set a mandatory target of 10 per cent by 2020 for renewable transport fuels (EC, 2009). Sustainability criteria were established, including a minimum GHG reduction of 35 per cent and prohibitions on biofuels grown in ecologically sensitive regions.

The EU markets may offer advantages to biofuel projects on the West Coast of Africa, due to lower transportation costs than those faced by landlocked countries in Southern Africa. In the longer term, India and China and other Asian countries may become significant renewable fuel importers, and the African East Coast would be a more favourable production location for Asian markets.

However, a focus on domestic markets has certain advantages; historical experience has illustrated the economic and social value of displacing imported oil, especially since many countries have no refineries and have to import the costlier refined petroleum products. With domestic renewable fuels increasing in the fuel mix and the growth of the national economy requiring more affordable transport, policies for domestic markets will gain in importance. Fuel distributors would then play an important role in supplying feedstocks and transporting fuels to processing facilities, as is already the case in some countries like Malawi. Existing oil distributors could engage and invest in biofuel distribution and even production in some cases, based on commercial opportunities that arise.

Section 4: government support and regulation

The role of national governments in managing growth in the biofuel sector arises from a variety of objectives, some of which may conflict with each other while others may reinforce each other:

- security of fuel supply;
- increased rural access to electricity;
- import substitution of vehicle fuels;
- cleaner and safer alternatives to the present wood fuels widely used for cooking;
- rural employment and empowerment;
- water availability must be balanced over regions and seasons; and
- productive land use must be prioritized – but not at the cost of displacing or marginalizing large segments of the population.

The national *bureaus of standards* work, among other things, with fuel standards that must be issued and enforced for imports, export, distribution, blending and use. Emission levels in both fuel production and use are monitored by *environmental ministries* and national *agencies*. Investments in the biofuel industry are typically promoted and monitored by national *investment centres* in association with the relevant government ministries, including ministries of *trade*, *transport* and *industry* and the ministries of *energy*. Notably, the ministries of *agriculture* or *water* are seldom directly involved in issuing investment permits or decisions. The national government, through state-owned *power companies*, can also be a country's main electricity producer and distributor, with responsibility for local and national grids, managing seasonal variations, etcetera.

Regional and local government bodies often have the major responsibility to manage land use, oversee and enforce water rights and make final decisions on the location of economic and social infrastructure. Thus, while they have a very important role, regional and local policy and implementation priorities are not always in harmony with policies at the national level.

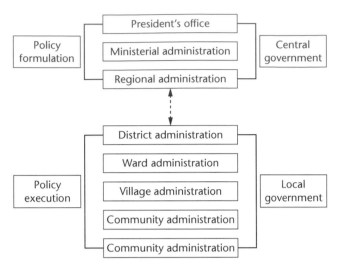

FIGURE 17.2 Example of institutional framework (Tanzania) (source: Habib-Mintz, 2010)

Conflicts between national and local/regional governments are especially evident and controversial in the context of land tenure and allocation of land and water resources to foreign investors. It is not uncommon that a number of partially overlapping land tenure systems co-exist and that national decisions to allocate these resources to investors are viewed by the local population as negating perceived traditional rights of access and use. Weak land rights can send subsistence farmers off their plots, but unfortunately stronger land rights may actually have similar effects, since it may be tempting or necessary to sell a land title at a time of need. Land ownership issues thus must be balanced to allow for certain flexibility, while creating stability for the large number of outgrowers that is the backbone of agriculture in many areas (see also Section 3 and Chapter 11).

Given their physical proximity, regional and local governments are likely to be under more pressure from the local population, particularly in relatively remote areas, than the national government. Foreign investors may have to deal with local governments directly concerning land issues and this brings with it the range of problems associated with communication and cooperation between two groups of actors so different in background and power.

As an example, the Kenya Energy Ministry sees a large potential in increasing agriculturally-based power production, including bagasse cogeneration. A potential of $300\,MW_{el}$ is identified considering the feedstock potential, of which bagasse potential is $38\,MW_{el}$ (F.O. Licht, 2010b). Tanzania, South Africa and other countries have also developed biofuel policies. Many agro-industries operating in Africa today could benefit from shifting part of their operations to biofuels and cogeneration. One of the important roles of government is to promote the exchange of experience and information in this area.

Section 5: investor types and financing options

While investors for biofuel projects may come from many different backgrounds, the arguments for reducing carbon emissions with the use of renewable energy sources carry a great deal of weight in the industrialized countries, while the production of biofuel feedstocks is often much more efficient in tropical regions and is cheaper in developing countries due to low labour costs. The focus for countries such as Tanzania and Ethiopia has thus been on foreign direct investment (FDI). These investors typically come with an industrial perspective (such as that of an energy company) and have a solid understanding of the processing technology, and although many have done business in the region before, they typically are equipped with a minimal understanding of agricultural production conditions for feedstock.

Overview of investment characteristics and experiences

Investors may expect and need substantial support from the government as well as regional/local authorities, and typically look for available grants and other subsidies to strengthen interest in investment. As a result, a spate of large-scale investors from developed countries and emerging economies, such as the EU and China respectively, have been acquiring large tracts of land in the South (primarily Africa), with promises of investment, job creation and energy supply that could contribute significantly to economic development in these countries (FAO, 2009). Notwithstanding, FDI can have a significant multiplier effect (often by a factor of ten) on local investment capacity.

While investors are often fairly inexperienced with sugar cane production, many countries have a long history of sugar plantations, often dating back to the colonial era. There is a wealth of technical and management expertise in sugar production: the problem lies in bringing this expertise to the attention of the foreign investor and also adapting it for the production and investment of renewable energy rather than sugar.

Key areas to address include connecting industrial investors to commercial agricultural production possibilities, and solving the question of who is responsible for regional infrastructure. Many investors, however, are opposed to contributing directly to regional and social infrastructure while at the same time the efficiency of using tax transfers for this purpose has been called into question. Other pressing aspects concern how the investors can get security for the longevity/sustainability of planned operations, related (but not limited) to: raw material sourcing, off-take, infrastructure maintenance and development and corporate conditions (reporting, taxation, etc.), including political stability.

Energy and agricultural sector investments in many African countries often include some degree of donor financing; funds may have been channelled to the government for use in supporting and regulating investment or are provided in the form of co-financing of economic and social infrastructure. While there is still limited involvement in the actual investment process, there have been widespread calls for donor financing to assist in such areas as technology transfer and adaptation and the analysis of environmental and social impacts (Parola, 2006).

Financial resources are of various types: managed funds, private capital, exchange traded funds and a number of others. The type of funding used by investors depends on the type of project. Typically, renewable energy finance may be segmented by the size of projects and type of debtor:

- consumer finance and microfinance for off-grid projects;
- corporate finance for small on-grid projects; and
- project finance for large-scale projects.

The funding of a larger project is typically done by a combination of different types, where FDI plays an important role. Domestic investors, when present, often act in consortium with foreign capital. Indeed, many countries require or reward some kind of local participation in energy and agricultural sector investments. At the same time, many foreign investors see advantages in associating themselves with local capital and expertise grounded in experience in Africa. The following sections review the different types of funding.

Domestic and regional private investors

There are wealthy domestic and regional individuals either acting in their own capacity or representing a family firm. These typically do not constitute a major source of funding, but this category could invest on other grounds than purely financial if the project aligns with the intentions of the individual or company. This category could potentially act more swiftly than an institutional investor such as a fund. An investment by a private investor would not necessarily be public or susceptible to the normal financial "due diligence" screening, including third-party ethical and environmental evaluation. These characteristics can be both a risk and an opportunity regarding public benefit and environmental and financial performance.

There is also the case of the local entrepreneur who develops the project-planning phase of the bio-energy project, which is a difficult and time-consuming task that needs institutional support in the form of feasibility/technical assistance grants. If the source of FDI can act in concert with a local entrepreneur who is well anchored in the local community, it can provide a framework for sustainable entrepreneurship. Issues complicating the participation of foreign investors in cultivating state-owned or traditionally owned lands can often be dealt with by involving domestic investors.

Established industry, including sugar cane-based sugar production, could play a part where waste and by-products are important inputs into cogeneration as part of an investment. Present industry would have advantages in a number of areas regarding biofuel production, and it would thus be important to include them in the planning for investment decisions.

To reduce risk, a large-scale investor places a high priority on ensuring reliable volumes of feedstock. Controlling the supply chain in a large-scale plantation is the most direct way of obtaining that security; however, the inclusion of commercial farmers as investors and owners could provide a good complement to outgrowers.

Domestic and regional institutional investors

For a project to go ahead, it may be wise to look also to national sources of funding. Local or domestic institutional investors, such as banks, insurance companies and pension funds, may open doors and give increased operability and credibility. A study on biomass cogeneration in Tanzania (Szogs and Wilson, 2008) shows that local companies are utilizing technologies that have been developed internationally, but financing them by using their own funds and thereby bearing all the project risks.

Private foreign direct investors

Foreign direct investment (FDI) inflows into the whole of Africa in 2009 amounted to a total of US$59 billion, down 16 per cent after almost a decade of uninterrupted growth. FDI into the region has been affected by the financial crisis, particularly from the oil-exporting and middle-income countries. Ironically, many low-income countries have been less affected since they tend to be less integrated into the global financial system, with new investors providing a buffer (UNCTAD, 2010)

Private capital, often channelled through funds, is a normal way to find finance for many types of projects. Many of these would undertake investments in a rather limited geographical area, such as Europe and/or the USA, suggesting that funds for bioethanol investments would be highly specialized towards Africa and/or the environment.

Such funds normally have an investment life-cycle, including exit strategy, spanning over five or ten years. The investment decision is preceded by a comprehensive due diligence of the proposed project, including the execution capability and experience of the management team. Other factors, such as industrial technology risk, market risk, political risk, infrastructure risk and ownership stability, are comprehensively evaluated. In the case of African bioenergy projects the infrastructure risk will receive additional focus. This includes road/rail/port infrastructure, local health and education infrastructure and electricity and water availability. These issues can add significant additional capital expense to a project.

A private investor can be expected to set performance measures for a given project, typically requiring EBITDA (earnings before interest, tax, depreciation and amortization) of 50–60 per cent, an IRR (internal rate of return) of 20–25 per cent, and EBIT (earnings before interest and tax) of 25 per cent. IRRs of 12 per cent have been calculated for an average plant in Brazil with an annual processing capacity of two million tonnes of sugar cane (BNDES, 2008); industrial consortia are more likely to invest in such projects, whereas private investors and established venture capital are less prone to invest under such conditions. Therefore the project developer must know which forms of finance are likely to be appropriate to his or her particular bioenergy opportunity.

Investors, either as private individuals, corporations, foundations or funds, typically manage a portfolio with a number of projects in different stages of the investment development life-cycle. The primary activity of the investor is thus to manage the portfolio, maximizing profit over the life of a fund.

Development finance institutions

Development finance institutions (DFIs) constitute the lynchpin in providing high-risk loans, equity positions, risk guarantees and other forms of credits to private investors (te Velde and Warner, 2007). Based in developed countries, the DFIs are state-owned or state-backed alternative financial institutions that may incorporate microfinance, community development and revolving funds (Levere *et al.*, 2006).

DFIs aim to bridge the gap between commercial investment finance and state aid. DFIs tend to invest in a partnership approach with a strong local business partner. They are meant to serve as a catalyst and to ensure that sustainable biofuel investments that otherwise would not be able to retrieve enough funding from the market can go ahead.

The DFIs typically have better back-up with additional capital than commercial banks and can also offer lower interest rates and longer payback periods, such as 10–15 years (te Velde and Warner, 2007). Examples of DFIs are EBRD (European Bank for Reconstruction and Development), EIB (European Investment Bank), NIB (Nordic Investment Bank) and Swedfund. The 15 members of the Association of European Development Finance Institutions (EDFI) are mandated by their governments to invest in developing countries and emerging markets. Together they had a consolidated investment portfolio in 2010 of approximately €21.7 billion across 4,088 projects (EDFI, 2010). Africa accounts for 28 per cent of the investment portfolio. The EDFI is providing approximately €4 billion in equity and loan investment on an annual basis. In comparison, the International Finance Corporation (IFC) has invested €25.1 billion in 1,560 projects (IFC, 2010).

A key value-adding component that DFIs have is their capacity to mobilize other investors by sharing knowledge, setting corporate investment standards, and so on. Their long-term presence gives a sense of security to other investors who may wish to co-invest in the region. It is also possible for a private investor to ask a DFI to co-invest in their projects for a period of time. This reduces the risk for both investor and DFI.

The DFIs have a significant direct and indirect impact on the developing countries in which they operate. A recent study (Dalberg, 2009) indicates that European DFIs together sustained close to two million direct and indirect full-time jobs through their investments in 2008. In addition, their investments generated around €2 billion in tax revenue for governments in developing countries.

International finance institutions

The international finance institutions are certainly playing an instrumental role in terms of providing finance to renewable energy projects. The multilateral development banks (the African Development Bank, the World Bank, the IFC) are running substantial funding programmes for renewable energy projects.

The World Bank (WB) has a particular emphasis on investments in climate mitigation measures, including renewable energy. In 2009 the WB Group financing of renewable energy and energy efficiency projects and programmes in developing countries increased by 24 per cent to reach US$3.3 billion. The lending is reserved for "new renewables" (solar, wind, geothermal, biomass and hydro below 10 MW) and energy efficiency projects. In the period 2005–2009 the WB in Africa has committed financing of US$552 million to energy efficiency projects; US$444 million to new renewable energy projects; and US$999 million to hydro projects greater than 10 MW. In 2009, financing in the renewable energy and energy efficiency area represented more than 40 per cent of the WB total energy-lending commitment.

In January 2004 the African Development Bank (AfDB) Board approved the new Bank Group Policy on the Environment, establishing policy on environmentally sustainable development in Africa, followed by an implementation plan. The work of the AfDB in renewable energy has been operationalized partly through the Dutch-funded Financing Small Scale Energy Users (FINESSE) programme. During 2008, FINESSE resources supported the integration of renewable energy into current Bank activities, as well as the preparation of stand-alone projects through technical and financial support to develop feasibility studies in several regional member countries. The overall goal of the FINESSE Africa programme is to assist countries in Africa to work through the Bank in formulating the appropriate policy and regulatory frameworks and developing capacity, in order to generate a pipeline of investment projects in renewable energy and energy efficiency. The AfDB also issues clean energy bonds. In May of 2010, two new issues (due November 2013) were offered to Japanese investors (AFDB, 2010).

Climate Investment Funds (www.climateinvestmentfunds.org) channels finance through the AfDB, the Asian Development Bank, the European Bank for Reconstruction and Development, the Inter-American Development Bank and the World Bank Group. Over US$6.4 billion has been committed to these funds. The funds are disbursed as grants, highly concessional loans and/or risk-mitigation instruments in the period up to 2012 through the multilateral development banks.

The Clean Technology Fund (CTF) has a more country-specific approach, and to date country investment plans in Africa have been approved for Egypt, Morocco, Nigeria and South Africa. The "Scaling-Up Renewable Energy Program In Low Income Countries" (SREP) is a targeted programme of the Strategic Climate Fund (SCF), which is within the framework of Climate Investment Funds (CIF). This programme has committed funds of $250 million; the Norwegian government is one of the significant programme supporters. Three of the six pilot countries for this programme are from Africa: Ethiopia, Kenya and Mali. The SREP aims to help low-income countries draw on new economic opportunities to increase energy access through renewable energy use.

The SREP is permitted to use a variety of financial instruments; equity and debt financing, credit enhancement, grants and loans, feed-in tariffs and technical assistance grants. The SREP programme is indeed multifaceted in its approach and allows for institutional capability building as well as financing of individual renewable projects, thus allowing sharing of knowledge between the African states over time. Institutions such as the African Development Bank (AfDB) may facilitate lending by arranging roundtables and bringing in different actors to reduce insecurity.

In order to instigate change we can also see that there is a "green banking" trend emerging, which will have a profound effect on capital availability over time. This is the approach adopted by the International Finance Corporation. The IFC provides financing in the form of loans and equity to climate-friendly projects and has financed considerable numbers of ethanol projects in Latin America, but in Africa other agro-industrial commodities receive the lion's share of the support (IFC, 2010).

A special type of fund is the African Biofuel and Renewable Energy Fund (www. faber-abref.org), which is also working on the basis of certified emission reductions (CERs). Focusing on West Africa, the fund works with a number of partners to contribute to the development of biofuels and renewable energy industries (Figure 17.3).

Debt financing

As mentioned, an investment typically includes a fraction of bank loans or debt in addition to equity. To give a loan that would match or surpass the equity stake, commercial banks would typically need to and want to reduce investment insecurity to a minimum and "ring fence" the profitability and performance of the plant.

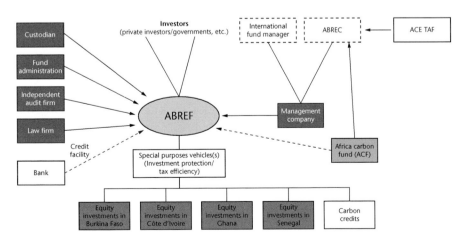

FIGURE 17.3 ABREF fund structure and partner outline (source: ABREF, 2010)

Commercial banks would thus undertake detailed client screening procedures and require substantial collateral as security against failure to repay the loan.

Aid organizations and donors

Aid organizations can help in many ways to facilitate biofuel investments where project plans align with development goals. Donors can complement more commercial investment to form a sustainable system for a given region. It has been suggested that rather than contributing to the investment directly with donor money, different types of training, infrastructure and equipment can be provided. By providing training in sustainable technologies for young businessmen, aid can interact with entrepreneurship. Additionally there can be deficits in terms of health, education and road infrastructure in the communities where projects are developed. Here the aid agencies can facilitate development in the local community. While the biofuel project can act as a substantial local stimulus, the involvement of an aid organization can provide support in terms of improving the social and community environment.

There is also the specialist form of aid organization that is focused on reducing poverty through the development of renewable energy as a means of corporate social responsibility. An example of this is the Koru foundation (KORU, 2011) of the United Kingdom where a group of European industrialists who are involved in the renewable energy industry in Europe assist communities that are energy poor and exposed to the impacts of climate change.

NGOs may play an important role in developing small-scale farming in many countries, in terms of access to credit, extension information and development of appropriate technological solutions to smallholder needs. NGOs and CSOs may also be important channels through which to communicate the needs and requirements of marginalized groups and the smallholder sector to the policy and investment process. They are equipped to carry out studies of the effects of different policy measures aimed at or affecting smallholders, and they are a cost-effective channel for disseminating information to smallholder agriculture; one example is "The Help Self Help Center" in Kenya.

Other factors/methods

If a project in sub-Saharan Africa includes imported goods, one way to ease the procurement process is through an export credit agency (ECA), when a private or quasi-governmental agency, typically from the country of the technology provider, takes the risk of the payment from the customer. ECA finance may offer competitive commercial terms, enhanced bankability and political risk cover for biofuels projects in Africa.

The carbon finance market is another channel, providing $6.5 billion a year directly to projects in developing countries to achieve GHG emission reductions. As of 2008, the Kyoto Protocol's Clean Development Mechanism (CDM) had registered over 1,000 projects, accounting for over one billion tonnes of GHG reductions in dozens of industries (Capoor and Ambrosi, 2008). The CDM has emerged as a $12 billion industry with a complex "business ecosystem" encompassing project developers, brokers, investment banks, rating agencies, consultants, lobbyists, regulators and the international trade shows that bring them all together. The CDM supported 183 bagasse cogeneration projects worldwide as of March 2009 (either pending or registered), of which 56 were in Brazil and six in Africa. There were 21 bagasse cogeneration projects rejected, as they could not prove additionality, i.e. that the project would not be carried out without the CDM (McNish *et al.*, 2009).

Companies such as Tricorona in Sweden and Climate Interchange AG in Germany develop projects that reduce carbon emissions and transfer technology from developed countries to developing countries. Certified emission reductions (CERs) are transferred to the developed country to meet required GHG reductions under the Kyoto Protocol (see also the case study in Section 6).

So far the bulk of carbon financing for renewable energy has gone to wind and solar power. There are no internationally accepted standards for GHG emissions in agricultural production, which constitutes a barrier to carbon financing for energy projects that involve crop cultivation.

National governments in developing countries can also co-invest in rural projects that will address development issues, such as housing, schools, hospitals, infrastructure, etcetera. Such investment can be direct and indirect, in the form of grants and soft loans or through legislation.

Different types of microloan schemes have developed through organizations such as Grameen Bank and Kiva. It can be argued that these schemes are generally not well suited for the medium to large-scale agricultural production associated with sugar cane; other means of funding should be sought.

Section 6: case studies from national and international investors

Africa is a vast continent with enormous differences in infrastructure and institutional capacity, and thus the three brief case studies presented below are not meant to be representative in any way. However, they do provide some concrete examples of the opportunities and challenges that arise in investment in sugar cane bioenergy projects. The first example concerns a multiple-product investment, the second relates more to ethanol production and the third relates only to cogeneration. It is also important to note that there can be synergies between ethanol and cogeneration investments, when combined appropriately within the economic market realities for sugar, ethanol and power production in the local and national context.

Sugar cane ethanol in Sierra Leone

Addax Bioenergy is a division of Addax & Oryx, a Switzerland-based energy group and a leading petroleum and mining company in Africa. Addax Bioenergy was established in 2008 to develop renewable energy projects. In 2008, AOG performed a feasibility study on the production of ethanol from sugar cane. Based on the study, the company established Addax Bioenergy with the intention to establish an integrated sugar cane plantation and ethanol distillery in the Makeni region in Sierra Leone: the planned agro-industrial complex will produce ethanol, electricity, biogas and food on 26,000 hectares of land. The project cost was estimated at about US$300 million, with planned production commencing in 2011. The plant will be capable of producing up to $170,000\,m^3$ of ethanol per annum, primarily for export to the European market. The biomass-fuelled power plant will achieve an excess capacity of up to 30 MW, which will be available to industry and consumers and will significantly boost electricity supply. Addax is co-funded in this venture by seven development financing institutions (DFIs): EU-EFP-EIB, the UK Emerging Africa Infrastructural Fund (EAIF) – UK; Deutsche Investitions- und Entwicklungsgesellshaft (DEG) – Germany, FMO – Holland, OEB – Austria, Swedfund – Sweden; and the AfDB – Africa. The project is designed to adhere to the EU sustainability standards in terms of the protection of the environment, greenhouse gas emissions and social responsibilities and also the IFC (World Bank) Performance Standards.

Sugar cane ethanol in Tanzania

Another example of an attempt at large-scale sugar cane ethanol investment was the project initially developed by the Swedish company SEKAB in Tanzania. SEKAB, a company owned by a number of municipalities in Sweden, had investigated the potential of cane ethanol for export from a number of African countries, and settled for Tanzania and Mozambique. Plans for a pilot processing plant and a surrounding 20,000 ha sugar cane plantation were drawn up for an area north of Dar es Salaam, as a first step towards an eventual expansion to some 400,000 ha. Funding came initially from the owners, with plans of funding the later phases of the project through FDI. SEKAB had similar but less advanced plans for Mozambique, through the majority-owned subsidiary, Ecoenergia de Mozambique Lda. The development of the pilot plant coincided with the establishment of a National Biofuels Taskforce in 2006, and subsequent capacity-building within the responsible branch of government, the Ministry of Energy and Minerals, funded by the Swedish and Norwegian states. At this time, several biofuel projects were under way in Tanzania (Sulle and Nelson 2009, referring to information mostly from 2007).

In the ensuing years, a number of difficulties emerged. The Tanzanian biofuels taskforce called for a moratorium in the wake of increasing food prices and international criticism of ethanol production, although ironically this criticism was mainly related to the food vs fuel debate for corn-based ethanol (in the US) that contributed to rising food prices (Rosillo-Calle and Schirley, 2010). The project also received criticism from NGOs regarding land rights, which sustained the call for a moratorium (ActionAid, 2009). Permits, leases and additional FDI proved difficult, partly due to limited water availability, and the international financial/economic downturn in 2008 did not help. The project stalled in 2008 when the public owners in Sweden refused to continue funding the Tanzanian branch of SEKAB operations; the Dar es Salaam office closed. After much negotiation, in 2009 the Tanzanian branch of SEKAB was sold to Ecodevelopment AB, whose main owner is a former CEO of SEKAB. An investment loan guarantee requested by SEKAB BT Tanzania from Sida did not materialize because Sida deemed the business plan to be "inadequate" (Västerbottens-Kuriren, 2009).

It is nevertheless possible that the tide may change again in terms of interest in expanded investment in sugar cane energy in Tanzania. Brazil has offered to help in fostering a biofuels industry by means of its deep experience in sugar cane ethanol, and would agree to discuss waiving Tanzania's debt owed to Brazil of US$240 million. Tanzania president Kikwete specifically mentioned that the country was ready to allocate close to 200,000 hectares for cultivating sugar cane for ethanol production as part of these initiatives (Africagoodnews, 2010).

Bagasse cogeneration CDM project in Kenya

Mumias Sugar Company (MSC) Limited and Japan Carbon Finance Limited have registered a project titled "35 MW Bagasse Based Cogeneration Project" by Mumias as a CDM Project (UNFCCC, 2008). The project intends to generate 35 MW, of electricity with 10 MW internal consumption by the factory and 25 MW export to the national grid. The proposed project is a power capacity expansion project involving the generation of electricity using sugar cane bagasse on site and consisting of installation of a five megavolt amp (MVA) transformer, a new 170 tonnes/hour high-pressure (87 bar) steam boiler and a 25 MW condensing extraction turbine alternator at MSC at different phases of the project, and finally the decommissioning of four 22 t/h and two 55 t/h bagasse-fired steam boilers currently operating there.

The technology to be employed for the Mumias Cogeneration Project is based on a conventional steam power cycle involving direct combustion of biomass (bagasse) in a boiler to raise steam, which is then expanded through a condensing extraction turbine to generate electricity. Some of the steam generated will be used in the sugar plant processes and equipment, while the power generated run the plant, with the excess (25 MW) exported to the national grid. Evaluation of the project revealed that it will reduce GHG emissions directly through the following means.

- Displacing grid electricity with GHG-neutral biomass (bagasse) electricity generation. This component of the project activity is expected to achieve GHG emission reductions of 1,245,652 tCO$_2$e over a ten-year period (2008–2017).
- Methane abatement through avoidance of dumping bagasse and instead using it to generate electricity. This component is expected to achieve GHG emission reductions of 50,262 t CO$_2$e over a ten-year period (2008–2017).
- The overall GHG emission reductions expected from the project are 1,295,914 t CO2e over the period 2008–2017.

The project will play an important role in the country's economic development, as more power will be available to offset the power supply deficits in the country. Expanded rural electrification, which could result from this project, would have far-reaching impacts on livelihoods in the rural community where the factory is located and where more jobs would be created (TÜV SÜD, 2008)

MSC, a limited liability company listed on the Nairobi Stock Exchange since 2001, is the project sponsor and operator. Company ownership is split between the government of Kenya (34 per cent), the farmers (30 per cent) and other stakeholders. MSC also funded the CDM registration process. The Japan Carbon Fund (JCF) acts as the purchaser of CERs. In addition, some of the CDM transaction costs (PIN and PDD preparation and validation) were funded by the JFC.

Conclusions

Providing financing for agriculture-based processing of bioenergy feedstocks on a large scale makes for a complicated investment, both from the point of view of those responsible for payback on financial capital and those entrusted to guide socio-economic development in an acceptable direction. Therefore, most investments in large-scale sugar cane production that include renewable energy, whether they are based on a plantation or an outgrower scheme, are likely to require some institutional or multilateral support. This is especially true in the present "early" phase of African biofuel project development. In this sense the development of "renewable energy development agencies" at the national level in several African countries is an encouraging sign in the process of supporting the introduction of funding programmes. However, funding is not the only consideration: equally important are the education, information and demonstration programmes. Investors need to adopt a holistic approach that is tailored to local and national needs and customs and must have at least a five to ten-year development timeframe in mind. Funders must have a long-term view and the managers of initiatives must be prepared to engage in a socially responsible manner in order for biofuel development to support the local community provide lasting benefits to the local and national economy.

References

ABREF (2010) African Biofuel & Renewable Energy Fund (www.faber-abref.org).

Africagoodnews (2010) "Tanzania and Brazil sign biofuel memo", 8 July, www.africagoodnews.com.

AFDB (2010) Press release 27 April, www.afdb.org.

ActionAid (2009) "SEKAB – ETANOL till varje pris? Hur SEKABs biobränsleprojekt i Tanzania drabbar lokalbefolkningen", Factwise on behalf of Actionaid, October, www.mynewsdesk.com/se/pressroom/actionaid/document/download/resource_document/5785.

BNDES (2008) "Sugarcane Bioethanol: Energy for Sustainable Development", Brazilian Development Bank.

Capoor, K. and Ambrosi, P. (2008) "State and Trends of the Carbon Market 2008", World Bank.

Cerin, P. and Dobers, P. (2008) "Editorial: the contribution of sustainable investments to sustainable development", *Progress in Industrial Ecology – An International Journal*, vol. 5, no. 3, pp. 161–179.

Dalberg (2009) "The Growing Role of the Development Finance Institutions in International Development Policy", Dalberg Global Development Advisors, Copenhagen (report funded by the Association of European Development Finance Institutions, EDFI).

EC (2003) Renewable Fuel Directive 2003/30/EC.

EC (2009) Directive 2009/28/EC of the European Parliament and of the Council of 23 April 2009 on the Promotion of the Use of Energy from Renewable Sources and Amending and Subsequently Repealing Directives 2001/77/EC and 2003/30/EC.

EDFI (2010) Association of European Development Finance Institutions, www.edfi.be/.

FAO (2008) "Biofuels, Prospects, Risks and Opportunities", Food and Agriculture Organization of the United Nations.

FAO (2009) "Land Grab or Development Opportunity? Agricultural Investment and International Land Deals in Africa", Food and Agriculture Organization of the United Nations (FAO), the International Fund for Agricultural Development (IFAD), the International Institute for Environment and Development (IIED).

F.O. Licht (2010a) "World Ethanol Price Report", June.

F.O. Licht (2010b) "World Biomass and Cogeneration Report", August.

Fulton, L., Howes, T. and Hardy, J. (2004) *Biofuels for Transport: An International Perspective*, International Energy Agency, Paris.

Goldemberg, J. (2008) "The Brazilian biofuels industry", *Biotechnology for Biofuels*, vol. 1, no. 6, BioMed Central.

Habib-Mintz, N. (2010) "Biofuel investment in Tanzania: omissions in implementation", *Energy Policy*, 38, pp. 3985–3997.

Haugen, H.M. (2010) "Biofuel potential and FAO's estimates of available land: the case of Tanzania", *Journal of Ecology and the Natural Environment*, vol. 2, no. 3 (March), pp. 030–037.

IFC (2010) "Telling Our Story: Renewable Energy".

KORU (2011) www.korufoundation.org.

Levere, A., Schweke, B. and Woo, B. (2006) *Development Finance and Regional Economic Development*, CFED, Washington, DC.

Mbohwa, C. (2003) "Bagasse energy cogeneration potential in the Zimbabwean sugar industry", *Renewable Energy*, 28, pp. 191–204.

McNish, T., Jacobson, A., Kammen, D., Gopal, A. and Deshmukh, R. (2009) "Sweet carbon: An analysis of sugar industry carbon market opportunities under the clean development mechanism", *Energy Policy*, 37, pp. 5459–5468.

Modi, V., McDade, S., Lallement, D. and Saghir, J. (2005) "Energy Services for the Millennium Development Goals", The International Bank for Reconstruction and Development/The World Bank and the United Nations Development Programme.

Mwamila, B.K., Kulindwa, O., Kbazohi, H., Majamba, H., Mlinga, D., Charlz, M., Chijoriga, A., Temu, G., John, R.P.C., Temu, S., Maliondo, S., Nchimbi-Msola, Z., Mvena, M. and Lupala, J. (2008) "Feasibility of Large-Scale Bio-Fuel Production in Tanzania", UDSM, SUA, Ardhi University, Morogoro and Dar es Salaam.

Parola (2006) "Technology Transfer Financing to Developing Countries in Climate Change, Carbon Offset and Environment: an Overview of Financiers and Instruments", Parola International Associates, Finland (funded by the Ministry of the Environment, Finland).

Rosillo-Calle, F. and Walter, A. (2006) "A global market for bioethanol: historical trends and future prospects", *Energy Sustainable Development*, 10, pp. 20–32.

Sulle, E. and Nelson, F. (2009) "Biofuels, Land Access and Rural Livelihoods in Tanzania", International Institute for Environment and Development, UK.

Swanström, L. and Cerin, P. (2006) "Management of Sustainability Issues in Industry – A Stakeholder Perspective", Centre for Environmental Assessment of Product and Material Systems, Chalmers University of Technology, Gothenburg (www.cpm.chalmers.se).

Szogs, A. and Wilson, L. (2008) "A system of innovation? Biomass digestion technology in Tanzania", *Technology in Society*, 30, pp. 94–103.

TÜV SÜD Industrie Service GmbH (2008) "Validation of the CDM-Project: '35 MW Bagasse Based Cogeneration Project', by Mumias Sugar Company Limited", Report No. 893334, 29 August, Kenya.

UNCTAD (2004) "Prospects for Foreign Direct Investment and the Strategies of Transnational Corporations, 2004–2007", United Nations Conference on Trade and Development, Geneva.

UNCTAD (2009) *Enhancing the Role of Domestic Financial Resources in Africa's Development: A Policy Handbook*, United Nations Conference on Trade and Development, Geneva.

UNCTAD (2010) Press release, 22 July.

te Velde, D.W. and Warner, M. (2007) "Use of Subsidies by Development Finance Institutions in the Infrastructure Sector", ODI Briefings 2, Overseas Development Institute, UK, www.odi.org.uk.

UNEP (2010) Cogen for Africa, http://cogen.unep.org.

UNDP (2007) "Asian Foreign Direct Investment in Africa: Towards a New Era of Cooperation among Developing Countries".

Västerbottens-Kuriren (2009) "Sida säger nej till Sekabs ansökan", *Västerbottens-Kuriren*, 16 October.

OECD (2008) *OECD Investment News*, 6 (March), The Investment Division of the OECD Directorate for Financial and Enterprise Affairs.

UNFCCC (2008) "A 35 MW Bagasse Based Cogeneration Project", Mumias Sugar Company Limited (MSCL), Version 12, August.

CONCLUSIONS

Francis X. Johnson and Vikram Seebaluck

Sugar cane is a global agricultural crop of great commercial significance, supporting developmental and societal needs in the many countries that grow it. It has become the world's most economically valuable bioenergy crop, with the potential for producing over 100 tonnes of biomass per hectare annually. Traditionally sugar cane has been exploited for the production of sugar as a sweetener but has more recently been demonstrated to support the generation of multiple products, particularly bioethanol and cogenerated electricity, on a large commercial scale. Although sugar production will continue to be the main commercial product from sugar cane in the near term, the crop is increasingly being recognized as a versatile resource that offers food, feed, fuel, fibre and various specialized products, which together can reduce dependence on fossil fuels, thereby favouring low carbon development.

Sugar cane's agricultural importance therefore extends into the energy and environmental spheres, as it provides a sustainable and flexible natural resource. This book has illustrated the suitability and the current socio-economic environment in Africa for sugar cane, which could sustain new developmental paths and contribute to a more dynamic and competitive economy in sub-Saharan Africa and other regions. The high resilience of the crop to adverse climatic factors, its multifunctionality and its efficiency in climate mitigation place it in a key position to support a sustainable agro-food security and bioenergy development strategy for the region. This chapter presents and synthesizes the key elements associated with developing the sugar cane resource in sub-Saharan Africa, ranging from cultivation and harvesting to impacts, policies, markets, technology investment and financing.

Agriculture

The sub-Saharan Africa region with its tropical and subtropical climate is well suited in many ways to expand the production of sugar cane. Possessing a high photosynthetic efficiency and biomass productivity, the crop has been genetically improved over the past decades to enhance both cane and sugar yields worldwide and to enable its expansion in a wider range of environments. Much effort has been directed towards the development of new sugar cane varieties that meet the specific requirements of growers as well as addressing emerging uses of the crop in relation to fibre content and resulting energy generation. Future advances to tap the enormous potential of the crop in both food and energy production will require a synergy across several fields of research, including traditional breeding, genetics, physiology and biotechnology to enhance breeding efficiency for sugar, energy and value-added products (Chapter 2). Cultivation of the crop in marginal areas for its total biomass potential beyond its ecological suitability is a further avenue to be explored.

The modern conceptions of bioenergy are challenging with respect to the risks associated with food security and environmental conservation. The most efficient use of natural resources, especially land, is a key element in harmonizing different agricultural demands such as food, feed, fibre and fuels. Agro-ecological zoning, an innovative approach to promote both food and fuel productions in synergy with the environment, could be used to develop bioenergy programmes in Africa, which has many regions with excellent natural conditions for biomass production. Key lessons from countries where agro-ecological zoning has been successfully applied, such as Brazil, could be adapted to the African context (Chapter 3). Such programmes will require the promotion of increases in agricultural efficiency as a whole, thereby identifying potentials and limitations for different regions and giving appropriate technical supports to explore these potentials.

One of the most important aspects for cane production in Africa is land availability for rain-fed cultivation, which appears to be significant in a number of regions. However, the estimates vary greatly given that the models use different key sugar cane growth determinants, integrate them differently, and exclude different land types as inappropriate for bioenergy feedstock production. Nevertheless, land suitability and availability mapping should be a leading tool in sugar cane bioenergy planning in Africa (Chapter 4). Such assessments should include analysis of long-term climatic data using water balance and crop models so that the major influence of radiation, temperature and water availability on plant growth can be incorporated in each unit area of land considered.

Harvesting and delivery of cane is a complex and costly operation for sugar production, which requires extensive coordination, especially for large-scale production and subsequent processing in centralized factories (Chapter 4). Rationalization of transport equipment and improved coordination can reduce production costs, and is a current focus area of industry development efforts, particularly in Southern Africa. Most of the cane grown in Africa is burnt prior to manual harvesting. However, environmental regulations together with the increasing importance of cane fibre for electricity generation suggests that reduced burning and more gathering and use of crop residues in Africa will become increasingly common in the coming decades. Such trends are already observed in Brazil, Mauritius and India, where bagasse cogeneration is now viewed as a strategic energy asset. Harvesting of cane residues also requires major changes to harvesting and transport systems, thus requiring cost assessments that include the economic value of residues and recognize local and regional differences. Assessments based on short-term economics may give different results from those that take long-term crop yield effects into account, but there is normally scope for harvesting of some portion of the non-stalk components of cane without adverse long-term effects.

Industry

Sugar cane is industrially processed into sugar and is marketed in different forms depending on cost, quality and customer requirements. The process of sugar manufacture from cane is fairly standardized worldwide and is a mature process, with only limited opportunities for efficiency and improved productivity in sugar production itself. However, it has been widely demonstrated that sugar cane can sustain a far more diverse and multifunctional role beyond its current primary use for sugar production, given its significant biomass potential. The viability of the industry can be increased through creation of a broader product portfolio that can improve profitability and competitiveness, especially through bioethanol from molasses/juice and cogenerated electricity from bagasse. Such developments have already been successfully demonstrated and implemented in Brazil, India and Mauritius; the challenge is now for other cane-producing countries to tap this wealth of experience and replicate, expand or adapt similar systems. Modern bagasse cogeneration plants operating at high pressures of 82–87 bars can export 130–140 kWh of electricity per tonne of cane processed, which is several times as much electricity as that available from typical systems, and this can be increased even more through further optimization of system performance and energy efficiency (Chapter 5). The use of cane agricultural residues, which are equivalent in volume to bagasse and which are usually left in fields, can double electricity production potential. In order for biomass cogeneration to be successful, supporting policy measures are needed along with the appropriate physical and institutional infrastructure to facilitate independent power production.

Ethanol production from molasses is practised worldwide while its production from cane juice is very common in Brazil, offering significant opportunities to increase its production and subsequent substitution for gasoline. Similar to the case for sugar, cane ethanol production is a mature commercial process, and its production from cane juice has some key synergies with sugar production given that several of the component processing workstations are shared by sugar and ethanol. Although further efficiency gains are minor in the Brazilian systems, reductions in production costs can still be achieved by producing clean microorganisms for fermentation and improved steam economy to increase surplus power generation (Chapter 6). Advanced technologies to produce ethanol from lignocellulosic materials are being developed, and upon commercialization cane fibres could be used for further increases in ethanol production. The effluents from a distillery, despite having a high polluting potential, can be purified and recycled to cane fields, thereby replacing part of the chemical fertilizer requirement. The use of ethanol as a transportation fuel, although currently concentrated in the USA and Brazil, is gradually spreading worldwide in developed and developing countries alike, driven by concerns about global warming and escalating oil prices. Blends of 10–20 per cent ethanol in gasoline have proven feasible in many countries, and the use of flexible-fuel vehicle technology now allows almost any blend to be used. Many cane sugar-producing countries can become cost-competitive ethanol producers, due to the lower cost of cane compared to other ethanol feedstocks and the fact that two-thirds of ethanol production cost is for the feedstocks. Sugar cane is widely recognized as the best alternative among first-generation biofuel feedstocks, because of its high productivity, high yield per hectare, potential for expansion, a very positive energy balance, the large potential for producing surplus electricity and the avoided life-cycle GHG emissions. Sugar factories are being transformed into "biorefineries" with multiple energy and non-energy products, which can be extended further in the future by second-generation biofuels technology based on cane fibres.

Besides ethanol and electricity production as the current key commercialized co-products from cane, exploitation of the sugar cane resource base in an integrated manner can facilitate recovery of many other value-added products and thereby contribute significantly to the creation of a sustainable sugar cane agro-industry in Africa and globally. Other co-products include speciality sugars, sucro-chemicals, alco-chemicals, pulp and paper, furfural and many other manufactured goods (Chapter 7). Although the technologies are available to produce literally hundreds of products from sugar cane and its co-products, key factors such as market access, returns on investments, volumes of production, availability of alternative resources and specific local conditions will dictate the pace of investments and the development of new products. However, it is clear that the full potential of cane is quite far from being tapped and more focused research is needed to look into the possibilities for substituting non-renewable energy and materials with products and services based on sugar cane feedstocks. The African sugar industry has so far concentrated on sugar production, with only a rather limited development of value-added products. There is wide scope for new co-product development in Africa, both on small and large industrial scales, given the availability of cane resources, the expansion of the industry, the market potential and development opportunities on the continent, and more importantly the need to improve the sustainability of the industry.

Markets

The world's physical trade of sugar is increasing at a faster rate than the growth in global output. Sugar produced in Africa is sold into growing domestic markets, with some African countries also enjoying sales into high-priced regional markets. There is also an export market, including sales to the EU, which under preferential trading arrangements typically offered higher prices than the world market but these have recently been subjected to bold reforms. In general, it can be concluded that in the case of preferential exporters, including those in Africa, the reform stimulates efficient producers to consolidate and increase production, which can in the short term be directed to the still lucrative EU market, with market access improving considerably (Chapter 8). The sugar industries of Africa will seek to reduce production costs and enhance their competitiveness in an increasingly crowded market place. Several industries are already targeting specific market opportunities arising from trade agreements with the EU and there is significant expansion of sugar production capacity under way in a number of African countries. Concurrently, renewable energy has become a strategic element in the expansion options for African sugar producers, based especially on emerging opportunities in the growing world ethanol market as well as in domestic and regional biofuel markets. Similarly, investment in bagasse cogeneration has already been recognized as a valuable addition to efficiency and low-cost production, simultaneously deriving additional revenue from the sales of surplus electricity.

The main driving forces for market diversification into renewable energy include uncertainties in oil and sugar prices, land and resources availability, power supply conditions, environmental policies and economic development policies. The challenges to diversification include lack of supporting national and international policies, difficulties with market access and technological constraints, as well as the more general barriers that stifle the development of renewable energy sources. The expansion of ethanol will depend on sugar prices and the willingness of producers to forego some sugar production; stronger policy incentives such as blending mandates as well as some financial incentives, such as carbon credits or loan guarantees, would be needed in order for national and regional markets to move towards a scale that is economically sustainable (Chapter 9). Similarly, regulatory frameworks for independent power production in combination with incentives for renewable energy in the power sector will be needed in exploiting bagasse cogeneration potential, as has already been occurring in Mauritius. The inherent economics of the sugar industry in Africa are fairly competitive by global standards and the question is more about the willingness to apply policy instruments to shift the industry towards renewable energy production and thereby improve the competitiveness and sustainability of both the sugar and energy sectors.

International trade in bioenergy is still at an early stage, and existing trade barriers have constrained market expansion, particularly in the case of ethanol fuel (Chapter 10). In addition, bioenergy trade is increasingly impacted by the imposition of sustainability criteria, established almost exclusively by organizations based in developed countries. In this respect, a strategy based only on exports could be risky for biofuels producers in some least developed countries of sub-Saharan Africa. It seems more prudent in such cases to start the production of ethanol in small to mid-scale, taking advantage of existing experience and establishing domestic markets for blending ethanol with gasoline; the benefits of such a strategy are especially clear in countries with high oil import dependency. Agricultural development and the expansion of a modern sugar cane industry are intertwined; a modern and dynamic agricultural sector is needed to facilitate an increase in sugar cane productivity and diversification of the sector into multiple products. In many countries this could be an excellent instrument for inducing regional economic development.

Impacts

Considering the level of poverty in Africa and the need to provide opportunities for income generation, there is a need to maximize the local benefits for small-scale stakeholders in the cane industry as well as addressing energy insecurity. Socio-economic and environmental impacts must be considered in relation to baseline conditions and available development alternatives (Chapter 11). Production of ethanol and bagasse electricity would pay national and regional economic dividends by mitigating the effects of rising oil prices and the impending power shortages in some regions. Sugar cane offers high socio-economic sustainability with respect to employment creation, due to the high labour requirements on the agricultural side. Other socio-economic indicators exhibit greater variation across countries and even within different regions of a given country; for example, the manual harvesting of cane can result in harsh conditions if labour regulations are not enforced, but at the same time in poorer regions there is also concern about increasing mechanization as this will reduce the quantity of jobs available. The production and processing of cane has significant environmental impacts, including water consumption, water pollution, biodiversity, soil impacts and air pollution. In comparison with many other commodity crops, pesticide use is relatively low. Apart from the application of chemical fertilizers, chemical application is mainly restricted to herbicides. Other impacts associated with expansion of land use to accommodate growth in cane production are highly location-specific. Changes in cogeneration technology can affect sustainability outcomes more than changes in ethanol production, due especially to the potential for significant reductions in GHG emissions through cogeneration, which leads to a decisive factor for sustainability.

Most of the schemes available for sugar cane certification worldwide are voluntary; sustainability certification schemes differ from technical standards such as ISO in that certification puts more emphasis on environmental and social impacts and somewhat less emphasis on technology management processes. The only sustainability scheme focusing exclusively on sugar cane is Bonsucro (previously known as the Better Sugarcane Initiative), a comprehensive scheme that is consistent with ISO Standard 65 and which includes a clause on efficiency that thereby addresses co-products (Chapter 12). Some sustainability schemes target more sustainable cane production based on consistency with local/national environmental regulations, while others focus on providing best practice guidelines and consider the socio-economics of cane production. There are increasing efforts in some African cane-producing countries to improve sustainability, efficiency and environmental and social regulations. The SADC region, where African sugar cane production is concentrated, has developed a region-specific framework that may serve as a model for other African regions. The regional scheme aims to provide a flexible enough system that can cater for the variations in agro-ecological conditions and policy objectives in SADC member states. The SADC efforts illustrate awareness of the risks and potentials related to bioenergy production from sugar cane and the region-specific scheme indicates proactive engagement to ensure that the developmental promises of biofuels are realized. The international schemes such as Bonsucro and the Roundtable on Sustainable Biofuels (RSB) are incorporating GHG assessments in order to respond to the EU Directive, which is the largest export market. Issues related to land use change (LUC) and High Conservation Value (HCV) areas are also incorporated in some of the international certification schemes, reflecting their growing importance in the sustainability discourse. Certification and/or standards cannot fully ensure sustainability, but can provide good guidance and appropriate market signals to steer bioenergy and sugar cane markets towards a more sustainable path.

Expansion in cane cultivation in African countries for bioenergy purposes is also closely linked to GHG emissions. Based on a case study in South Africa, it has been found that irrigation has significant impact on GHG emissions of cane production due to the consumption of carbon-intensive electricity (Chapter 13). Analysis of different scenarios for ethanol production from sugar cane at South African mills suggests that ethanol could mitigate emissions in comparison to gasoline. The use of coal to supplement the deficient energy needs from bagasse burned inefficiently in conventional boilers would represent a significant emission source in the ethanol life-cycle. However, such a practice could be easily altered by using commercially available modern bagasse cogeneration systems, which would completely eliminate the use of coal and result in significant GHG savings. Additional emissions reductions could even be reached through the export of excess electricity generated from bagasse, which would contribute to diversifying the national electricity mix, thereby favouring exploitation of renewable energy sources. The direct emissions from LUC (land use change) are expected to be zero or negative, while for the conditions proposed there will be no indirect LUC involved in the expansion of sugar cane in Southern Africa.

Strategic issues and comparisons

African countries face a broad spectrum of challenges in their pursuit of progress, among which the continued lack of access to modern energy services and lack of opportunities for income generation affect development and contribute to perpetuating poverty in the region. The continued dependency on traditional biomass energy remains one of the biggest challenges to African economies. But given the natural endowment of the region, development of a viable bioenergy industry is an interesting prospect, and one that could potentially assist many African countries to solve many of their pressing needs. As a natural resource-based and land-intensive activity, biomass energy has courted controversy amongst analysts and various stakeholders. Despite the divergent opinions, indications are that if properly implemented and guided by sustainability criteria, biofuels and other modern forms of bioenergy can play an important role in reducing energy poverty and energy import bills while promoting health, economic growth and environment protection. Co-products from an efficient domestic resource like sugar cane can add local and regional value to agricultural activities and thereby overcome the tendency, which dates to colonialist times, for exports to be composed of exploitative cash crops that add little local value. Biofuels could boost energy security via direct reductions in oil imports (physical energy security) or through increased economic opportunities, leading to a greater ability to pay for energy (economic energy security) (Chapters 14 and 15).

Important lessons can be learnt from the bioenergy experiences of Mauritius and Brazil, countries that have demonstrated that political commitment to the exploitation of indigenous resources can achieve far-reaching economic benefits with the necessary policy framework. The widespread use of ethanol in Brazil and bagasse-cogenerated electricity in Mauritius have proven that bioenergy from sugar cane is technically and economically feasible on a large scale, and furthermore that valuable environmental and energy security benefits are thereby obtained. Countries such as Mozambique, Tanzania and Zambia, which are totally dependent on foreign oil, at a huge cost to their economies, could significantly benefit from cane bioenergy exploitation (Chapter 14). African countries could in principle cover their domestic energy demand with indigenous biofuel resources and still export surplus production to the world market. Whether countries will opt to export some of the produced biofuels will depend on national policy priorities, favouring physical or economic energy security. However, before realizing their biofuels production potential, sub-Saharan Africa countries need to address a number of political, technological and financial obstacles.

The agricultural sector in many African countries needs modernization, not only to produce more food but also because their industries require a dynamic market push, and their populations need jobs, income and electricity. Despite the agricultural and bioenergy opportunities close at hand, they often depend too much on international donors, particularly with regard to development assistance to reform institutions and make the initial infrastructure investments. However, donors have not necessarily prioritized such developments. At the same time, the countries need to put enabling policy mechanisms in place to create the market opportunities for cane energy production. This includes cutting subsidies on oil, improving the regulatory framework to favour domestic over imported energy, and creating fiscal incentives for sugar cane farmers and private investors to explore the energy generation potential. Thus proper institutional mechanisms and coordination amongst concerned stakeholders, including both private and public sectors, are required for the production and commercialization of such energy products (Chapter 15). It is very important for the international community to acknowledge this potential and provide the necessary support in the form of technical, institutional and economic assistance to help such countries realize their bioenergy potential, which is attractive both in terms of national development and climate change mitigation. Often the global dimension of the climate change problem is emphasized and policy makers approach it on the basis that the global issues are all-encompassing. However, there are important connections to local, national and regional impacts, especially in LDCs. Thus, climate mitigation strategies need to address multiple scales, considering both top-down and bottom-up approaches given that the costs and benefits are distributed differently. Bottom-up actions can favour diversity and innovation that is suited to local or national needs and conditions (Chapters 15 and 16). Technology transfer across countries and regions also requires careful consideration of the public–private interface and the appropriate level of technological maturity in combination with bringing the key stakeholders into a constructive process that empowers them as resource managers.

Finally, the design and implementation of financing mechanisms and the ability to attract foreign and domestic investment are indispensable elements in the realization of sugar cane-based bioenergy projects (Chapter 17). Although sugar cane has a long history in some parts of Africa, investment in its energy side remains quite limited but has received greater attention in recent years due to concerns over climate change, energy security and rural development. Providing financing for agriculture-based processing of bioenergy feedstocks on a large scale makes for a complicated investment, both from the point of view of those responsible for payback on financial capital and those entrusted to guide socio-economic development in an acceptable direction. Therefore, most investments in large-scale sugar cane production, whether on a plantation or through an outgrower scheme, require institutional or multilateral support. This is especially true in the present "early" phase of African biofuel project development. In this sense the development of "renewable energy development agencies" at the national level in several African countries is an encouraging sign in the process of supporting the introduction of funding programmes. However, besides funding needs, education, information and demonstration programmes are equally important. Investors need to adopt a holistic approach and must often consider a a five to ten-year timeframe as a minimum. The sourced funding must take a long-term view and must be accompanied by responsible social engagement, in order for biofuel development to support a sustainable transformation in the local and national economy.

INDEX

Page numbers in *italics* denote tables, those in **bold** denote figures.